Historical Acoustics

Historical Acoustics

Relationships between People and Sound over Time

Special Issue Editors

Francesco Aletta
Jian Kang

MDPI • Basel • Beijing • Wuhan • Barcelona • Belgrade • Manchester • Tokyo • Cluj • Tianjin

Special Issue Editors
Francesco Aletta
University College London
UK

Jian Kang
University College London
UK

Editorial Office
MDPI
St. Alban-Anlage 66
4052 Basel, Switzerland

This is a reprint of articles from the Special Issue published online in the open access journal *Acoustics* (ISSN 2624-599X) (available at: https://www.mdpi.com/journal/acoustics/special_issues/historical_acoustics).

For citation purposes, cite each article independently as indicated on the article page online and as indicated below:

LastName, A.A.; LastName, B.B.; LastName, C.C. Article Title. *Journal Name* **Year**, *Article Number*, Page Range.

ISBN 978-3-03928-526-6 (Pbk)
ISBN 978-3-03928-527-3 (PDF)

Cover image courtesy of Elena Bo.

© 2020 by the authors. Articles in this book are Open Access and distributed under the Creative Commons Attribution (CC BY) license, which allows users to download, copy and build upon published articles, as long as the author and publisher are properly credited, which ensures maximum dissemination and a wider impact of our publications.

The book as a whole is distributed by MDPI under the terms and conditions of the Creative Commons license CC BY-NC-ND.

Contents

About the Special Issue Editors vii

Francesco Aletta and Jian Kang
Historical Acoustics: Relationships between People and Sound over Time
Reprinted from: *Acoustics* **2020**, *2*, 128–130, doi:10.3390/acoustics2010009 1

Zühre Sü Gül
Acoustical Impact of Architectonics and Material Features in the Lifespan of Two Monumental Sacred Structures
Reprinted from: *Acoustics* **2019**, *1*, 493–516, doi:10.3390/acoustics1030028 5

Alicia Alonso, Rafael Suárez and Juan J. Sendra
The Acoustics of the Choir in Spanish Cathedrals
Reprinted from: *Acoustics* **2019**, *1*, 35–46, doi:10.3390/acoustics1010004 29

Zorana Đorđević, Dragan Novković and Uroš Andrić
Archaeoacoustic Examination of Lazarica Church
Reprinted from: *Acoustics* **2019**, *1*, 423–438, doi:10.3390/acoustics1020024 41

Lidia Álvarez-Morales, Mariana Lopez and Ángel Álvarez-Corbacho
The Acoustic Environment of York Minster's Chapter House
Reprinted from: *Acoustics* **2020**, *2*, 13–36, doi:10.3390/acoustics2010003 57

Rupert Till
Sound Archaeology: A Study of the Acoustics of Three World Heritage Sites, Spanish Prehistoric Painted Caves, Stonehenge, and Paphos Theatre
Reprinted from: *Acoustics* **2019**, *1*, 661–692, doi:10.3390/acoustics1030039 81

Nikos Barkas
The Contribution of the Stage Design to the Acoustics of Ancient Greek Theatres
Reprinted from: *Acoustics* **2019**, *1*, 337–353, doi:10.3390/acoustics1010018 113

Braxton Boren
Acoustic Simulation of Julius Caesar's Battlefield Speeches
Reprinted from: *Acoustics* **2019**, *1*, 3–13, doi:10.3390/acoustics1010002 131

David E. Witt and Kristy E. Primeau
Performance Space, Political Theater, and Audibility in Downtown Chaco
Reprinted from: *Acoustics* **2019**, *1*, 78–91, doi:10.3390/acoustics1010007 143

Dario D'Orazio and Sofia Nannini
Towards Italian Opera Houses: A Review of Acoustic Design in Pre-Sabine Scholars
Reprinted from: *Acoustics* **2019**, *1*, 252–280, doi:10.3390/acoustics1010015 157

Dario D'Orazio, Anna Rovigatti and Massimo Garai
The Proscenium of Opera Houses as a Disappeared Intangible Heritage: A Virtual Reconstruction of the 1840s Original Design of the Alighieri Theatre in Ravenna
Reprinted from: *Acoustics* **2019**, *1*, 694–710, doi:10.3390/acoustics1030041 185

Pamela Jordan
Historic Approaches to Sonic Encounter at the Berlin Wall Memorial
Reprinted from: *Acoustics* **2019**, *1*, 517–537, doi:10.3390/acoustics1030029 203

About the Special Issue Editors

Francesco Aletta, AFHEA, is a Research Associate at the Institute for Environmental Design and Engineering, The Bartlett, University College London from 2018. He was a Post-Doctoral Research Fellow at the Department of Information Technology of Ghent University from 2016 to 2018, and Research Associate at the School of Architecture of the University of Sheffield, working on the soundscapes of both indoor and outdoor spaces. He has been Visiting Lecturer at Birmingham City University and Visiting Professor at both the Polytechnic Institute of Turin and the University of Roma Tre. His work on soundscape descriptors and predictive models has informed policy documents and international standards. He is the Secretary of the Technical Committee for Noise of the European Acoustics Association. He serves on the Editorial Board of several peer-reviewed international journals in the fields of building and environmental acoustics, environmental psychology, and urban studies. Francesco Aletta has worked in soundscape studies for 10 years, authoring 70+ publications. He is the recipient of several research awards and grants for research equipment. He serves as reviewer for 50+ journals and works as referee for research projects for the Italian Ministry for Research and University.

Jian Kang, Professor, FREng, FIOA, FASA, FIIAV, CEng, holds Chair in Acoustics at the Institute for Environmental Design and Engineering, The Bartlett, University College London, from 2018. He has been Professor of Acoustics (now Visiting Professor) at the School of Architecture of the University of Sheffield since 2003, and previously worked at the University of Cambridge, the Fraunhofer Institute of Building Physics in Germany, and Tsinghua University in China. He chairs the Technical Committee for Noise of the European Acoustics Association, and EU COST Action on Soundscape of European Cities and Landscapes. He was awarded the IOA Tyndall Medal 2008, the Peter Lord Award 2014, the NAS Lifetime Achievement Award 2014, and CIBSE Napier Shaw Bronze Medal 2013. He is Fellow of the Royal Academy of Engineering. Jian Kang has worked in environmental and architectural acoustics for 30+ years, with 70+ research projects, 800+ publications, 90+ engineering/consultancy projects, and 20+ patents. His work on acoustic theories, design guidance, and products has brought about major improvements to the noise control in underground stations/tunnels and soundscape design in urban areas. He is a recipient of the prestigious Advanced ERC Grant Award, and currently working internationally on developing Soundscape Indices.

Editorial

Historical Acoustics: Relationships between People and Sound over Time

Francesco Aletta * and Jian Kang *

UCL Institute for Environmental Design and Engineering, The Bartlett, University College London (UCL), Central House, 14 Upper Woburn Place, WC1H 0NN London, UK
* Correspondence: f.aletta@ucl.ac.uk (F.A.); j.kang@ucl.ac.uk (J.K.); Tel.: +44-(0)20-3108-7338 (J.K.)

Received: 18 February 2020; Accepted: 20 February 2020; Published: 23 February 2020

The Special Issue "*Historical Acoustics: Relationships between People and Sound over Time*" was the inaugural collection of the recently established journal "Acoustics (MDPI)", so it felt appropriate to give it a focus to history, places and events of historical relevance, seeking to explore the origins of acoustics, and examining the relationships that have evolved over the centuries between people and auditory phenomena. Sounds have, indeed, accompanied human civilizations since the beginning of time, helping them to make sense of the world and to shape their cultures. While the establishment of "acoustics" as the science of the "production, transmission and effects of sound" in our contemporary understanding could be located approximately two-hundred years ago, scientific acoustical studies date back to the 6th century BCE, with the ancient Greek philosophers, and were developed later on by Roman architects and engineers. In fact, the interest human communities have expressed towards acoustical phenomena goes back much further than that, regarding which recent research outcomes from the emerging field of archaeoacoustics (investigating the auditory and acoustic environment of prehistoric sites and monuments) have been very fruitful. Societies and cultures have been more or less aware of the importance of "sound" and the science underpinning it, and acoustics have always played a central role for our lives and evolution.

Submissions were invited for research dealing with, for instance, acoustic characterization of prehistorical and historical spaces and buildings, acoustics of worship spaces (e.g., temples, mosques, churches, etc.) and ancient theatres, auralization of soundscapes of the past, soundscape of heritage sites and sound as cultural heritage, and literature reviews about acoustic treaties. Considering the relatively broad spectrum that an arbitrary definition such as "historical acoustics" could cover, the contributions gathered in this Special Issue seem to cluster around a few main themes.

The acoustics of historical worship buildings was a theme common to several contributions. Sü Gül [1] investigated the acoustic environments of the Hagia Sophia and the Süleymaniye mosque by carrying out measurements of room acoustics parameters and comparing the results with previous measurement datasets retrieved from the literature. The work included acoustical simulations to compare unoccupied and occupied conditions and to inform the discussion of the original materials. Other contributions related to the theme of worship spaces dealt with Christian churches. Alonso et al. [2] analysed the acoustic evolution of the choirs of several Spanish cathedrals by means of both on-site measurements and simulation models, which confirmed that room acoustics conditions typically provide suitable intelligibility of sung text. Đorđević et al. [3] studied the sound field of the Orthodox medieval church of Lazarica (Serbia); also, in this case, room acoustics measurements were used to calibrate a computer model and perform an acoustic simulation of the church to investigate the effects of the space occupancy, the central dome and the presence of the iconostasis on a number of reverberation- and speech-related parameters. Álvarez-Morales et al. [4] studied the Gothic York Minster cathedral; the authors used measured and simulated room impulse responses to better understand how its architectural features contribute to its highly reverberant acoustic field and reflected on the implications these had on the evolution of the site from a meeting place of the cathedral's Chapter to its contemporary use for a variety of cultural events.

Several contributions addressed the theme of the acoustics of sites of archaeological interest, with these cases typically dealing with open-air sites. Till [5] reported on the acoustics of three UNESCO World Heritage Sites (the La Garma cave complex in Spain, Stonehenge stone circle in the UK, and the Paphos Theatre in Cyprus), covering a time span of almost 40,000 years. The author reflects on the evolving acoustics of ritual sites in human civilizations, highlighting the role of sound in defining the character of such spaces. The acoustics of ancient Greek theatres is a broad research topic, per se; in his paper, Barkas [6] collates data from twenty ancient theatres in Greece and shows the positive effect of the scenery in contemporary performances of ancient drama to improve the acoustic comfort, concluding that, in spite of considerable alterations (e.g., historical changes, accumulated damage, etc.), most of those theatres are still fit for purpose (i.e., they are theatrically and acoustically functional). Boren [7] used acoustic simulation to confirm historical records of Julius Caesar giving a speech to 14,000 soldiers after the battle of Dyrrachium and another one to 22,000 soldiers before the battle of Pharsalus during the Roman Civil War. Results seem to indicate that, under reasonable background noise conditions, Caesar could, indeed, have been heard plainly by 14,000 soldiers in the speech at Dyrrachium; on the other hand, even in favourable environmental conditions, it is realistic to estimate that Caesar could not have been heard by more than some 700 soldiers in the case of Pharsalus. Likewise, Witt and Primeau [8] used acoustic modelling tools to investigate an Ancestral Puebloans site in Downtown Chaco in the United States that served as an open-air performance space for both political theatre and sacred ritual, focusing, in particular, on the inter-audibility between various locations within the performance space of this site.

The acoustics group of the University of Bologna proposed two contributions on the acoustics of historical opera houses. In the first paper, D'Orazio and Nannini [9] reviewed publications about theatre design written by pre-Sabinian Italian scholars (treatises, essays, etc.), pointing out elements of consistency among some 19th century minor Italian opera houses, to explore to what extent acoustics-related scientific and empirical knowledge would have been part of the construction practice during the golden age of the Italian opera. In the second paper, D'Orazio et al. [10] focuses on the Alighieri theatre in Ravenna, designed by the Meduna brothers and, through acoustic measurements and simulations, the authors compare the current condition with the original one, before the proscenium of the stage was removed in order to open an orchestra pit during a refurbishment in the late 1920s.

Finally, Jordan [11] brought forward the topic of soundscapes as cultural intangible heritage. The author applied both qualitative and quantitative methods to a case study at the Berlin Wall Memorial in Germany, investigating both the past and present soundscape of the site (as understood and perceived), with the support of binaural recordings, psychoacoustic analysis, and soundscape surveys based on standardized soundscape protocols.

A common methodological trait for most of the research that deals with "historical acoustics" is the presence of both acoustic measurements and acoustic simulations in the investigated cases. For the measurements, the researchers' work should always be commended, for the considerable challenges they face in implementing standardized measurement protocols in locations that often present serious accessibility and operability issues. For the acoustic simulations, it is interesting to observe that, while acoustic software was originally conceived (mainly) for the design of spaces that are yet to be built, it is also a powerful tool to investigate acoustic environments that no longer exist (or at least not in pristine condition).

The works gathered in this Special Issue show a vibrant research activity around the "acoustics of the past", which should be the foundation layer to inspire the discipline's future paths. During these months of interaction with authors, reviewers and editors, we realized that there is an increasing interest in these themes and, for this reason, the Journal agreed to establish a permanent topic collection on "Historical Acoustics", to which we invite all interested authors to contribute.

Author Contributions: Conceptualization, F.A. and J.K.; methodology, F.A. and J.K.; writing—original draft preparation, F.A.; writing—review and editing, F.A. and J.K.; funding acquisition, J.K. All authors have read and agreed to the published version of the manuscript.

Funding: This work was funded through the European Research Council (ERC) Advanced Grant (No. 740696) on "Soundscape Indices" (SSID).

Acknowledgments: The Editors would like to thank all authors for their submissions and all reviewers for their thorough work on the manuscripts.

Conflicts of Interest: The authors declare no conflict of interest.

References

1. Sü Gül, Z. Acoustical Impact of Architectonics and Material Features in the Lifespan of Two Monumental Sacred Structures. *Acoustics* **2019**, *1*, 493–516. [CrossRef]
2. Alonso, A.; Suárez, R.; Sendra, J.J. The Acoustics of the Choir in Spanish Cathedrals. *Acoustics* **2019**, *1*, 35–46. [CrossRef]
3. Đorđević, Z.; Novković, D.; Andrić, U. Archaeoacoustic Examination of Lazarica Church. *Acoustics* **2019**, *1*, 423–438. [CrossRef]
4. Álvarez-Morales, L.; Lopez, M.; Álvarez-Corbacho, Á. The Acoustic Environment of York Minster's Chapter House. *Acoustics* **2020**, *2*, 13–36. [CrossRef]
5. Till, R. Sound Archaeology: A Study of the Acoustics of Three World Heritage Sites, Spanish Prehistoric Painted Caves, Stonehenge, and Paphos Theatre. *Acoustics* **2019**, *1*, 661–692. [CrossRef]
6. Barkas, N. The Contribution of the Stage Design to the Acoustics of Ancient Greek Theatres. *Acoustics* **2019**, *1*, 337–353. [CrossRef]
7. Boren, B. Acoustic Simulation of Julius Caesar's Battlefield Speeches. *Acoustics* **2019**, *1*, 3–13. [CrossRef]
8. Witt, D.E.; Primeau, K.E. Performance Space, Political Theater, and Audibility in Downtown Chaco. *Acoustics* **2019**, *1*, 78–91. [CrossRef]
9. D'Orazio, D.; Nannini, S. Towards Italian Opera Houses: A Review of Acoustic Design in Pre-Sabine Scholars. *Acoustics* **2019**, *1*, 252–280. [CrossRef]
10. D'Orazio, D.; Rovigatti, A.; Garai, M. The Proscenium of Opera Houses as a Disappeared Intangible Heritage: A Virtual Reconstruction of the 1840s Original Design of the Alighieri Theatre in Ravenna. *Acoustics* **2019**, *1*, 694–710. [CrossRef]
11. Jordan, P. Historic Approaches to Sonic Encounter at the Berlin Wall Memorial. *Acoustics* **2019**, *1*, 517–537. [CrossRef]

© 2020 by the authors. Licensee MDPI, Basel, Switzerland. This article is an open access article distributed under the terms and conditions of the Creative Commons Attribution (CC BY) license (http://creativecommons.org/licenses/by/4.0/).

Article

Acoustical Impact of Architectonics and Material Features in the Lifespan of Two Monumental Sacred Structures

Zühre Sü Gül

Department of Architecture, Bilkent University, 06800 Ankara, Turkey; zuhre@bilkent.edu.tr

Received: 14 April 2019; Accepted: 3 July 2019; Published: 16 July 2019

Abstract: Hagia Sophia and Süleymaniye Mosque, built in the 6th and 16th centuries, respectively, are the two major monuments of the İstanbul World Heritage Site. Within the context of this study, sound fields of these two sacred multi-domed monumental structures are analyzed with a focus on their architectonic and material attributes and applied alterations in basic restoration works. A comprehensive study is undertaken by a comparative analysis over acoustical field tests held in different years and over an extensive literature review on their material and architectural characteristics. Initially, the major features of Hagia Sophia and Süleymaniye Mosque are presented, and later, basic alterations in regard to function and materials are provided. The methodology includes the field tests carried both within the scope of this research as well as the published test results by other researchers. Acoustical simulations are utilized for comparison of unoccupied versus occupied conditions and also for discussion on original materials. The impact of historical plasters on the acoustics of domed spaces is highlighted. Common room acoustics parameters as of reverberation time and clarity are utilized in comparisons. The formation of multi-slope sound energy decay is discussed in light of different spiritual and acoustical needs expected from such monumental sacred spaces.

Keywords: Hagia Sophia; Süleymaniye Mosque; room acoustics; historical structures; restoration

1. Introduction

This research aims to investigate the effects of the main architectural features of monumental sacred structures on their interior sound fields. With their large volumes and multi-domed upper shell typology in the form of main dome spaces coupled to sub spaces sheltered either by domes or vaults, these historical structures have an outstanding and particular acoustical environment or so called "interior soundscape". In architectural literature, domed structures are mostly observed in religious buildings. Sacred spaces have a specific role of emphasizing spirituality. Acoustics is one means of recognizing the indoor environment and a critical element to augment the divine character of sacred mega-structures. The basic architectural features of such monuments i.e., the central dome and huge volumes, result in the long and shallow reverberant decay, contributing to the key role of holy spaces.

1.1. Acoustics of Sacred Spaces

The literature on the acoustics of sacred spaces is extensive. Within this framework the acoustics of mosques [1–6], churches [7,8], basilicas [9] and cathedrals [10–13] have always been an attractive research topic for acousticians. In relation to this research there are mainly two groups of studies. The first group includes virtual re-construction of historical structures [2,3,11], renovation of pre-designed and built structures [1] for enhancing their indoor acoustical qualities, and acoustical design of new generation mosques [6]. The prominent methods in sound field analysis are field tests, scale modelling, and acoustical simulations [14]. The second group of studies focuses on non-exponential decay formation in multi-volume mega-structures [7,9,10,15]. The acoustical studies on sacred spaces search for the desired interior sound field in relation to function, which differs slightly in different religions. Worshippers sometimes need solitude, while at other times they want to feel in absolute unity with the others in attendance.

There are three basic acoustical requirements for mosques in relation to clarity as well as the spiritual effects of sound; audibility of the namaz—prayer orders of the Imam (prayer leader)—, recognizable sermon of the preacher, and listening to or joining in the recital of the musical versions of the Holy Quran. In mosques, in order to overcome the large sound attenuation of his voice over the rows of worshippers on the floor, the imam delivers his sermon at minbar and müezzin delivers the commands of namaz from müezzin's mahfili. This scheme was specifically developed in the time of historical mosques when there was no electronic sound reinforcement. It was a positive effect that the sound source is above the receiver locations for providing more sound strength to the audience in mosques' original schemes. In the acoustical design of mosques the male voice must be augmented by the architectural features of the space for the envelopment and spaciousness. Thus, the architecture must be incorporated within the expected formal language, which satisfies the spiritual aspects of a worship space. The common use of carpet as a floor finish in mosques instead of the stone floors prevalent in Christian spaces results in faster sound decay rates, thus shorter reverberation times. In Christian sacred spaces, intelligibility of sermons by a priest is one acoustical necessity [14]. On the other hand, recitals of liturgical music are much more frequent in churches and cathedrals when music as an activity is compared to those in mosques. For that reason, it is much more appropriate that the reverberation times are longer in churches or cathedrals when compared to those in mosques.

1.2. Room Acoustics Coupling in Multi-Domed Superstructures

Most od the time, the historical sacred spaces with multi-domed upper-structures have even longer reverberation times than classical venues for music. At this point, the basic challenge in the design of sacred spaces is optimizing reverberance. It is well-known that very high reverberation negatively affects speech intelligibility, while very short reverberation causes a dry acoustical environment that would reduce the envelopment and spaciousness in a religious space. These two contradictory requirements, clarity and reverberance, are hard to reconcile with classical decay properties of a single space room. Some previous studies highlight that multi-domed or vaulted monumental spaces are composed of multiple sub-spaces that are connected to each other with apertures in the form of arches. These arches behave as coupling apertures and depending upon the size of the apertures and as well as the volume of individual sub-spaces, a convex decay curve is observed in large cathedrals, basilicas [7,10,11], and in mosques [15]. Non-exponential or multiple sound energy decay with regard to the inherent properties of early and late decays can maintain the sense of intimacy and clarity that is mostly observed in coupled spaces. This outcome was further discussed in a previous study within a broader context of this research [16] and acoustical indicators in multi-domed superstructures were re-evaluated [17].

In order to contribute to the previous literature on acoustics of sacred structures, this study investigates two historically significant monuments of the İstanbul World Heritage Site. The first monument is Hagia Sophia, which was originally built in the 6th century and functioned as a church and later converted to a mosque, while currently serving as a museum. The second monument Süleymaniye Mosque is a 16th century structure. Their historical significance and religious uses in relation to acoustics have motivated many researchers to test and discuss characteristics of their interior sound fields. A comprehensive study is necessitated to assess the acoustical conditions of both structures in relation to their architectural features and interior finish materials and to examine the acoustical changes that have occurred due to the repairs undergone within their lifetime. Initially, the major architectural features of Hagia Sophia and Süleymaniye Mosque are presented, and later, basic alterations with regards to function and materials are provided. The methodology includes the field tests carried both within the scope of this research as well as the published field test results by other researchers. Acoustical simulations are applied for comparison of unoccupied versus occupied conditions, reflecting major activity patterns. The simulations also provide basis for a discussion on the acoustical outcomes of original or historical plasters versus the current condition after restorations. Finally, the formation of multi-slope sound energy decay in case structures is discussed briefly in light of different spiritual and acoustical needs expected from such monumental sacred spaces.

2. Hagia Sophia

Hagia Sophia, the Holy Wisdom of Christ, was constructed as a church between 532 and 537 in Constantinople (Istanbul) during the reign of the Byzantine Emperor Justinian. Hagia Sophia was both the center of religious life and had been a legend for the new wave of church buildings in the West in the 12th and 13th centuries. After the Ottoman conquest of 1453, during the rule of Mehmet II, it was converted from a church to a mosque. In 1932 upon order from Atatürk, Hagia Sophia (Figure 1) started to function as a museum. Hagia Sophia is a masterpiece of not only Byzantine art but also of the world's historical heritage. Its architectural success and carried messages affected both the Ottoman Empire and architects such as Sinan the Architect [18–21].

Figure 1. Hagia Sophia exterior view (www.ayasofyamuzesi.gov.tr).

The influences of Hagia Sophia's form have revitalized many other domed spaces both in churches and mosques. On the other hand, the building physics aspects of the structure and materials has been a subject for many other researches [22]. The interior sound field of such an immense monumental building has also inspired acousticians. In some acoustical studies on Hagia Sophia, the room acoustical parameters were analyzed for its current condition, as further detailed in Section 5.1. By some other researchers, the acoustical history of the structure was virtually generated for different time periods and activity patterns [23]. One study focused on recreating Hagia Sophia's acoustics within other venues by the use of electronic architecture [24].

Hagia Sophia is a large space with many coupled sub-spaces. High reverberance within such a large interior, lends unique properties to the music and demands suitable repertoire and singing ability in terms of liturgical music. The sound field of such a volume with dominating geometric and material attributes also inspired this study in terms of acoustical coupling investigations. In a broader framework, the non-exponential energy decay formation in Hagia Sophia for its current state was previously discussed [17].

2.1. Major Architectural Features

The major figure of Hagia Sophia is an expanded dome basilica: a rectangular building covered by a central dome between two half domes and integrating longitudinal and centralized planning

(Figure 2). This structure has an interior length of 73.5 m and a width of 69.5 m, excluding the narthex and the apse. The length of the entire interior from the exonarthex to the edge of the apse is 92.3 m. The central nave is built on an east-west axis, and a large dome is constructed right in the center to cover up the space. The central dome that rises 55 m above the pavement of the nave is not exactly round but slightly elliptical today, with a diameter of 31.2 m on one axis and 32.8 m on the other [18]. The domed central space is skirted by two large hemicycles covered by half-domes to the east and west. The diameter of these half-domes roughly equals to that of the central dome. From the crown of each of the main piers and between the four great arches, the pendentives fan out on the interior and rise to a roughly circular projecting cornice [25,26]. The side aisles are separated from central nave by columns and arches sheltered with vaults. There are galleries over the side aisles and inner narthex, creating a U-shape that reinforces the centralizing tendency of this expanded dome basilica. The central oval vessel of enclosed space is further expanded by barrel-vaulted spaces that terminate along the building's longitudinal axis at the east and the west ends of its nave [18]. Overall, Hagia Sophia has an approximate interior acoustical volume of 150,000 m^3, which creates an outstanding visual and acoustical environment.

Figure 2. On the left: section view of Hagia Sophia from the central axis by Salzenberg [25] (p.280), on the right: interior view—an old painting [25] (p.284).

Stone, brick and mortar are the main materials that compose the above-ground structure including the piers, columns, arches, vaults, and domes. The stone is either limestone or green stone—local granite. Above the springing level of the gallery vaults, most of the piers are built of brick. The surfaces of all the walls as well as the large supportive piers, bordered by sculptured moldings, are covered with polished slabs of veined marble and other colored stone. Internally most of the original marble revetment of the walls remains in place and most of the original non-figural mosaic decoration of the vaults has remained undamaged at ground level [18,20]. Some of the non-figural decorations in the galleries and above are distinguished partly by a much greater use of tesserae cut from natural stones and terracotta. All surviving figural mosaics are of a later date. The floor is paved today with large rectangular marble slabs, not the original one which was crushed in 1346 [20,26].

2.2. Basic Repairs and Alterations

Over the 1400 years of its existence, Hagia Sophia has suffered much damage essentially due to the major earthquakes. Three main phases of structural repair and strengthening are recorded. The first repairing phase was in 1317, in the reign of Andronicus Palaeologus; the second was in 1573 under the architect Sinan; and the third repair was in 1847 under the Swiss architect Gaspare Fossati, assisted by his brother Giuseppe. The principal work undertaken in 1317 was the construction of new buttresses. In 1573, Sinan built a new minaret in place of one that was to be demolished. In 1847 major works were the rectification of a number of columns in the gallery exedra, the installation of new ties for critical locations, and other repairment at the level of the dome base [20,21].

Hagia Sophia has also undergone many alterations due to the changes in its activity patterns. In conversion from a church to a mosque, some Christian elements were removed, and Islamic additions were introduced. Since pictorial representations are traditionally not permitted in Islam, after 1453 the mosaics were gradually covered up, whitewashed, or plastered over and hence preserved. After the Ottoman conquest, all the Christian furnishings were removed [20]. In 1847 the Sultan commissioned a pair of Swiss architects, Gaspare and Giuseppe Fossati, to restore both the fabric and the decoration of the building. During these works all the surviving mosaics were uncovered and copied in order to provide visual record. Unfortunately, many of the mosaics recorded by Fossati's had disappeared by this date, most probably lost in the great earthquake of 1894. The original altar, screen, and ambo were badly damaged in the first collapse, and rebuilt in the subsequent restoration [18].

Among the Islamic additions, apart from four minarets on the exterior, a mihrab was added on the kiblah axis (the church is oriented to the east) The minbar was constructed in the same direction. Prayer carpets and banners of victory were hung on the walls during the mosque period. The Müezzin's mahfili in the center of the structure and four more galleries in the narthex were added. On the left side of the central nave there is the preaching pulpit, that still exists. Imperial Pavilion and Imperial Loge (Figure 3) are some other additions which did not exist during the Christian period. Another significant addition of the Islamic period are the huge disks of inscriptions/calligraphies over wooden roundels with diameter of 7.6 m, which are acknowledged to be largest inscriptions in the entire Islamic world [26].

Figure 3. On the left: imperial pavilion and loge, on the right: preaching pulpit [25] (p.176,179).

In 1932 upon order from Atatürk, Prof. Thomas Whittemore—founder of the Byzantine Institute of America—had been given the permission to uncover and clean the mosaics [19]. In 1992, a major restoration and consolidation of the mosaics in the dome was started by the Central Laboratory for Restoration and Conservation of Istanbul in collaboration with an international team of experts funded by UNESCO [18]. Not all listed here but many other Hagia Sophia restorations were held in its lifespan. Even today, the scaffoldings, which are used for the repairment of current mosaic damages at the central dome due to the water leakage and humidity, occupy a huge place within the nave of Hagia Sophia.

3. Süleymaniye Mosque

Being the largest of the Ottoman building enterprises of the time, Süleymaniye Mosque and Complex (Figure 4) was sponsored by Kanuni Sultan Süleyman in his ruling and designed by Sinan-the architect laureate of the Ottoman Empire-. The main construction was held between 1550 and 1557, while the whole structure with the fine works of tombs was completed in 1568 [27]. The complex is currently located in the historical island of Istanbul [28]. Süleymaniye Complex has a plan of three adjacent squares: the mosque itself, the courtyard with the last prayers section, and backyard, where the tombs of Sultan and his wife Hürrem are located. The site plan of the complex takes its references from the topography. The complex is composed of 22 different structures with different functions surrounding an interior courtyard [29–31]. Süleymaniye Mosque, which is the central figure of the whole complex, has always been an inspiration source for many fields including architectural aesthetics, structure and construction, acoustics, and material science. The acoustics of the mosque in this respect is an important subject of research.

Figure 4. Süleymaniye complex exterior view [28].

3.1. Major Architectural Features

Süleymaniye Mosque's main structural elements are domes, arches, and flying buttresses. The mosque is covered centrally by a single dome which is supported on two sides by semi domes. The two semi domes align with the direction of the mihrab. Side aisles are sheltered by five smaller domes which complete the upper structure. Pendentives are utilized to smooth the central dome, secondary half dome, and arch connections. The inner plan of the mosque measures approximately 63 by 69 m. The main dome rests on four elephant feet and 32 footings on a circular wheel with a diameter of 26.2 m. The height from the foundation to the impost is 33.7 m. The inner rise of the dome is 14.0 m, and thus the height of the dome from the ground to the keystone is 47.8 m (Figure 5). The middle and corner smaller domes on aisles have diameters of 9.90 and 7.20 m [32,33]. Süleymaniye Mosque's interior acoustical volume is approximately 75,000 m^3.

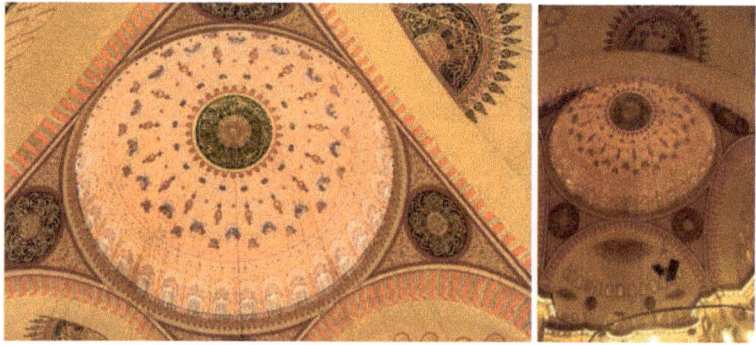

Figure 5. Interior views of domes and pendentives in Süleymaniye Mosque (photos by the author).

The uniqueness of the building comes from abundant sources of stone supplies delivered from various ruins of ancient cities all over the world [34]. For its original state, other basic interior materials are plaster and paint on brick, tile, ceramic pots, glass, and wood. The interior walls are faced with stone revetments. The ceilings of the pulpits and the royal box, the domical superstructure, and the pendentives are painted (Figure 5). In contrast to lavishly painted domes and pendentives in lower zones the stone revetments left relatively bare. The prominent architectural features of the interior are historical columns, marble panels, porphyry discs, great arches, mihrab, minbar (pulpit) and royal box, stained glass windows, and inscriptions.

Limestone is the main structural stone as well as the facing stone for interior walls and wall footings. Piers carrying the main dome and suspension arches are of cut limestone while the inner faces are painted as of Hereke conglomerate and Proconnesian marble. Columns are of Egyptian porphyry (red sparrow eye). Different bricks were utilized as the core material of domes due to its lightness. Painted brick domes are then decorated with gold foiled pen paintings. The mihrab and the minbar are of carved white marble and have stained glass windows on the sides [35–37]. Lime, fine sand, gypsum, linen, and straw are the basic ingredients of plaster layers and seams. Linen is applied in dome plasters of Süleymaniye Mosque in its original state. Wood in the interior is mostly used for flat ceilings, doors, window frames and furniture. Only the two windows on each side of mihrab have kündekari wooden work shutters. The floor finish of the mosque is carpet with straw backing—which were collected from the finest straws grown in Nile delta—as stated in original documents. Carpets had originally been woven in Egypt and Aydın-Tire [35,36]. Another feature of Süleymaniye Mosque of its original state is the use of Sebu's (clay pots), that are essentially applied to lessen the weight of the dome, but also believed to function as cavity resonators for the control of excessive low frequency sound (Figure 6).

Figure 6. Muqarnas details (photos by the author).

In accounting registers of Süleymaniye Mosque construction (account book number 108), it is declared that for plaster finish of dome (Beray-i sıva-i kubbeha-i cami'-i şerif) 134 kantar of linen was purchased [35]. This is significant information in terms of acoustical assessment of historical plasters that is detailed in following sections. As a final acoustical point, within the mosque muqarnas/stalactites are used as a transition element in column heads, in the skirting of half domes. These elements do not have any structural function and are out of gypsum. Muqarnas helps to enhance the sound diffusion in mostly curvilinear and concave transition planes by fragmenting these surfaces (Figure 6).

3.2. Basic Repairs and Restorations

Since the 16th century, Süleymaniye Mosque has not experienced any major structural or volumetric/spatial changes. Aside from some small extensions, the structure as a whole has kept its integrity and original form. However, there have been a couple of major interior material modifications. The initial restorations were held in 19th century successively in 1840, 1844, 1845, 1847, 1870, and 1873. In these interventions basically pen-carved paintings and plasters were modified, which ended up being substantially different—in terms of chemical and physical ingredients—interior finish characteristics. In the 1840s and 1880s restorations, held by Italian experts, it is recorded that the clay pots were covered and closed, and the original dome plasters were modified with gypsum plaster [36–38].

Since 1958, the restorations have been held within the control of T.R. Grand Commission of Memorial'. In 1959–1969 restorations basic interventions were the renewal of oil paints on elephant feet, removal of wooden cabinets at some locations, painting of door and window jambs, and renewal of dome and arch paintings. In these restorations, some of the 19th century paintings were removed in order to uncover the original paintings [38,39]. Prior to 2007–2011 restorations, on stone and wooden surfaces different types of material deteriorations were detected. In particular, the pen-paintings on dome facings and on ceilings of exterior side galleries were corroded due to moisture. Damage reported prior to 2007–2011 restorations also included structural damage, the damage due to cement-based plastering or seam fills, and damage due to some other inappropriate use of material [36,38]. The cement-based plasters and application of pen-wall paintings on these plasters are significant in assessing the changes in acoustical field of the Mosque. During 2011 restorations, the samples of original Horasan plasters in the dome were collected. Tests and analyses were held for obtaining cement free plasters that are compatible with original ones, then applied on renewed plastering and pen-wall paintings (nakış). The paintings within the Mosque took their final shape after 1957–1959 and 2007–2011 restorations [38]. Besides, it was also declared that the mouths of in total 256 clay pot (Sebu) voids were opened and cavities were repaired [40].

4. Methodology

4.1. Room Acoustics Measurements

In a broader scope of this study [16], acoustical field tests were held and results specifically in relation to multiple-decay formation were discussed previously for Hagia Sophia [17] and Süleymaniye Mosque [15,17]. This particular paper aims to discuss the findings in relation to different restorations together with acoustical measurements taken in different years by various research teams. As the impulse responses collected by other researchers are not available to this research for post-processing, results of T30 and C80, which are the common parameters available in all field tests, are presented in comparative analysis.

Field tests, within this study's scope, were held in Süleymaniye Mosque on 23 February 2013, in between the hours of 19:30 p.m. and 3:00 a.m. at the main prayer hall, and in Hagia Sophia on 25 August 2014, in between the hours of 09.00 a.m. and 12.00 a.m. at the ground floor, for both in unoccupied conditions [17]. In assessing room acoustic parameters, field measurements were held in accordance with ISO 3382-1,2009 [41]. B&K (Type 4292-L) standard dodecahedron omni-power sound source was used in acoustical signal generation with B&K (Type 2734-A,) power amplifier. The impulse responses at various measurement points were captured by B&K (Type 4190ZC-0032) microphone, incorporated into the hand held analyzer (B&K-Type 2250-A). Sampling frequency of the recorded multi-spectrum impulse was 48 kHz, covering the interval of interest between 100 and 8000 Hz. DIRAC Room Acoustics Software Type 7841 v.4.1 was used for both generating different noise signals and for post-processing of the measured impulse response data for each receiver position. For reliable decay parameter estimations, it was aimed to obtain signal that was at least 45–50 dB higher than the noise in all octaves. Tested signals were E-sweep, MLS, MLS-pink, balloon pop, and wood clap. Up to five pre-averages were applied over multiple measurements, with impulse response length of 21.8 s. E-sweep provided the highest PNR values, thus those samples were mostly utilized in post-processing. As an example, comparison of some sample source signals is provided in Figure 7.

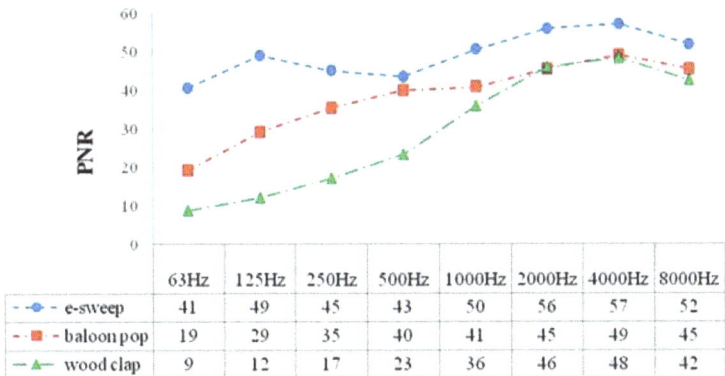

Figure 7. Hagia Sophia PNR measurements in 1/1 octave bands in Hz; comparison of e-sweep, balloon pop, and wood clap source signals.

The measurement positions of field tests are represented in Figure 8, in which the sizes of each structure in relation to one another can also be compared on the plan scheme. Major source locations in Süleymaniye Mosque are one in front of mihrab (S1) and the other at müezzin mahfili (S3). The positioning of sound source is important to assess the acoustics of Süleymaniye Mosque for its traditional use. For acoustical coupling or multi-decay rate investigation, additional source locations including one underneath the main dome (S4) and one underneath side corner dome (S2) were tested [42]. Eight receiver locations (R1–R8) were coupled with four source locations (S1–S4) providing measured source–receiver configurations. In order to observe multi-slope decay formation for different locations and to discuss the effects of spatial variations—such as main central space versus space underneath side galleries—in Hagia Sophia, three source (S1–S3) and six receiver (R1–R6) positions were tested in numerous configurations (Figure 8).

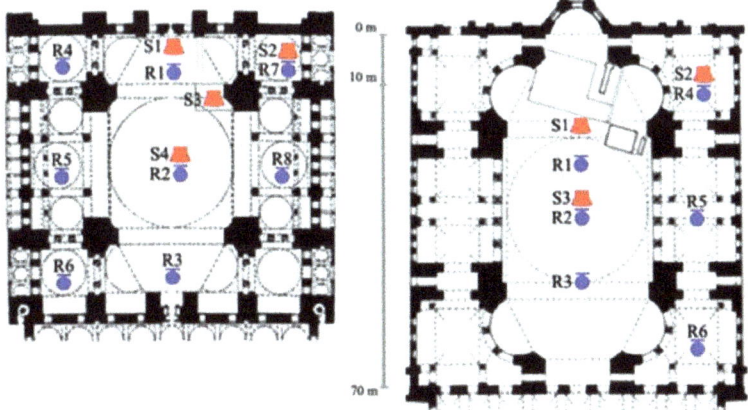

Figure 8. Sound source (**in red**) and receiver (**in blue**) locations used in field tests of Süleymaniye Mosque—METU-2013 tests (**on the left**) and Hagia Sophia—METU-2014 tests (**on the right**).

Other measurements following different restoration periods are also summarized in this section. The first set of measurement in Süleymaniye Mosque was held after 1959–1969 restorations by Gazi University [43] in 1988 (GU-1988); the measurements were taken in 1/3 octaves for the frequency spectrum in between 100 and 8 kHz. The following field tests were taken by Middle East Technical University (METU) in 1996 (METU-1996); broadband noise signal was emitted at two source locations including one in front of mihrab and one over müezzin's mahfili. Impulse responses were collected for six receiver locations at the prayer's area [44]. The final group of measurements after 1959–1969 restorations, as can be found in the academic literature, were held in 2000 within the scope of Conservation of the Acoustical Heritage by the Revival and Identification of the Sinan's Mosques CAHRISMA project [3]. Ferrara University (UNIFE-2000) and Denmark Technical University (DTU-2000) held two different measurements within that scope. Sweep signal was used as a source signal, and impulse responses were collected for different configurations of three source and six receiver locations. UNIFE used test signal possessing of the 80–18,000 Hz spectrum range and DTU used test signal possessing of the 35–11,500 Hz spectrum range. DTU recorded impulses for 10 s of capture length [3]. The first measurements after 1959–1969 restorations were held within the context of this thesis research [16] and abbreviated as METU-2013 in following discussion for the ease of comparisons.

Following the major alterations due to the functional changes, the basic restoration and consolidation of the mosaics in the dome of Hagia Sophia started in 1992, as previously summarized. There is no available measured acoustic data in the literature before that period. All of the available field data of Hagia Sophia were collected after 2000. Previous to Hagia Sophia measurements held in the scope of this research [16] in 2014 (METU-2014), two other set of data were recorded. Ferrara University (UNIFE-2000) and Denmark Technical University (DTU-2000) held two different measurements in the context of the CAHRISMA [3] project in 2000, similar to those explained for the Süleymaniye Mosque.

4.2. Acoustical Simulations

The acoustical simulations of historical structures are carried out in a broader frame of this research by two main methods that are ray-tracing simulations and diffusion equation model (DEM) analysis in a finite element scheme. For applying different estimation methods, the acoustical models of both structures were generated, utilizing the latest building survey registers (röleve) for Hagia Sophia provided by the Turkish Ministry of Culture and Tourism—General Directorate of Turkish Cultural Heritage—and for Süleymaniye Mosque provided by the General Directorate of Pious Foundations. For highlighting the different volumetric relations and major form of the encapsulated interior sound fields, in Figure 9 the ray-tracing models of both structures are presented.

Figure 9. Ray tracing models of Hagia Sophia (on the left) and Süleymaniye Mosque (on the right).

Ray tracing simulations were carried out by ODEON Room Acoustics Software v.14.00 (Odeon A/S, Lyngby, Denmark) [45]. For field tests there were limited time within permitted dates, due to the historical and touristic significance of both structures. For that reason, before field tests, ray-tracing simulations were held for obtaining a general overview regarding to multi-slope decay formation for most probable positions. After field tests, the models were tuned and further utilized for experimentation of different materials of the original state and fully-occupied conditions. The DEM method was utilized for in depth-analysis and understanding of the probable reasons of multi-slope decay formation in such multi-domed monumental structures by energy flow vector, energy flow decay, and spatial sound energy distribution analysis and results are presented in previous papers [15,46,47].

In this study, main objective was to discuss the acoustics timeline of the case structures. For briefness, only the Süleymaniye Mosque's ray-tracing simulation results are presented for different materials and occupancy states to represent its original condition. The source and receiver positions applied in simulations are identical to those used in field tests. Yet again, similar to the field tests, omni-power sound sources are utilized in simulations. Materials are defined with associated sound absorption and scattering coefficients of the interior surfaces by tuning the model with the present field test results (Table 1). Later, the full occupancy of the Mosque during a Friday's sermon is estimated and compared to its unoccupied condition.

Table 1. Sound absorption coefficient and scattering data of applied materials in acoustical simulations.

Materials/Locations	Sound Absorption Coefficients over 1/1 Octave Bands						Scattering Coefficient
	125 Hz	250 Hz	500 Hz	1 kHz	2 kHz	4 kHz	
Ornamented stone piers, arches columns, and column heads	0.05	0.05	0.06	0.09	0.11	0.11	0.20
Current carpet [3]	0.02	0.12	0.25	0.38	0.55	0.58	0.10
Carpet with prayers [48]	0.18	0.22	0.41	0.58	0.69	0.72	0.70
Historical plasters tested in 30% humidity [49]	0.10	0.17	0.23	0.29	0.32	0.32	0.10
Current plasters on brick	0.13	0.09	0.07	0.05	0.03	0.04	0.10
Large pane of glass	0.18	0.06	0.04	0.03	0.02	0.02	0.05
Marble slabs	0.01	0.01	0.01	0.01	0.02	0.02	0.05
Wooden doors and furniture	0.10	0.07	0.05	0.04	0.04	0.04	0.10
Gypsum moldings for muqarnas and similar decorations	0.29	0.10	0.05	0.04	0.07	0.09	0.50
Copper decorative elements	0.12	0.08	0.02	0.01	0.01	0.01	0.50

The sound absorption coefficient data for standard materials as of single layer gypsum, marble on concrete, stone cladding, large pane or multi-layer glass surfaces are well defined in various sources that are more or less identical and mostly reflective as well. The critical issue in this case is assignment of comparatively absorptive materials to those listed in the former. The absorptive materials can significantly change the sound energy decay rates within such immense volumes depending upon their applied surface areas. For the optimization of sound absorption coefficients of current carpet two different sources were visited. Among those, one sample was tested under the scope of the CAHRISMA project [3] and the other is a previously tested carpet sample [50], which is widely used in Turkish Mosques. The sound absorption coefficient data of people on the prayer area applied in simulations—reflecting occupied state of the mosque—are taken from previous laboratory tests of mosque congregations [48]. The attained scattering coefficients on different surfaces are based on the amount of surface irregularities, in small or large scale. In the calibration process of acoustical models, field test data were utilized and T30 values were compared to simulations for different receiver locations. Just noticeable difference (JND) for T30 was kept under 5%, when simulated results were tuned to measured ones. Basically, the unknown sound absorption performances of plasters applied today -and modified in previous restorations-, has given direction to the tuning.

The original acoustical features of the Mosque were examined based on an assumption that the interiors, before restorations, were lime-based plasters with pozzolanic additives which are compatible with the historical ones belonging to the same era. Specifically, the sound absorption coefficients of interior plasters in current state were estimated to be in the range of 0.05 to 0.07 for the mid frequency range, while the sound absorption coefficients of the historical lime-based plasters were previously documented to be in between 0.20 to 0.30 in the mid frequency range [49]. Such a simulation analysis is expected to offer an insight into the probable acoustical conditions in respect to historical lime-based plasters versus today's repair plasters, as discussed in Section 5.3.

5. Results and Discussion

5.1. Acoustics of Hagia Sophia after the 1990s

The data presented for METU-2014 measurements were initially collected for different source signals including wood claps and popping of a balloon. Hagia Sophia was too large to be excited with either a balloon pop or wood clap, and measurements did not provide the sufficient signal (or peak) to noise ratio in the overall frequency spectrum, while neither of those source signals were perfectly omni-directional (Figure 7). Among MLS and e-sweep excitations, e-sweep provided the highest peak-to-noise (PNR) values ranging in between 45 to 60 in the overall spectrum, which were found more reliable to be used especially in multi-slope investigations. For that reason, in Figures 10 and 11 only the e-sweep generated impulse response results are presented for METU-2014 tests.

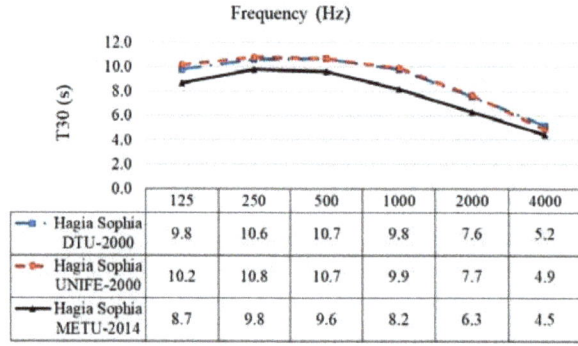

Figure 10. Comparison of T30 results over collected impulse responses in Hagia Sophia by Denmark Technical University (DTU)-2000, Ferrara University (UNIFE)-2000, and METU-2014 in field tests, in 1/1 octave bands from 125 to 4000 Hz.

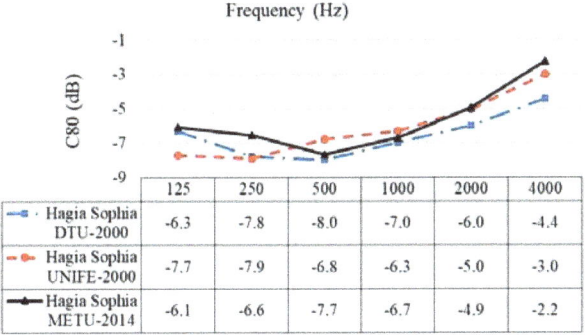

Figure 11. Comparison of C80 results over collected impulse responses in Hagia Sophia by DTU-2000, UNIFE-2000, and METU-2014 in field tests, in 1/1 octave bands from 125 to 4000 Hz.

In Figure 10, DTU and UNIFE indicate measurements taken in 2000 under the supervision of the CAHRISMA project [3]. Hagia Sophia field tests held in 2000, highlight that T30 values in overall are higher than field test results of METU-2014 measurements. The drop of T30 values in METU-2014 measurements below 250 Hz in Hagia Sophia could be the result of the barrier wall separating one side aisle from main volume over the whole longitudinal length, which was present within the space during field tests due to ongoing restorations. Out of single layer gypsum board connected to back studs/profiles in certain intervals, it is probable that the system behaved as an effective low frequency membrane absorber. As another outcome, the mid to high frequency drop in T30 values of METU-2014 measurements in comparison to pre-2000 restorations are one probable result of present scaffolding in on-going restorations that were re-built after 2012. Thus, the presence of additional architectural elements/constructions within the space during METU-2014 measurements resulted in a drop of 1 to 2 s, in the overall frequency spectrum, of reverberation times. On the other hand, the trend of the sound decay over the frequency spectrum is similar for all Hagia Sophia field tests.

C80 is a parameter that is very much dependent on source-receiver positions. Therefore, the deviations in average values of different field test data is expected. It is well known that, subjectively, clarity is the hearing or sensation of individual pieces within music or speech. In objective terms, clarity is the ratio of early to late energy and it is a competing or reverse parameter to the reverberation. Thus, the trend of C80 values for overall field tests is in the opposite manner of T30 results.

It is known that there were no major architectural or form modifications between 2000 and 2014 in Hagia Sophia restorations. Field recorded data is not available for before 1992 restorations, when the consolidation of the mosaics in the dome started. Thus, it is not possible to compare the basic alterations of Hagia Sophia, especially in regard to changes in its function, and to discuss on the success of cleaning of the mosaics, and their acoustical results through field tests.

Overall, in the current state of Hagia Sophia, the T30 results obtained in recent years by different research groups indicate very high reverberation times within the mega-structure. Mid and low frequency T30 averages are around 9 s and high frequency averages are around 5 s. This superb and extraordinary aural environment is unique and has the potential to provide acoustical field conditions in relation to coupled spaces for specific locations. Non-exponential sound energy decay is one outcome of Hagia Sophia's architectonic language, a brief discussion of which is presented in Section 5.4.

5.2. Acoustics of Süleymaniye Mosque for before and after 2007–2011 Restorations

The architectural and material modifications on Süleymaniye Mosque in different time periods are briefed in Section 3.2. With an aim of assessing the acoustical effects of previous restoration works, the field measurements taken in 2013 (METU-2013) are compared with previous field test data. The common acoustical parameter measured and assessed in all field tests for Süleymaniye

Mosque is the reverberation time (T30). Thus, measured T30 values over octave bands are the basis of comparisons (Figure 12). Some previous literature recommends reverberation times of 4-4,6 s for mid frequencies as an optimum range for mosques with similar volume [3]. According to Figure 12, all of the field tests indicate very long reverberation times, higher than recommended ranges for the mosques for unoccupied condition. Especially, 125 Hz is very problematic considering the intelligibility of speech; which will be even worse when the electro-acoustic system is on, as in today's applications. The T30 values over 2000 Hz are comparatively more suitable for the Mosque's function in comparison to mid to low frequencies. The main reason of this drop in higher bands is the absorption of air of such an immense volume.

In METU-2013 measurements below 100 Hz, reverberation times reaching 20 s are observed. It is also noted that there are some measurement spots reaching 20 s at 125 Hz in GU-1988 measurements that are excluded from the average as considered to be the highest deviation [43]. The rise in 250 Hz of METU-2013 measurements are due to the very high values of T30 at 200 Hz in the 1/3 octave band. One reason for the deviations among different measurements might also be due to the averaging of 1/3 octaves or only plotting 1/1 octaves. In 1996 and 2000 field tests, T30 values at 125 Hz were measured in between 9 and 11 s, which are much lower than 1988 and 2013 measurements. This result is most probably caused by the insufficient capture/impulse response lengths of the other field tests. For instance, the DTU-2000 measurement set-up uses 10 s of impulse response length, in the real case, which could easily have missed the chance of estimating a longer reverberation time. On the contrary, both of the first measurements taken by GU-1988 with analogue equipment and the final measurements taken by METU-2013 with digital equipment and with an impulse length of 22 s, indicate 15–17 s of T30 values at 125 Hz. Thus, it could be stated that 2007–2011 renovations have not significantly affected the T30 values at 125 Hz, which are still very high and above the acceptable limits for mosque function.

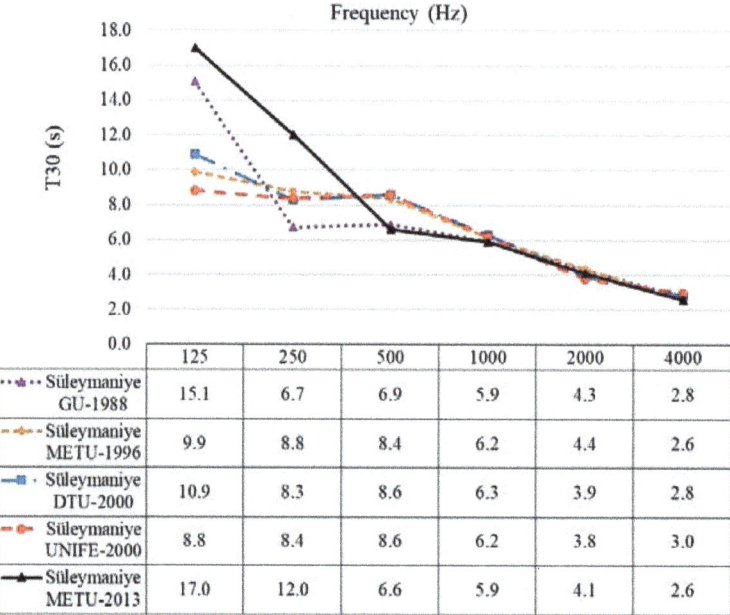

Figure 12. Comparison of T30 results over collected impulse responses in Süleymaniye Mosque by Gazi University (GU)-1988, METU-1996, DTU-2000, UNIFE-2000, METU-2013, and AU-2013 in field tests, in 1/1 octave bands from 125 to 4000 Hz.

According to METU-2013 measurements, reverberation times for octave bands in between 500 and 4000 Hz have been lowered by different ratios compared to previous years' test results. The major change is the 2 s of drop at 500 Hz. When T30 values at 250 and 500 Hz are compared for 1988 measurements and 1996–2000 measurements, an increase of 1.5 to 2 s is observed in the latter. In order to be able to explain this increase, the probable undocumented minor-restorations or material changes in between the years of 1988 and 1996 should be known. Another point is that the measurement positions vary in different field tests, so only average values could be compared.

The comparative analysis of field test results indicate that the 2007–2011 restorations resulted in a slightly positive decrease in reverberation time at 500 Hz, whereas overall, the values are still higher than the recommended ranges. The attempts for removing cement-based plasters, and application of plasters that are compatible with historical lime-based plasters, as so declared, are constructive but not yet efficient acoustical interventions; especially in control of low frequency sound content. The effects of so stated intrusion on Sebu (clay pot) voids—opening the mouths and repair of inside cracks—are still vague in acoustical terms. In order to be able to scientifically comment on acoustical effects of Sebu repairs, the present geometrical and dimensional properties of these elements should be identified, which can only be possible by their systematic inspection on site.

Reverberation time (RT, T30, T20) is one of the key parameters affecting intelligibility of speech. Another important acoustical parameter that affects intelligibility is the level of background noise. High background noise can mask the speech subjectively and can drastically lower the speech transmission index (STI) values. Background noise is a frequency dependent parameter, so its spectrum differs for different noise sources. METU-2013 measurements were taken at night, when the Mosque is closed for visit or prayer and intriguing environmental/traffic noise was at minimum. The measured background noise level was LAeq 39.3 dBA, which is higher than the recommended—NC5 (35 dBA)—upper limit of background noise levels for religious spaces [51]. The high background noise levels detected are caused by the cooling fan units of electrical panels located within the Mosque at the first level right across the mihrab wall. It is evident that one of the basic reasons behind the claims for acoustical or intelligibility problems within Süleymaniye Mosque after the final restorations is the presence of fan noise, which can easily distract speech related activities such as hutbe's by imam.

5.3. Interpretation of Süleymaniye Mosque's Acoustics for Its Original State

Süleymaniye Mosque has gone through various restorations in years, and the acoustical conditions within the Mosque after specific restorations are compared and assessed in previous sections by the data out of field tests. However, none of the field test results reflect the acoustical conditions of the Mosque in its original state. On the other hand, the acoustical expectations from the Mosque in the years of its construction and expectations from a mosque in these contemporary years are quite different. All of the historical mosques today incorporate the use of sound-reinforcement systems. For historical mosques, which were designed for natural sound in their time, it is evident that problems would occur in application of electro-acoustic systems unless necessary precautions are taken.

One of the basic architectural parameters in room acoustics affecting natural speech/sound are the volume of the main space and its geometry. It is known that the dimensions and basic geometrical features of Süleymaniye Mosque have not been altered to date. The dominating form of the Mosque is the central dome which is supported by two semi-side domes. Acoustical focusing effects of the dome can be prevented to some extent, in the case that the lower end of the diameter/circumference of the dome section is located much higher than the receiver/prayer ear height [6]. In Süleymaniye Mosque, even the focusing zone of the biggest central dome is located 20 m above the prayer plane/floor. By the help of this, the first order reflections might have been minimized. However, the dome form causes sound foci at its focal points and negatively affects the sound scattering.

Sinan the Architect in his mosques applied the Sebu (clay pot) technique, which may have enabled acoustical asymmetry within the dome by scattering the sound and enlarging the dome reflection zone. By that, much even distribution of sound within the prayer zone could be provided. Sebu forms are similar to that of amphoras, which have a short neck and a backing volume. The Sebu technique in the structure enables to lessen the weight of the dome, while in acoustics they can easily function as Helmholtz resonators. These elements can scatter sound. Especially at low frequencies (63–250 Hz) they can behave as narrow band volume absorbers. In the account books (D.88. Yp. 19/a) of Süleymaniye Mosque's construction work, it is recorded that each for 2 akçe's (the chief monetary unit of the Ottoman Empire, a silver coin) total of 255 "Sebu" were purchased [35]. Several investigations state that 64 of these pots are located on a circular disk at the central dome [43]. On the other hand, after 2007–2011 restorations it was declared that 256 pots open towards the interior space were detected [40]. According to these declarations, the numerous applications of these clay pots (Sebu's) with various sizes could widen the frequency bandwidth that they are effective in. Thus, it could be predicted that such an application in the original state of the Mosque, to some extent, had healed the excessive low frequency sound content. Together with Sebu voids, the fragmentation of parallel surfaces in both sections and plan schemes of the Mosque by architectural elements such as mahfil's, niche's, and surface treatments such as muqarnas, kündekari, and glazed ceramics have provided sound scattering in a wide frequency spectrum, so that an even distribution of sound throughout the prayer zone is obtained.

In spaces with excessive volume as in Süleymaniye Mosque, the expected long reverberation times have to be controlled by increasing the sound absorptive surface area. For an even distribution of reverberation over frequency, the sound absorption of materials in octave bands should be well balanced. Carpet is an absorptive material for mid and high frequency range only, unless it has at least 5–10 cm height platform, or air space, underneath. In the records of construction documents of Süleymaniye Mosque such a platform is not mentioned [35]. This means that the original carpet's sound absorption performance would be similar to that of today's carpet for low frequencies, and not much of a difference is expected from high frequencies in reference to some other research on carpet effects' in mosques [50]. One other significant piece of information is that in its original state, straw is laid underneath carpet of Süleymaniye Mosque [35], which would have provided a slight improvement in absorption of mid frequency sound content. Carpet is still significant in terms of providing positive absorption area for intelligibility (of consonants). Apart from those, in a prayer scenario the presence of people and their compactness improve sound absorption as well as scattering within the Mosque, as observed in full-occupancy simulation results (Figure 13).

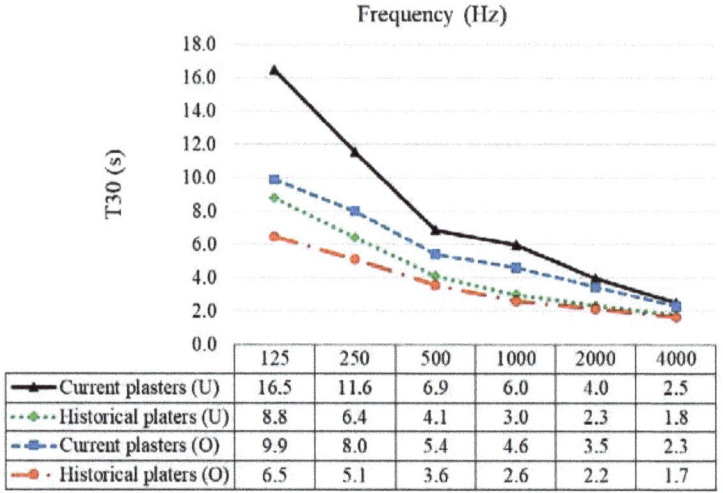

Figure 13. Simulated T30 results of Süleymaniye Mosque Hz in 1/1 octave bands from 125 to 4000 Hz for its current state and trial with historical plasters for unoccupied (**U**) and unoccupied (**O**) conditions.

The absorptive carpet floor surface is still not sufficient to tolerate the long reverberations of this immense volume. On the other side, in Süleymaniye Mosque shiny and tight stone walls, column, and elephant feet surfaces compose the reflective areas. The sound absorption coefficients of applied stones within Süleymaniye Mosque can be predicted for the current and original states in the range of 0.05–0.10. The unknown is the sound absorption coefficient data of original/historical plasters with or without paintings that are renewed in several restorations. These plasters are applied at dome, arch, and mostly upper wall surfaces, and compose a very large surface area of approximately 19,000 m^2. The influence of these plasters on reverberation times would be noteworthy [52].

In the account books (D.108) of Süleymaniye Mosque's construction work [35], it is recorded that 134 scale linen were purchased to be used in plastering of dome. In some other account books, it is also recorded that 524 kantars (Ottoman weight unit corresponding to 56.5 g) of linen were purchased for plastering of the structure in general. Use of linen within the composition of interior surface plasters in Süleymaniye Mosque is an important acoustical element. Linen is commonly used to improve the tensile strength of mortar and/or plaster materials; on the other hand, in acoustic terms it can also increase the sound absorption performance of the plaster at low frequency range. Linen can also augment the mechanical strength of plasters [53,54]. It can be predicted that by the use of linen Sinan the Architect increased sound absorption area serviced by plastered surfaces, which then provided a more controlled reverberation within the mosque in both low and mid frequencies.

Another significant point is that the previous literature highlights the acoustical significance of porous and soft horasan mortar with linen and hemp fiber ingredients that are applied on dome and wall surfaces in Sinan's mosques [43]. The replacement of historical mortar and plaster, which are good absorbers at low to mid frequency range, with tight and stiff cement-based plasters in specific restorations are accused of causing excessive reverberation times observed especially at low to mid frequencies at most of the historical mosques. The use of natural fibers as an ingredient in historical plasters, in fact, is considered to be significant information in terms of acoustic performance of historical plasters, and needs to be investigated with further studies.

In another study, the historical Turkish Baths of the same era were investigated in terms of the acoustical performances of plasters applied in their interior wall surfaces. The sound absorption coefficients of historical multi-layered pozzolanic lime plasters were tested by the impedance tube method [49]. According to that, historical Turkish Bath plasters have sound absorption coefficients of (α) 0.25–0.30 in the mid frequency range and (α) of 0.10–0.17 in the low frequency range, which are higher than the current plasters estimated to have (α) of 0.03–0.09 in the mid frequency range. In Süleymaniye Mosque the current plastered surfaces cover a significant area that should have a drastic impact on effective sound absorption within the Mosque.

For testing the above argument, the acoustical model of Süleymaniye Mosque (Figure 9) was utilized in a simulation study that compares today's plasters, after multiple restorations, with historical multi-layered pozzolanic lime-based plasters. The procedure of model tuning for current state with field tests are presented in Section 4.2. The results are initially attained for unoccupied condition as in field tests, later the results are obtained for occupied condition. The simulations are done for both current state of the Mosque and for modified acoustical model with historical plaster data [49]. Occupied condition represents the full-occupancy as in Friday's Sermon when clarity of speech is more important.

The plot of the simulations tuned by the acoustical field tests in Süleymaniye Mosque's current condition is quoted as "Current plasters (U)" in Figure 13. The measurement results used in this tuning are given in Figure 12 indicated by "Süleymaniye METU-2013." According to the simulation results for unoccupied condition, the replacement of cement-based repair plasters with historical ones ended up with a drop of 2.8 s in T30 value at 500 Hz and almost a drop of 8 s at 125 Hz (Figure 13). T30 values for the fully-occupied mosque in its current state is 6.6 s lower than that of its unoccupied condition for 125 Hz, and 3.6 s lower than unoccupied condition for 250 Hz. However, the low frequency reverberation times are very high and distracting for intelligibility in its current state.

This situation causes even worse acoustical defects, in fact, when the electro-acoustic system is on. In the fully-occupied state of the Mosque with the application of representative historical plasters, the reverberation times are lowered by 2 to 3 s in mid and low frequencies in comparison to the same occupancy state with current plasters. The reverberation time values in between 5 and 6.5 s in the low octaves and 2.6–3.6 s in the mid frequency range indicate that with this trial of historical plasters, Süleymaniye Mosque's interior sound field is much suitable and closer to the recommend limits for mosques with similar volume [3,55]. Although there were some recent attempts in removing cement-based plasters (2007–2011 restorations), the 2013 field measurements show that the problems are not thoroughly solved. Results indicate that if the historical plasters could have survived till now or if the Mosque underwent repairs with the plasters totally compatible with the historical ones, the acoustical conditions would be much more suitable for the function of a mosque today, as well.

As a final note, being located in a courtyard of a big complex (külliye) surrounded by walls together with a thick masonry wall construction—widths ranging in between 65 and 200 cm—and strong exterior shell of domes [29], Süleymaniye Mosque was very well isolated from any environmental noise that might be present in its time. Vehicle traffic, industrial noise, or mechanical equipment noise, as in today's, did not exist in the past. Thus, the intelligibility of speech within the Mosque, had been at least masked, or distorted, in its original state.

5.4. Interior Sound Fields of Hagia Sophia and Süleymaniye Mosque in Relation to Major Architectural Parameters

The aim of this section is initially to assess the dissimilar sound field of Hagia Sophia in comparison to Süleymaniye Mosque, due to volumetric, geometric, and material input. The joint interpretation of results over data obtained at field measurements in 2013 (or METU-2013) for Süleymaniye Mosque and in 2014 (or METU-2014) for Hagia Sophia are presented. The results are provided for T30 in Figure 14 and for C80 in Figure 15, over octave bands from 125 to 4000 Hz and with average, maximum, and minimum values for overall source-receiver positions (Figure 8). As can be observed from Figure 14, the deviation in between measured receiver positions are mostly less than 1 s, and reach up to 2 s in a few octave bands. It is evident that when different receiver positions are compared over minimum and maximum values, the difference is noticeable due to the immense volume. However, the discussion on subjective response evaluation should not rely on single decay metrics but rather be considered under the view of multi-slope sound energy decay formation.

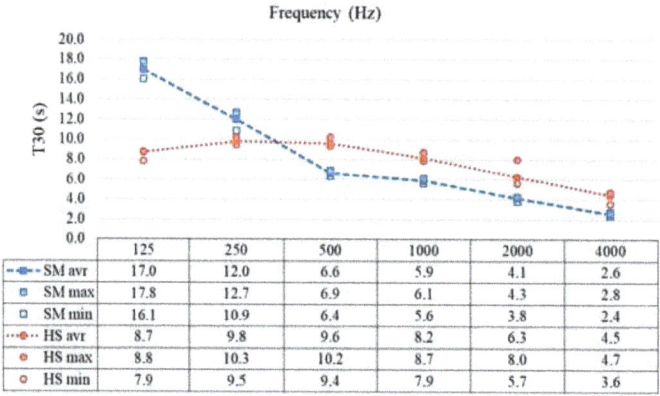

Figure 14. Comparison of T30 results over collected impulse responses in field tests of Süleymaniye Mosque by METU-2013 and in Hagia Sophia by METU-2014, in 1/1 octave bands from 125 to 4000 Hz.

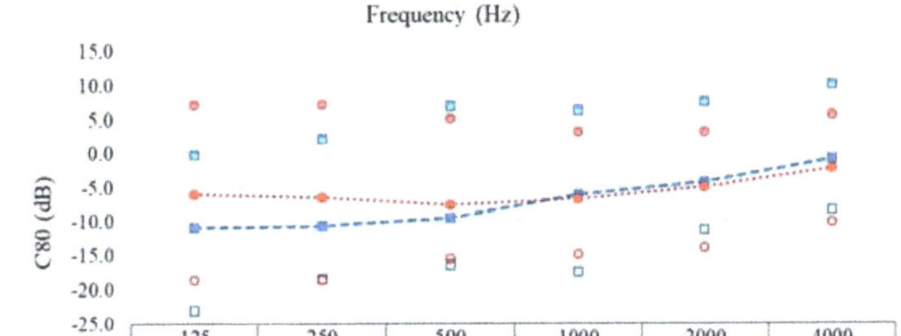

Figure 15

	125	250	500	1000	2000	4000
SM avr	-10.9	-10.6	-9.5	-6.1	-4.2	-0.9
SM max	-0.3	2.2	6.9	6.3	7.5	10.1
SM min	-23.1	-18.4	-16.4	-17.5	-11.2	-8.3
HS avr	-6.1	-6.6	-7.7	-6.7	-4.9	-2.2
HS max	7.2	7.2	5.1	3.1	3.2	5.7
HS min	-18.5	-18.5	-15.5	-14.9	-13.9	-10.1

Figure 15. Comparison of C80 results over collected impulse responses in field tests of Süleymaniye Mosque by METU-2013 and in Hagia Sophia by METU-2014, in 1/1 octave bands from 125 to 4000 Hz.

Due to volumetric differences, dimensional proportions and variable sound absorption characteristics of interior finish materials Süleymaniye Mosque have expectedly lower reverberation times in the frequency range over 250 Hz in comparison to Hagia Sophia. This result was expected considering the doubling of acoustical volumes, which is roughly around 75,000 m^3 for Süleymaniye Mosque, whereas roughly around 150,000 m^3 for Hagia Sophia. The plan schemes of two structures are compared in Figure 8, indicating the considerable differences in size, even in plan layout. Another significant aspect in relation to reverberation, aside from the interior volumes, is the interior finish materials. Marble floor surface in Hagia Sophia in comparison to the carpet floor finish in Süleymaniye Mosque is also very much effective in the increase of reverberation times, specifically over mid to high frequencies. Whereas, the carpet is not that effective under 250 Hz, and this time the reverberation is higher in Süleymaniye Mosque.

In Süleymaniye Mosque, T30 values at 125 Hz are observed to rise to very high levels, which are even much longer than T30 values obtained at 125 Hz in Hagia Sophia. That indicates a probable occurrence of resonances within the main space of Süleymaniye Mosque within that particular frequency band interval. Overall, the distribution of reverberation times over source-receiver positions are smooth and there are no big differences between the average and minimum-maximum values. This indicates a reasonably even distribution within both structures with such great volumes. The diffuseness is due to the surface fragmentations provided by efficient scattering of sound by means of arches, muqarnases or stalactites and mahfili's. It should also be noted that high reverberance inside Hagia Sophia provides suitable repertoire and singing ability in terms of liturgical music, while in Süleymaniye Mosque music is not practiced that much, but only ritual sermons. Thus, the interior sound field of Süleymaniye as a mosque and Hagia Sophia as a church are practically much more suitable for their authentic function of use.

As shown in Figure 15, the average trend of C80 values over frequency spectrum obtained for Hagia Sophia and Süleymaniye Mosque are similar, while values for both structures are much lower than those typically obtained in other venues such as concert halls. It is not surprising that such high reverberation times in overall frequency spectrum would cause low values of clarity. Another reason for very low clarity is the major geometric attributes of both mega-structures. The great sizes and heights of central domes and the lack of close-by surfaces to the receiver zone—such as overhead reflectors in concert halls—cause late reflections to dominate early reflections. The C80 parameter is very much dependent on receiver position, so the deviations in the data that can be observed from average, minimum, and maximum values obtained in both structures are as expected. The C80 information is provided not essentially for comparison but mostly for archiving the measured data obtained within the case historical structures over years.

Another significant finding is multi-slope decay formation as an outcome of the measurements in case structures through a detailed analysis by Bayesian decay parameter estimations. The number and level of decay rates/slopes and their locations were discussed in a previous study for both structures [17]. This discussion is not included in detail within the content of this paper, as the main objective is to discuss the material and geometric changes over years and analyze their possible acoustical effects from their past to present.

The available and comparable data in previous field tests are mostly for T30, and in a few for C80. The multi-slope decay analysis, on the other hand, do not solely rely on a 20 or 30 dB drop of energy levels, but rather investigates the sound energy decay curve for all portions within Bayesian probabilistic inference framework [15]. In regard to the non-exponential energy decay formation, it can be stated that the overall number of decay slopes is greater in Süleymaniye Mosque than in Hagia Sophia, especially in the main prayer zone underneath central domes. The absorption introduced by carpet floor finish in Süleymaniye Mosque versus a reflective upper structure provides un-diffused sound field. Consequently, the sound energy density fragmentation in Süleymaniye Mosque is much more obvious in comparison to Hagia Sophia, due to the relatively diffused sound field of Hagia Sophia.

On the other hand, the span of arches which separate the main prayer hall in Süleymaniye Mosque and the arches connecting the side galleries to the central nave in Hagia Sophia differ in dimension (Figure 16). Due to the larger proportions of arches in Süleymaniye Mosque, the side aisles behind those major arches supporting the central dome are not restricted as much as arcades separating side aisles of Hagia Sophia. While, the smaller arches of Hagia Sophia create much more defined coupling apertures and occasionally cause triple slope decays underneath side aisles. To summarize, interior finish material distribution and/or geometric attributes of the two mega-structures provide sound energy flows inside, among sub-volumes. All these geometric and material attributes cause sound energy concentration and fragmentation within the spaces that augment the non-exponential or multiple sound energy decay formation [17].

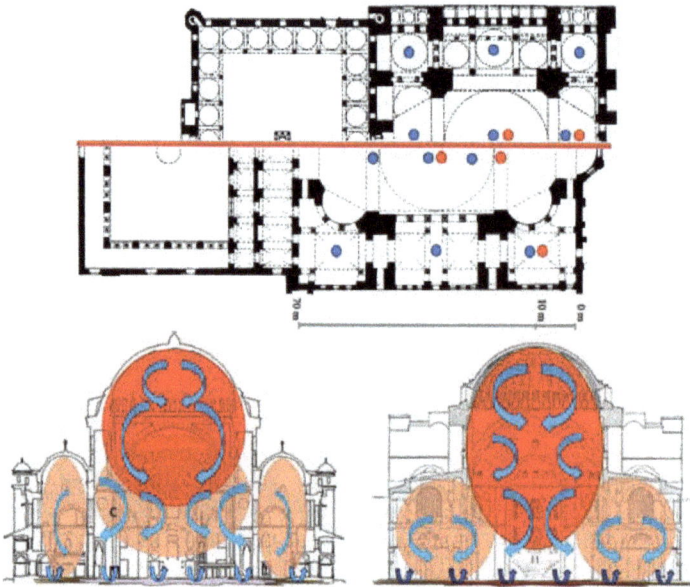

Figure 16. Above: partial plans of Süleymaniye Mosque (**above**) and Hagia Sophia (**below**); below: conceptual section view of sound energy flows of Süleymaniye Mosque (**on the left**) and Hagia Sophia (**on the right**).

6. Conclusions

Within the context of this study, sound fields of two sacred multi-domed historical structures, namely Hagia Sophia and Süleymaniye Mosque, are analyzed with a focus on their architectonic and material attributes, and applied alterations in basic restoration works. The historical significance of the monuments attracted the attention of different research groups, ending up with extensive data in the literature both in acoustics and also in many other fields such as structure, material science, restoration, or art history. A comprehensive study on the acoustics of these sacred spaces is necessitated, by a comparative analysis over acoustical field tests held in different years and over wide-ranging literature on their material and architectural features. The methodology included the field tests carried both within the scope of this research as well as the previously published test results by other researchers. Acoustical simulations were utilized for comparison of unoccupied versus occupied conditions and also for discussion on original materials.

The major difference in between the two monuments is that Hagia Sophia was originally designed to function as a church, while Süleymaniye is a mosque. The basillican layout of Hagia Sophia in comparison to the square and central plan of Süleymaniye Mosque resulted in volumetric differences. Both structures have a highly reflective upper structure, with stone and plaster claddings, while also have many sound scattering elements as of muqarnases or stalactites and space dividing arches. Another major difference in effect of acoustics is that Süleymaniye Mosque has a carpet floor finish whereas Hagia Sophia has a marble floor cladding.

The major acoustical outcomes of this study can be summarized as below:

- Having a larger volume (almost double) and reflective floor finish, Hagia Sophia resulted in higher reverberation times in comparison to Süleymaniye Mosque in mid to high frequencies. This result is appropriate for the original function of Hagia Sophia considering the liturgical music held in churches and cathedrals, in comparison to mosques where liturgical music is seldom.
- In lower octave bands, Süleymaniye Mosque has greater reverberation values in comparison to Hagia Sophia. The excessive reverberation in low frequencies of Süleymaniye Mosque is discussed for its original state by a trial on applying historical lime-based plasters through simulations. Inherently good sound absorption characteristics of historical multi-layered lime-based plasters have resulted in much proper acoustical parameter results within Süleymaniye Mosque.
- Due to the effective surface area, the slight difference on the plasters' sound absorption performance significantly changes the overall acoustical character of sacred spaces. Thus, the physical and chemical characteristics of plasters used in restorations and renovations are highly important, not only for other building physics aspects, but as well for acoustics science.
- The high reverberance of such mega-structures may negatively affect speech intelligibility due to its masking effect. However, a very short reverberation rating would cause a dry acoustical environment and reduce the envelopment and spaciousness in a religious space. Thus, the optimization studies of reverberation times in sacred spaces should not disregard the spiritual aspects.
- The recent restoration work on Süleymaniye Mosque when compared to previous years' field test results has ended up in lower and much more optimal reverberation times in mid to high frequencies, on average. This is most probably due to the removal of cement-based plasters, but is still not totally effective. However, the low frequencies are even higher. One reason for lower decay rates of previous field tests in low frequencies is thought to be due to the insufficient impulse response length.
- A concave dome is a representation of the wholly world, while at the same time this geometry is the primary reason for sound focusing and all relevant acoustical defects as of echoesBurnt-clay pots (Sebu's) as discussed in Süleymaniye Mosque, might had cured the low frequency sound energy built-up within the Mosque, as well as provided acoustical asymmetry in overcoming acoustical defects as of sound foci or echo formation. This approach should be a motivation for today's scientific research with much developed acoustical technology.

- On the other hand, the particular geometrical configuration of multi-domed case structures and interrupted plan-layout with arches, and elements such as mahfili's, elephant feet's, columns, and piers, help to overcome fluttering echoes in between parallel wall surfaces and also help an even distribution of sound.
- All of the field tests in Hagia Sophia were held after the major architectural revisions in relation to the function of this sacred space. The most recent field tests held within the scope of this research indicates that the reverberation times in overall frequency spectrum is lowered by 1–2 s.
- Another discussion point, briefly mentioned in this study, is that the sacred monuments provide not a single decay with a single reverberation time but multiple sound energy decays, which is rarely observed in modest sized single-space structures. The significance of multiple-decay, in form of early and late energy, is that the early decay enhances clarity or definition of sound, while the late decay contributes to the reverberance that complements spiritual needs.

To conclude, the comparative analyses as done in this research are beneficial for acquiring insight to the acoustical outcomes of repairs and restorations. However, the discussion on the basic acoustical metrics that have been studied for years on historical sacred mega-structure is a challenge, especially for the cases where there are multiple-sound energy decays. It should be noted that the unusual sensation (hearing) of sound in such monumental sacred spaces may be explained by overlapping decay curves and their acoustical outcomes, which also might have contributed to them being called out as the world's significant architectural master-pieces, for both the selected case studies as well as the other sacred edifices of the world.

Acknowledgments: The author would like to express her gratitude for the given permissions by General Directorate of Pious Foundations for Süleymaniye Mosque field tests, and to Turkish Ministry of Culture and Tourism, and General Directorate of Turkish Cultural Heritage and Museums for Hagia Sophia field tests. Mehmet Çalışkan, Ayşe Tavukçuoğlu and Ning Xiang are gratefully acknowledged for their guiding and support during different phases of this research.

Conflicts of Interest: The authors declare no conflict of interest.

References

1. Abdelazeez, M.K.; Hammad, R.N.; Mustafa, A.A. Acoustics of King Abdullah Mosque. *J. Acoust. Soc. Am.* **1991**, *90*, 1441–1445. [CrossRef]
2. Suárez, R.; Alonso, A.; Sendra, J.J. Virtual acoustic environment reconstruction of the hypostyle mosque of Cordoba. *App. Acoust.* **2018**, *140*, 214–224. [CrossRef]
3. CAHRISMA, *Conservation of the Acoustical Heritage by the Revival and Identification of Sinan's Mosques' Acoustics*; Project report No: ICA3-CT-1999-00007, Work package 2, Deliverables; Yıldız Technical University: Istanbul, Turkey.
4. Abdou, A.A. Measurement of acoustical characteristics of mosques in Saudi Arabia. *J. Acoust. Soc. Am.* **2003**, *113*, 1505–1517. [CrossRef] [PubMed]
5. Sü, Z.; Yılmazer, S. The acoustical characteristics of the Kocatepe Mosque in Ankara, Turkey. *Arch. Sci. Rev.* **2008**, *51*, 21–30. [CrossRef]
6. Gül, Z.S.; Çalışkan, M. Impact of design decisions on acoustical comfort parameters: case study of Doğramacızade Ali Paşa Mosque. Technical Note. *App. Acoust.* **2013**, *74*, 834–844. [CrossRef]
7. Magrini, A.; Magnani, L. Models of the influence of coupled spaces in Christian churches. *Build. Acoust.* **2005**, *12*, 115–139. [CrossRef]
8. Luigi, E.; Martellotta, F. Acoustics as a cultural heritage: the case of Orthodox churches and of the "Russian church" in Bari. *J. Cult. Heritage* **2015**, *16*, 912–917.
9. Martellotta, F. Identifying acoustical coupling by measurements and prediction-models for St. Peter's Basilica in Rome. *J. Acoust. Soc. Am.* **2009**, *126*, 1175–1186. [CrossRef]
10. Anderson, J.S.; Anderson, M.B. Acoustic coupling effects in St Paul's Cathedral, London. *J. Sound Vib.* **2000**, *236*, 209–225. [CrossRef]
11. Alonso, A.; Suárez, R.; Sendra, J.J. On the assessment of the multiplicity of spaces in the acoustic environment of cathedrals: The case of the cathedral of Seville. *App. Acoust.* **2018**, *141*, 54–63. [CrossRef]

12. Suárez, R.; Alonso, A.; Sendra, J.J. Intangible cultural heritage: the sound of the Romanesque Cathedral of Santiago de Compostela. *J. Cult. Heritage* **2015**, *16*, 239–243. [CrossRef]
13. Pedrero, A.; Ruiz, R.; Díaz-Chyla, A.; Díaz, C. Acoustical study of Toledo Cathedral according to its liturgical uses. *App. Acoust.* **2014**, *85*, 3–33. [CrossRef]
14. Kleiner, M.; Klepper, D.L.; Torres, R.R. *Worship Space Acoustics*, 1st ed.; J. Ross Publishing: Plantation, FL, USA, 2010; pp. 227–271.
15. Sü Gül, Z.; Xiang, N.; Çalışkan, M. Investigations on sound energy decays and flows in a monumental mosque. *J. Acoust. Soc. Am.* **2016**, *140*, 344–355. [CrossRef] [PubMed]
16. SüGül, Z. Assessment of Non-Exponential Sound Energy Decays within Multi-Domed Monuments by Numerical and Experimental Methods. Ph.D. Thesis, Middle East Technical University (METU), Ankara, Turkey, January 2015.
17. Sü Gül, Z.; Çalışkan, M.; Tavukcuoglu, A.; Xiang, N. Assessment of acoustical indicators in multi-domed historic structures by non-exponential energy decay analysis. *Acoust. Aust.* **2018**, *46*, 181–192. [CrossRef]
18. Klenbauer, W.E.; White, A.; Matthews, H. *Hagia Sophia*; Scala Publishers: London, UK, 2004.
19. Oyhon, E.; Etingü, B. Hagia Sophia, Church of Divine Wisdom. In *Churches in İstanbul*; YKY: Istanbul, Turkey, 1999.
20. Mark, R.; Çakmak, A.S. *Hagia Sophia from the age of Justinian to the Present*; Cambridge University Press: New York, NY, USA, 1992.
21. Kahler, H.; Mango, C. *Hagia Sophia*; Frederick A. Praeger: New York, NY, USA, 1967.
22. Mainstone, R.J. *Hagia Sophia: Architecture, structure and liturgy of Justinian's Great Church*; Thames and Hudson: London, UK, 1988.
23. Weitze, C.A.; Rindel, J.H.; Christensen, C.L.; Gade, A.C. *The Acoustical History of Hagia Sophia Revived through Computer Simulation*; Forum Acusticum: Seville, Spain, 2002.
24. Abel, J.S.; Woszczyk, W.; Ko, D.; Levine, S.; Hong, J.; Skare, T.; Wilson, M.J.; Coffin, S.; Lopez-Lezcano, F. Recreation of the Acoustics of Hagia Sophia in Stanford's Bing Concert Hall for the Concert Performance and Recording of Cappella Romana. In Proceedings of the International Symposium on Room Acoustics, Toronto, ON, Canada, 9–11 June 2013.
25. Akgündüz, A.; Öztürk, S.; Baş, Y. *Kiliseden Müzeye Ayasofya*; Osmanlı Araştırmaları Vakfı: Fatih/İstanbul, Turkey, 2006; pp. 174–180, 280–285.
26. Eyice, S. *Ayasofya*; Yapı Kredi: Istanbul, Turkey, 1984.
27. Cantay, T. *Süleymaniye Camii*; EREN Yayıncılık: Istanbul, Turkey, 1989.
28. T.R. Prime Ministry Directorate General of Foundations Archive. 2011.
29. Saatçi, S. Temelden Aleme İnşaat Süreci. In *Bir Şaheser Süleymaniye Külliyesi*; Mülayim, S., Ankara, T.C., Eds.; Kültür ve Turizm Bakanlığı: Ankara, Turkey, 2007; pp. 57–74.
30. Yılmaz, Y. *Kanuni Vakfiyesi Süleymaniye Külliyesi*; Vakıflar Genel Müdürlüğü: Ankara, Turkey, 2008; pp. 87–134.
31. Necipoğlu-Kafadar, G. The Süleymaniye Complex in İstanbul: an interpretation. *Muqarnas* **1985**, *3*, 92–117. [CrossRef]
32. Mungan, I. Strüktür Çözümü. In *Bir Şaheser Süleymaniye Külliyesi*; Mülayim, S., Ankara, T.C., Eds.; Kültür ve Turizm Bakanlığı: Ankara, Turkey, 2007.
33. Kuban, D.A. Symbol of ottoman architecture: The Süleymaniye. In *Ottoman Architecture*; Antique Collectors' Club: Suffolk, UK, 2010; pp. 277–294.
34. Kolay, I.A.; Çelik, S. Ottoman stone acquisition in the mid-sixteenth century: The Süleymaniye Complex in Istanbul. *Muqarnas* **2006**, *23*, 251–272. [CrossRef]
35. Barkan, L.O. *Süleymaniye Camii ve İmareti İnşaatı (1550-1557)*; Türk Tarih Kurumu Matbaası: Ankara, Turkey, 1972; Volume 1, pp. 14–45.
36. Eyüpgiller, K. Restitüsyon ve Renovasyon. In *Bir Şaheser Süleymaniye Külliyesi. Mülayim. Mülayim, S.*; Mülayim, S., Ankara, T.C., Eds.; Kültür ve Turizm Bakanlığı: Ankara, Turkey, 2007; pp. 193–232.
37. İrteş, S. Kalemişi, Cam ve Revzen. In *Bir Şaheser Süleymaniye Külliyesi. Mülayim. Mülayim, S.*; Mülayim, S., Ankara, T.C., Eds.; Kültür ve Turizm Bakanlığı: Ankara, Turkey, 2007; pp. 293–328.
38. Ersen, A.; Nilgün, O.; Akbulut, S.S.; Yıldırım, B.S. Süleymaniye Camii 2007–2010 yılları restorasyonu ve restorasyon kararları. *Restorasyon* **2011**, *3*, 6–27.
39. Acar, S. Süleymaniye'nin düşündürdükleri. *Tasarım* **2000**, *102*, 108–117.

40. A Discussion on the Acoustics of Süleymaniye Mosque for its Original State. Available online: http://www.radikal.com.tr/turkiye/256_bos_kupun_sirri-1028387 (accessed on 17 July 2013).
41. ISO 3382-1. *Acoustics-Measurement of Reverberation Time of Rooms with Reference to Other Acoustical Parameters*. ISO: Geneva, Switzerland, 2009.
42. Sü Gül, Z.; Çalışkan, M.; Tavukcuoglu, A. On the acoustics of Süleymaniye Mosque from past to present. *Megaron* **2014**, *9*, 201–216. [CrossRef]
43. Kayılı, M. Mimar Sinan'ın Camilerindeki Akustik Verilerin Değerlendirilmesi. *Mimarbaşı Koca Sinan: Yaşadığı Çağ ve Eserleri*. **1988**, *1*, 545–555.
44. Topaktaş, L. Acoustical properties of classical ottoman mosques, simulation and measurements. Master's Thesis, Middle East Technical University (METU), Ankara, Turkey, July 2003.
45. Naylor, G.M. ODEON—another hybrid room acoustical model. *Appl. Acoust.* **1993**, *38*, 131–143. [CrossRef]
46. Sü Gül, Z.; Xiang, N.; Çalışkan, M. Diffusion equation based finite element modeling of a monumental worship space. *J. Comput. Acoust.* **2017**, *25*, 1–16.
47. Sü Gül, Z.; Odabaş, E.; Xiang, N.; Çalışkan, M. Diffusion equation modeling for sound energy flow analysis in multi domain structures. *J. Acoust. Soc. Am.* **2019**, *145*, 2703–2717. [CrossRef] [PubMed]
48. Ahnert, W.; Feistal, S.; Behrens, T. Speech Intelligibility Prediction in very Large Sacral Venues. In Proceedings of the Meetings on Acoustics ICA2013, Montreal, QC, Canada, 2–7 June 2013.
49. Tavukçuoğlu, A.; Aydın, A.; Çalışkan, M. Tarihi Türk Hamamlarının Akustik Nitelikleri: Özgün Hali ve Bugünkü Durumu. In Proceedings of the TAKDER 9th National Congress, Ankara, Turkey, 26–27 May 2011.
50. SüGül, Z.; Çalışkan, M. Acoustical Design of Turkish Religious Affairs Mosque. In Proceedings of the 21st International Congress on Acoustics ICA2013, Montreal, QC, Canada, 2–7 June 2013.
51. *ASHRAE Handbook HVAC Applications*, SI ed.; American Society of Heating, Refrigerating, and Air Conditioning Engineers: Atlanta, GA, USA, 2011.
52. SüGül, Z.; Tavukcuoglu, A.; Çalışkan, M.A. Discussion on the Acoustics of Süleymaniye Mosque for its Original State. In Proceedings of the 9th International Symposium on the Conservation of Monuments in the Mediterranean Basin (MONUBASIN), Ankara, Turkey, 3–5 June 2014.
53. Bos, H.L.; Van Den Oever, M.J.A.; Peters, O.C.J.J. Tensile and compressive properties of flax fibres for natural fibre reinforced composites. *J. Mater. Sci.* **2002**, *37*, 1683–1692. [CrossRef]
54. Dalmay, P.; Smith, A.; Chotard, T.; Sahay-Turner, P.; Gloaguen, V.; Krausz, P. Properties of cellulosic fibre reinforced plaster: Influence of hemp or flax fibres on the properties of set gypsum. *J. Mater. Sci.* **2010**, *45*, 793–803. [CrossRef]
55. Orfali, W.A. Sound Parameters in Mosques. In Proceedings of the 153rd Meeting of ASA, Salt Lake City, UT, USA, 4–8 June 2007.

 © 2019 by the author. Licensee MDPI, Basel, Switzerland. This article is an open access article distributed under the terms and conditions of the Creative Commons Attribution (CC BY) license (http://creativecommons.org/licenses/by/4.0/).

Article

The Acoustics of the Choir in Spanish Cathedrals

Alicia Alonso *, Rafael Suárez and Juan J. Sendra

Instituto Universitario de Arquitectura y Ciencias de la Construcción, Escuela Técnica Superior de Arquitectura, Universidad de Sevilla, Seville 41012, Spain; rsuarez@us.es (R.S.); jsendra@us.es (J.J.S.)
* Correspondence: aliciaalonso@us.es; Tel.: +34-954-559-51

Received: 13 November 2018; Accepted: 4 December 2018; Published: 6 December 2018

Abstract: One of the most significant enclosures in worship spaces is that of the choir. Generally, from a historical point of view, the choir is a semi-enclosed and privileged area reserved for the clergy, whose position and configuration gives it a private character. Regarding the generation and transformation of ecclesial interior spaces, the choir commands a role of the first magnitude. Its shape and location produce, on occasions, major modifications that significantly affect the acoustics of these indoor spaces. In the case of Spanish cathedrals, whose design responds to the so-called "Spanish type", the central position of the choir, enclosed by high stonework walls on three of its sides and with numerous wooden stalls inside, breaks up the space in the main nave, thereby generating other new spaces, such as the trascoro. The aim of this work was to analyse the acoustic evolution of the choir as one of the main elements that configure the sound space of Spanish cathedrals. By means of in situ measurements and simulation models, the main acoustic parameters were evaluated, both in their current state and in their original configurations that have since disappeared. This analysis enabled the various acoustic conditions existing between the choir itself and the area of the faithful to be verified, and the significant improvement of the acoustic quality in the choir space to become apparent. The effect on the acoustic parameters is highly significant, with slight differences in the choir, where the values are appropriate for Gregorian chants, and suitable intelligibility of sung text. High values are also obtained in the area of the faithful, which lacked specific acoustic requirements at the time of construction.

Keywords: worship acoustics; Spanish cathedrals; choir space

1. Introduction

Throughout history, the development and construction of worship spaces in Europe has been adapted to the different artistic styles of each era. Although the same style can be dated for different centuries in each European country, there is a number of common characteristics as well as formal and constructive aspects that define each typology. It is evident that from the acoustic point of view, the materials, the geometry, and the variety of particularities belonging to each style can exert a significant effect on the evaluation of the sound field of the temple. In this regard, several studies have analysed the acoustic evolution of worship spaces and have focused on the main qualities of each era and characterised the sound field through the analysis of various acoustic parameters [1,2]. However, not only do dimensions, materials, and stylistic considerations determine the acoustic behaviour of a space, but the spatial conception of its interior also constitutes another main aspect that determines the acoustics of the temple. The variation of the use of the temple in terms of the requirements of each epoch has given rise to certain spatial changes undergone over the years, that is to say, the organisation of its component zones: Naves, transept, ambulatory, and choir.

Occasionally, acoustic needs have influenced the formal, artistic, and spatial configuration of the temples. Knowledge on the origin and situation of elements that have influenced the acoustics of the temple allow the way of solving the interior liturgical space to be understood more clearly, thereby enabling the various configurations adopted in each area to be outlined in accordance with the time or type of event developed. The variation in the location of the choral space, apart from its stylistic, constructive, and chronological considerations, has changed the configuration of worship spaces throughout history [3]. Due to its ability to create and transform the indoor space, the choir would become the main enclosure to configure and articulate the spatial relationship of the temple and the location of the clergy and the faithful. Therefore, it is confirmed that the position of this semi-enclosed and privileged area reserved for the clergy occasionally produces major modifications of these indoor spaces.

In Europe, there are several types of churches and cathedrals that establish the choir space in different positions. In this article, the evolution and acoustic impact of the case study of Spanish cathedrals is analysed, in which the central position of the choir stalls fragments the space in the main nave where, unlike other models elsewhere in Europe, the congregation is not permitted to participate in all types of ceremonies. This work concerns the reconstruction of the process by which Spanish cathedrals have a strong personality within the European panorama, by virtue of the location of the choir.

2. The Choral Space in Churches and Cathedrals

From the architectural point of view, the choir enclosure commands a role of the first magnitude in the configuration of ecclesial indoor spaces and thereby constitutes one of the elements that configure the acoustic issue of worship spaces. The cardinal element of the liturgy can also be considered in its spatial development, and the conception of the choir can only be understood in the space for which it was conceived. The hierarchy of the choir constitutes the most faithful portrait of the clergy and responds to a different personality in each diocese [4]. In the case of cathedrals, the choir position has a significant condition in the architectural floor plan, since liturgical functions are developed around the choir. The fact that it is enclosed by high stonework walls on three of its sides, with numerous wooden stalls inside, has a noticeable effect on the sound field.

In the following sections, the historical evolution and the main typologies of European choirs are defined according to how their location affects the interior of the ecclesial space.

2.1. European Typologies of European Cathedrals

Ever since the construction period of the first Christian churches, the presence of a space reserved for the clergy has stood out, which would be used for the recital of the word in the form of singing. At first, it was located at the head of the temple and later, in its central nave, facing the presbytery, which then led to the schola cantorum, or choir enclosure, in early Christian basilicas [1].

The first position of the choir was later altered in European cathedrals, whereby that central area disappeared and the grouping of spaces for the clergy, altar, and choir were established, thereby creating a deep head in the temple. In Europe, various models and typologies, which are shown in Figure 1, were established, depending on the location of the choir:

- French typology. At first, the choral space was established at the head of the temple around the High Altar, changing its name to choeur. In this regard, the French model had a deep head of the temple to lodge the High Altar and the choir, which, in certain cases, was later moved to the centre for functional reasons. This model corresponds to the initial state of great French cathedrals, such as those of Reims, Amiens, Chartres, and Notre Dame. This opacity and the choir–presbytery represented, to a certain extent, an enclosed canonical church within the cathedral space, and hence the solemn liturgy could not be seen but could be heard by the faithful. As mentioned previously, this typology was an innovation with respect to the Romanesque churches, whose position of the choral space was in the centre of the nave following the model of Cluny Abbey [5]. In the central nave, the sequence followed was that of altar–choir–faithful.

- English typology. English choirs, such as those in the cathedrals of York and Lincoln, created a large private enclosure and were totally closed off to the faithful, and were located within a very long header. Among many other aspects in relation to the arrangement of the floor plan, form, and use of the cathedral, the processional entrance to the choir led through the door that opened into the retrochoir (trascoro). The sequence followed was that of altar–choir–faithful.
- Italian typology—Trentino model. The Schola cantorum of early Christian basilicas preserved throughout the Middle Ages has its origin in Italy. In accordance with major reforms enacted in the Council of Trent (16th century), decisive actions affecting the interior space and the liturgy of the cathedral were carried out. For this reason, renovation work based mainly on the transfer of choral space was promoted in certain European cathedrals, in order to strengthen this estranged relationship between the clergy and the faithful. This typology responds to the solution funded by Carlos Borromeo in the cathedral of Milan, in which the choir is located behind the main altar, taking this intermediate position between the choral space and the faithful. In the central nave, the sequence followed was that of choir–altar–faithful.

These European models (Figure 1) allow the congregation to be accommodated throughout the central nave in all kinds of celebrations.

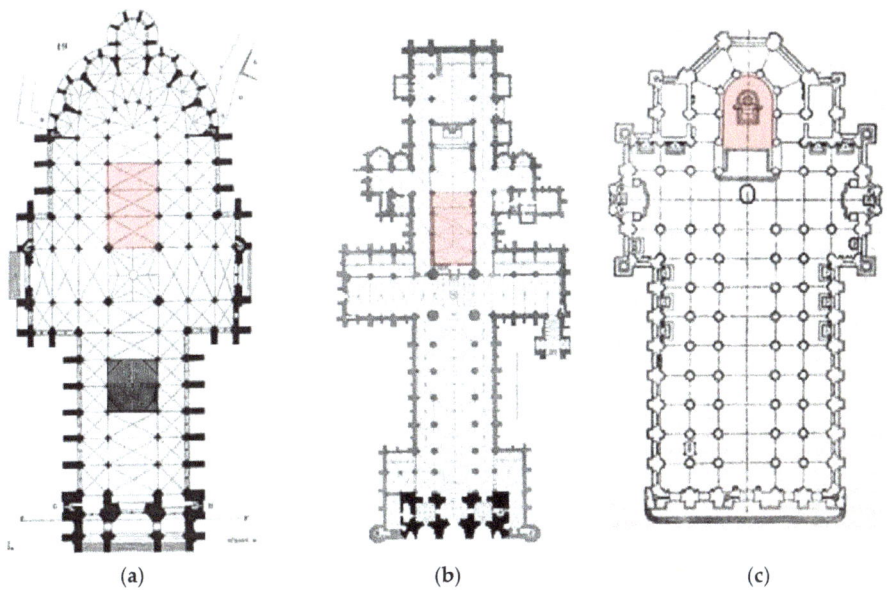

Figure 1. European models. (a) French typology, cathedral of Reims (13th century); (b) English typology, cathedral of Lincoln (14th century); (c) Italian typology—Trentino model, cathedral of Milan (16th century). Red zone: Choir stalls.

2.2. Historical Evolution of the Choir Space in Spanish Cathedrals

In Spain, the choir is usually placed in the middle of the central nave, thereby responding to the old tradition of monasteries, and these, in turn, respond to the schola cantorum of the early Christian basilicas [1,3]. Thus, the chancel and the choir, closely united, became a church for the clergy within the cathedral itself, where the congregation was excluded. In the Spanish typology, the position of the choir generated a division of the central nave into two differentiated zones: On the one hand, the solemn space of the great liturgy and, on the other hand, the space where the ordinary religious ceremonies were developed. This configuration delimited a space for preaching in front of the presbytery, to which the voice could be reached from the pulpits. Moreover, the central position of the choir stalls fragmented the space in the main nave and generated other new spaces, such as the trascoro. As previously indicated, this arrangement of the choir responds to the old tradition of Early Christian basilicas in which the schola cantorum had a similar disposition. In this case, the sequence adopted in the central nave was that of altar–faithful–choir–trascoro–faithful. In the Spanish model, the development of both the word and the musical liturgy is supported by a series of celebratory sources, such as the choir. Longitudinally, the choir space has the dimension of two intercolumniations. Throughout history, the celebration of certain events has required the transfer of this space to other areas of the interior, such as the trascoro, a space that becomes a new stage for music, thanks to its spatial amplitude.

A brief description is made of the course of the evolution of the choir space in Spain, whereby a description is given of the spatial configuration of some of the main cathedrals, both for their outstanding heritage character and the influence of their architectural configuration, transferred to other Spanish and New World cathedrals. Figure 2 shows various floor plans of Spanish cathedrals. Table 1 summarises the data of the main dimensions of the cathedrals under study and their choir spaces.

The tour of historical evolution of the choir space starts with the Latin cross floor plan of the cathedral of Santiago de Compostela (12th century), which is based on monastic buildings such as Cluny Abbey, with a shallow presbytery and a wide stone choir in the central nave [5,6]. This choir space is located in the first three sections of the nave from the transept, with a fourth section assigned to the leedoiro, a high tribune used for sermons in ceremonies. It was in the cathedral of Santiago where the location of the choir between the altar and the faithful in the centre of the nave was first made in Spain. The choir typology of this cathedral played a role of obligatory reference for future Spanish temples. It should be noted this typology generates a problem that has been analysed in other studies, which is the need to look for a configuration that accommodates a large congregation, since its position prevents the use of the depth of the central nave [7]. Although in many Spanish cathedrals, the architectural project raised the location of the choir in the main nave, it should be noted that in certain cathedrals, such as the cathedrals of Burgos and of León, the French model was initially followed. However, later work was carried out in order to restructure the model and move the choir to the central location. These proposals for the transfer of the choir to the nave obey no aesthetic reasoning, but rather a desire for ecclesial renewal.

In the cathedral of Toledo (14th century), the model of the cathedral of Santiago is followed, where the choir, named compostelano [4], is positioned in the centre of the temple, in this case at a double height. Its artistic style is varied and cannot be associated to a single period, since it was built in various stages.

In the case of the cathedral of Seville (15th century), following the previously established model, the choir space is also located in the middle of the central nave, facing the presbytery. It should be noted that the floor plan of the cathedral of Seville was the cathedral model exported to Ibero–America, that of a rectangular church with a straight ambulatory, thus ensuring the continuity of the processional space.

Unlike the other cathedrals, in the cathedral of Granada (16th century), there have been major interventions that have involved the modification of the situation of the choral space. At first, the choral space was located at the centre of the main nave, used mainly for the performance of instrumental music and singing. In the early 20th century, as in the case of the cathedral of Santiago, the choir was moved to the head of the temple, thereby enabling an optimal space-functional relationship by recovering its position at the High Altar. The last intervention, carried out at the end of the 20th century, completely suppressed the choir by distributing the stalls throughout the temple.

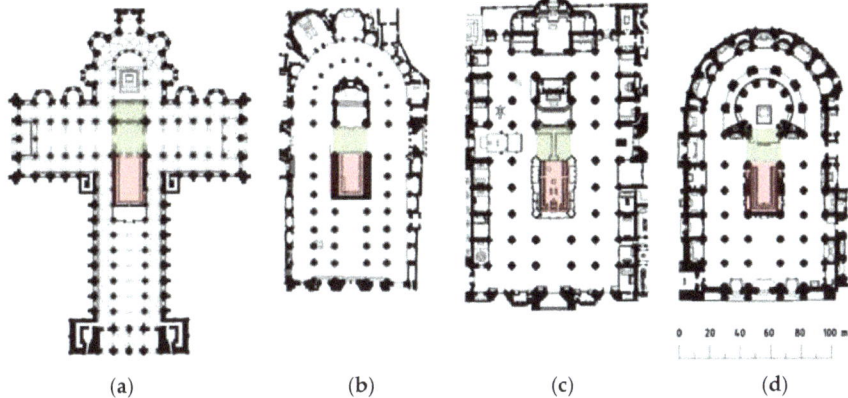

Figure 2. Spanish typology. Floor plans of: (**a**) Cathedral of Santiago de Compostela (12th century); (**b**) cathedral of Toledo (13th century); (**c**) cathedral of Seville (15th century); (**d**) cathedral of Granada (16th century). Red zone: Choir stalls; Green zone: Area of the faithful.

Table 1. Dimensions of Spanish cathedrals under study and their choir spaces.

Cathedral	Dimensions			
	Long (m)	Width (m)	Height (m)	Volume (m^3)
Santiago	174	40 (130 transept)	20 to 43	48,000
Choir	30	16	6.5	-
Toledo	108	55	19 to 31	125,000
Choir	19	11.5	6	-
Seville	116	76	35 to 40	205,000
Choir	22	13	8	-
Granada	106	63	25 to 32	160,000
Choir	20	11.5	7	-

3. Methodology

In order to analyse the acoustic environment of the Spanish cathedrals when the sound source is located in the choir space, two different techniques were carried out, depending on the case study: On the one hand, onsite acoustic measurements were carried out in two of the cathedrals that are preserved as they were built centuries ago (Toledo and Seville); and on the other hand, an archaeoacoustics procedure was developed, that is to say, a virtual acoustic reconstruction of the configuration of the spaces in the past using acoustic simulation tools (Santiago and Granada). However, the cathedral of Granada was also acoustically measured, since this procedure enabled the calibration of the virtual model that faithfully represents the current indoor acoustic environment, a procedure that is studied in Alonso et al. [8]. Once the model was calibrated, spatial transformations were incorporated that reproduce the acoustic sound field of the cathedral in the past with a different ancient choir. This technique, in which onsite measurements and computer simulation are used as a method of evaluation of the acoustic quality of worship spaces, is evidenced by other previous studies [2,7–9].

3.1. Onsite Measurements

An experimental technique was carried out based on onsite measurements to obtain room impulse responses (RIRs) of the cathedrals of Toledo, Seville, and Granada [7–9]. The measurements were carried out within two research projects, each of which lasted for 3 years, financed by the National Plan, in order to meet the objectives that were set out in the project, in which the recovery of the intangible heritage was proposed, as was the addition of possible solutions for the improvement of the acoustic behaviour of the cathedrals. The results of these research projects are published in several studies [7–10]. The RIR reflects the behaviour of a space between the point of emission and that of reception and can be used to calculate the acoustic parameters. Measurements were conducted at night in the unoccupied temple and followed the standard ISO 3382 [11] and guidelines established by Martellotta [12]. Temperature values and relative humidity were monitored, and the values of each cathedral are shown in Table 2.

Table 2. Data monitored during measurements in each cathedral.

Cathedral	Temperature (°C)	Relative Humidity (%)
Toledo	15.8–18	53–66.7
Seville	25–25.5	55–58
Granada	11.7–12.5	36–43

High impulse-to-noise ratio (INR) values require the adjustment of the level and duration of the measured signal and determine a suitable quality of the signal recorded, which provides reliable results of the main acoustic parameters. Since background noise during the measurements was low, it was possible to obtain INR values above 45 dB.

The measurement process started with the excitation response of the cathedrals by issuing sine sweep signals, emitted by an omnidirectional dodecahedron sound source (AVM do-12), raised up to 1.5 m from the floor or support surface, with frequency increasing exponentially over time. The duration of the sweep was set to 20 s and covered the octave bands from 63 to 16 kHz with a power amplifier (B&K 2734). In the case of the cathedral of Seville, a self-amplified Beringher Eurolive B1800D-Pro subwoofer was incorporated in order to improve the low-frequency results. A single output channel was used, since the signal is sent to the subwoofer, which is self-powered and has a high-pass filter (cut-off frequency at 90 Hz) that allows the signal to be sent up to the dodecahedron [13].

The signal generation, acquisition, and analysis were performed using the WinMLS2004 software tool together with an audio interface EDIROL UA-101 in the case of Seville and Granada, and the commercial software EASERA v1.2 with AUBION x8 multichannel sound card in the case of Toledo. Consequently, RIRs were acquired using a multipattern microphone Audio-Technica AT4050/CM5 that can be easily changed from omnidirectional to figure-of-eight in order to simulate the spatial impression of the listener through the lateral acoustic energy perceived, and whose amplifier and polarisation source is Earthworks Audio LAB1. In this study, acoustic parameters were considered when obtained only from omnidirectional directivity characteristics. The microphone was placed at a height of 1.2 m from the floor, which is the approximate ear level of a seated person, at each of the reception points throughout the audience area, with direct vision from the sound source. For the binaural RIRs, a Head Acoustics HMS III (Code 1323) dummy head was used. A moment of the measurement procedure in the choir space of the cathedrals of Seville and Toledo is captured in Figure 3a,b.

3.2. Simulation

Acoustic simulation is a highly useful tool, both for the prediction of the acoustic behaviour of an enclosure in the design stage and for the recovery of the sound of the past in spaces of worship that no longer exist. In this regard, simulation software also enables the characterisation of ancient acoustic conditions by offering the possibility of listening to the acoustics of the space of another era [13]. This is the case of the cathedral of Santiago since the old stone choir space, executed by Master Architect Mateo, was removed from the cathedral in the 17th century to be replaced by a wooden choir in the central nave. Since in the 20th century, it was also decided to dismantle the wooden choir, there is a need to use acoustical prediction tools to analyse the acoustic environment of the temple in past eras. Virtual acoustic models of the cathedral of Santiago and the cathedral of Granada were created in order to reconstruct these spaces from the past.

In the case of Santiago, documented floor plans researched by Conant [6] were used and the model contained 5600 planes and a volume of 48,202 m^3, and in the case of Granada, almost 5000 planes were used to create a volume of 160,000 m^3. After having created the models using SketchUp 3D modelling software, they were then exported to CATT-Acoustic v9.0c [14]. The predicted values were provided using the calculation engine TUCT v1.1a, incorporated in the acoustic software. On the one hand, the data collection of sound absorption/scattering coefficients was carried out by categorising the existing materials in the space after a visual inspection [15]; and on the other hand, coefficients of materials of inexistent enclosures, such as the choir space, were based on documented bibliography. The sound source was located in the choir space, corresponding to the liturgical use that is the object of study, and the receivers were positioned according to occupation. The acoustic model simulations provided the RIRs, which led to the calculation of the most significant acoustic parameters of the cathedral. Virtual models of the cathedral of Granada and Santiago, into which the inexistent choir space was incorporated, are shown in Figure 3c,d.

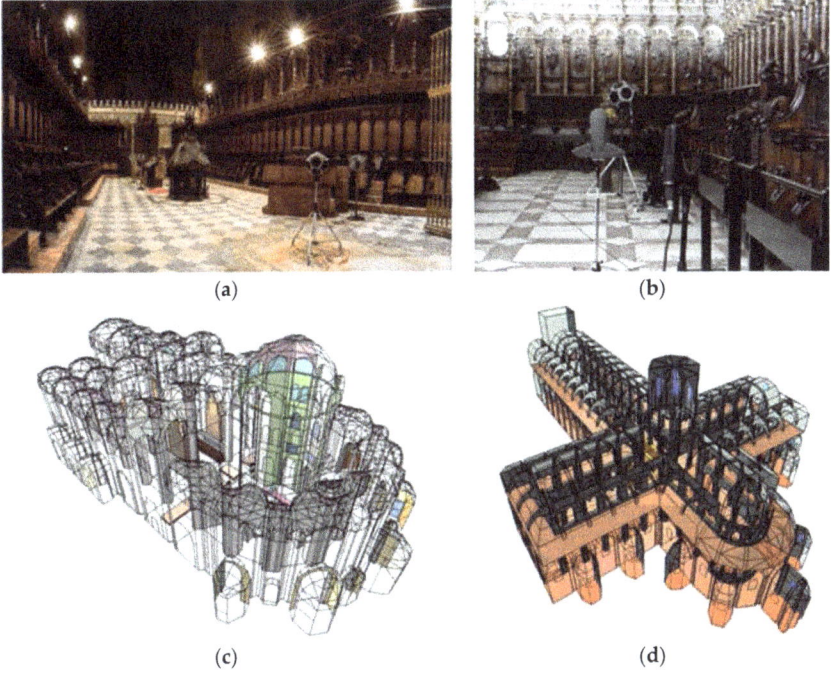

Figure 3. Acoustic measurements conducted inside cathedrals: (**a**) Choir stalls of the cathedral of Seville; (**b**) choir stalls of the cathedral of Toledo. Virtual models in which the choir space was incorporated: (**c**) cathedral of Granada; (**d**) cathedral of Santiago.

4. Acoustical Assessment of the Choir Space in Spanish Cathedrals

This study analysed the acoustic environment of Spanish cathedrals when the sound source is located in the choir stalls. The sound field of the Spanish cathedrals under study was evaluated through the analysis of the RIRs measured or predicted at the receiver points with direct sound from sources located in the choir and in the main altar. In this regard, a comparison of the values obtained was established in order to evaluate the acoustic impact of locating the sound source in the choir space, while considering different groups of receiver positions: One group located in the choir space itself, and another group located in the area of the faithful. This latter group was located in the area between the main altar and choir stalls, with a distribution of between 3–6 receiver points, depending on the cathedral studied. In order to correctly obtain the acoustic parameters, it was ensured that all the receiver points analysed guaranteed the arrival of direct sound. As previously stated, in relation to the acoustics, it is known that the choir constitutes one of the elements that configures the acoustic issue of cathedrals. This analysis enables the different acoustic conditions existing between the choir itself and the area of the faithful to be verified, while also noting the significant improvement of the acoustic quality in the choir space. The experimental results are shown in the various forms explained below.

Figure 4a–d depicts the acoustic parameter values, spatially averaged, versus frequency octave-band when the sound source is located in the choir and the receiver points are located in both positions of the choir and the area of the faithful. Table 3 provides a summary of spatial average values (500–1000 Hz) of the acoustic parameters analysed in the study, in accordance with Table A.1 ISO 3382 [11]. On the other hand, Table 4 shows the comparison in terms of the Just Noticeable difference (JND) of acoustic parameters between values of points located in the choir space and in the area of the faithful, when the sound source is located in the choir.

It can be observed from Figure 4a that T_{30} values are independent of the position inside the cathedral, since the reverberation time is obtained by calculating the time the signal needs to drop by 30 dB. In terms of early decay time (EDT), the calculation is performed using the first 10 dB, thereby raising the importance of the variation of the first reflections. This fact means that there is probably a difference in the results when the sound receivers vary the position and, therefore, vary the existence of the closest surfaces. Moreover, the choir enclosure is surrounded by high stonework walls on three of its sides and with numerous wooden stalls inside, which varies both the absorption and scattering coefficients. This study also centres on the values provided by parameters such as clarity (C_{80}), which measures the range of the listening perception for musical use, the centre time (Ts), which is related to the balance between early and late energy reaching the receiver, and definition (D_{50}), which shows the capacity of the enclosure to generate precision in the articulation of vocal sounds. All these factors lead to a noticeable effect in the sound field when the receiver points are located inside the choir space, as can be seen in Figure 4b–d and Table 4. It becomes apparent that the spectral values of EDT at mid-frequencies, spatially averaged, present a significant difference, greater than 10 JNDs in all cases, and more than 2 JNDs for C_{80}.

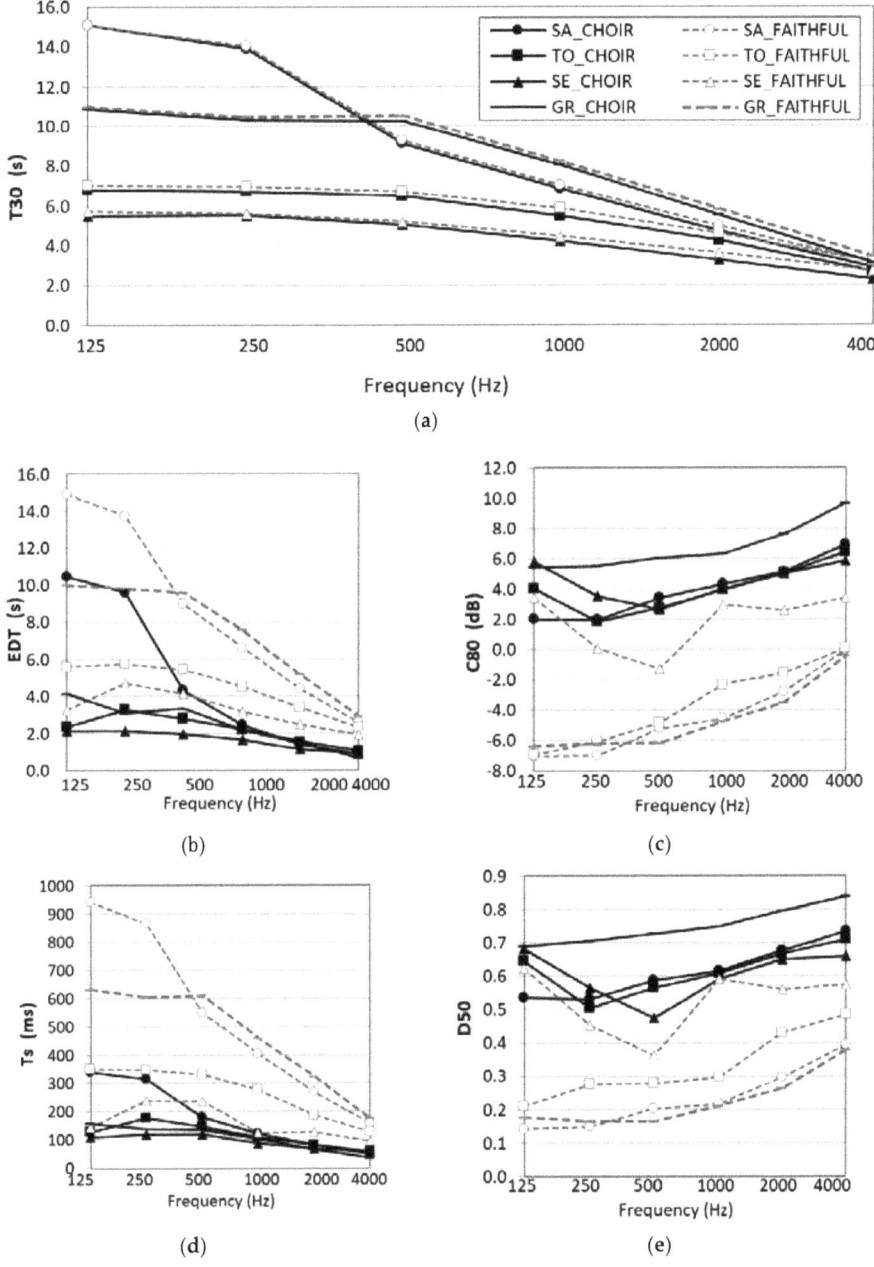

Figure 4. Acoustic parameter values, spatially averaged, versus frequency octave in the unoccupied cathedrals studied when the sound source is located in the choir and in the altar: (**a**) T30 (s); (**b**) EDT (s); (**c**) C_{80} (dB); (**d**) Ts (ms); (**e**) D_{50}. SA = Santiago, TO = Toledo, SE = Seville. EDT: Early decay time.

Table 3. Summary of average values (500–1000 Hz) of acoustic parameters analysed, in accordance with Table A.1 ISO 3382 [11]. Spatial average values by zone: Choir space and area of the faithful.

	T_{30} (s)	EDT (s)	C_{80} (dB)	T_S (ms)	D_{50}
SA_CHOIR	7.99	3.37	4.28	150	0.60
SA_FAITHFUL	8.15	7.81	−4.21	477	0.21
TO_CHOIR	6.01	2.50	3.89	128	0.59
TO_FAITHFUL	6.31	4.97	−2.93	302	0.29
SE_CHOIR	4.65	1.83	3.88	102	0.53
SE_FAITHFUL	4.84	3.66	1.42	181	0.48
GR_CHOIR	9.14	2.75	6.67	119	0.74
GR_FAITHFUL	9.37	8.59	−4.83	535	0.19

Table 4. Differences in JND of acoustic parameters between values of receiver points located in the choir space and in the area of the faithful, when the sound source is located in the choir.

	T_{30} (s)	EDT (s)	C_{80} (dB)	T_S (ms)	D_{50}
SA	0.4	26.3	5.7	29.2	3.9
TO	1.0	19.8	3.8	18.2	3.0
SE	0.8	11.2	2.6	6.4	0.6
GR	0.5	42.4	46.7	7.7	5.5

The variations of the EDT parameter are highly significant, with slight differences in the choir, where the values are appropriate for the Gregorian chant, and high values in the area of the faithful. The Gregorian chant is characterised as being a prayer sung with a prosodic rhythm, centred on intonation. This is a plainchant and a purely melodic, monodic song into which more complex developments, such as organum parallelum, have been incorporated. The relevance of the Gregorian chant is justified due to the importance of ecclesiastic chant at the liturgy and owing to the significant role played by choirs in cathedrals. In the case of D_{50}, the situation is similar, whereby adequate intelligibility is attained of the text sung in the choir.

Another influential factor is that of the acoustic quality of the cathedral, since the differences observed are lower when the acoustic conditions are better and the reverberation is lower. In the case of the cathedral of Seville (4.79 s), whose values of T_{30} are drastically lower than those of Granada (9.14 s), the differences in certain parameters are extremely small when the source is located on the altar and in the choir. In the cathedral of Seville, mostly rough and highly porous limestone, as well as the ornate decoration, exert considerable influence on the sound absorption by the venue, and, by extension, on its acoustic parameters [7]. It is interesting to observe that with the exception of the cathedral of Seville, the results reflect very similar sound conditions in the choir area, identified as the church of the clergy, with greater acoustic requirements for the chant. In the same way, the significant differences between the results of the two areas are shown, as influenced by the conditions of the enclosure. The area of the faithful represents the church without the many current acoustic requirements.

In addition, it should be noted that the JNDs in the cathedral of Granada are highly significant; however, this fact is justified by the great reverberation and the diffuse sound of the cathedral in the area of the faithful, which contrasts with the excellent acoustic conditions of its choral space. With the exception of the cathedral of Santiago, the subjective reverberation values in the choir are below 3 s, and the clarity values are in the range 2–4 dB, which is considered optimal for large-scale spaces [16].

5. Conclusions

The choir enclosure is one of the elements that structures and influences the acoustic issue of worship spaces, since it plays a role of the first magnitude in the configuration of ecclesial indoor spaces. From a historical point of view, the choir is a private, semi-enclosed, and privileged area reserved for the clergy. The spatial complexity of cathedrals responds to liturgical requirements. In this regard, in Europe, different models and typologies are established, depending on the location of the choir. In the case of most Spanish cathedrals, the central position of the choir stalls fragments the space in the main nave.

In this article, a grand tour is developed regarding the acoustic impact of the case study of Spanish cathedrals: Santiago, Toledo, Seville, and Granada. A noticeable effect is perceived in the sound field when the source is located inside the choir space since, in relation to the acoustics, it is known that the position of the choir is one of the elements that influences the acoustic issue of cathedrals.

When analysing the evolution of the acoustic impact of the choir in the Spanish cathedrals, it is not possible to establish an improvement in acoustic quality over time, since factors, such as the dimensions of the enclosure and the height of the walls, can influence the results. This analysis has successfully confirmed the various acoustic conditions existing between the choir itself and the area of the faithful, whereby the significant improvement of the acoustic quality in the choir space, identified as the church of the clergy, can be noted. The variations in the acoustic parameters are highly significant, with slight differences in the choir, where the values are appropriate for the Gregorian chant, with EDT values in the range of 1.8–3.3 s, and suitable intelligibility of sung text, with D_{50} values higher than 0.60 and C_{80} in the range of 3.8–6.6 dB. On the other hand, high values are obtained in the area of the faithful, which represents the congregation of the church, without the many current acoustic requirements.

Author Contributions: Conceptualization, A.A., R.S. and J.J.S.; Methodology, A.A., R.S. and J.J.S.; Software, A.A.; Validation, A.A., R.S. and J.J.S.; Investigation, A.A., R.S. and J.J.S.; Resources, A.A., R.S. and J.J.S.; Data Curation, A.A.; Writing-Original Draft Preparation, A.A.; Writing-Review & Editing, R.S. and J.J.S.; Visualization, A.A.; Supervision, R.S. and J.J.S.; Project Administration, R.S. and J.J.S.; Funding Acquisition, R.S. and J.J.S.

Funding: This work has been financially supported by FEDER funds and the Spanish Government, with reference BIA2014-56755-P.

Conflicts of Interest: The authors declare that there are no conflicts of interest. The founding sponsors played no role in the design of the study; in the collection, analyses, or interpretation of data; in the writing of the manuscript; and in the decision to publish the results.

References

1. Suárez, R.; Sendra, J.J.; Alonso, A. Acoustics liturgy and architecture in the Early Christian Church. From the Domus ecclesiae to the Basilica. *Acta Acust. United Acust.* **2013**, *99*, 292–301. [CrossRef]
2. Pedrero, A.; Ruiz, R.; Díaz-Chyla, A.; Díaz, C. Acoustical study of Toledo Cathedral according to its liturgical uses. *Appl. Acoust.* **2014**, *85*, 23–33. [CrossRef]
3. Navascués, P. *La Catedral en España: Arquitectura y Liturgia*, 1st ed.; Lunwerg: Barcelona, Spain, 2004; ISBN 8497851080.
4. Franco-Mata, A. El coro de la cathedral de Toledo. *Abrente* **2010**, *42*, 113–165.
5. Suárez, R.; Alonso, A.; Sendra, J.J. Archaeoacoustics of intangible cultural heritage: The sound of the Maior Ecclesia of Cluny. *J. Cult. Herit.* **2016**, *19*, 567–572. [CrossRef]
6. Conant, K.J. *Cluny, les Églises et la Maison du Chef D'ordre*, 1st ed.; Mâcon: Cambridge, MA, USA, 1968.
7. Alonso, A.; Sendra, J.J.; Suárez, R.; Zamarreño, T. Acoustic evaluation of the cathedral of Seville as a concert hall and proposals for improving the acoustic quality perceived by listeners. *J. Build. Perform. Simul.* **2014**, *7*, 360–378. [CrossRef]
8. Alonso, A.; Suárez, R.; Sendra, J.J. Virtual reconstruction of indoor acoustics in cathedrals: The case of the Cathedral of Granada. *Build. Simul.* **2017**, *10*, 431–446. [CrossRef]
9. Girón, S.; Álvarez-Morales, L.; Galindo, M.; Zamarreño, T. Acoustic evaluation of the cathedrals of Murcia and Toledo, Spain. In Proceedings of the International Congress of Sound and Vibration (ICSV24), London, UK, 23–27 July 2017.
10. Martellotta, F.; Álvarez-Morales, L.; Girón, S.; Zamarreño, T. An investigation of multi-rate sound decay under strongly non-diffuse conditions: The crypt of the cathedral of Cadiz. *J. Sound. Vib.* **2018**, *70*, 261–274. [CrossRef]

11. ISO 3382-1. *Acoustics—Measurement of Room Acoustic Parameters. Part1: Performance Rooms*; International Organization for Standardization: Geneva, Switzerland, 2009.
12. Martellotta, F.; Cirillo, E.; Carbonari, A.; Ricciardi, P. Guidelines for acoustical measurements in churches. *Appl. Acoust.* **2009**, *70*, 378–388. [CrossRef]
13. Suárez, R.; Alonso, A.; Sendra, J.J. Intangible cultural heritage: The sound of the Romanesque cathedral of Santiago de Compostela. *J. Cult. Herit.* **2015**, *16*, 239–243. [CrossRef]
14. Dalenbäck, B.I.L. *CATT-Acoustic v9 Powered by TUCT. User's Manual*; CATT: Gothenburg, Sweeden, 2011.
15. Vorländer, M. *Auralization*; Springer: Berlin, Germany, 2008.
16. Gade, A.C. Acoustics in halls for speech and music. In *Springer Handbook of Acoustics*, 2nd ed.; Rossing, T.D., Ed.; Springer: New York, NY, USA, 2007; pp. 301–350. ISBN 978-1-4939-0755-7.

© 2018 by the authors. Licensee MDPI, Basel, Switzerland. This article is an open access article distributed under the terms and conditions of the Creative Commons Attribution (CC BY) license (http://creativecommons.org/licenses/by/4.0/).

Article

Archaeoacoustic Examination of Lazarica Church

Zorana Đorđević [1],*, Dragan Novković [2] and Uroš Andrić [3]

1. Institute for Multidisciplinary Research, University of Belgrade, 11000 Belgrade, Serbia
2. The School of Electrical and Computer Engineering of Applied Studies, 11000 Belgrade, Serbia; dragannovkovic72@gmail.com
3. Front Element Media, 11000 Belgrade, Serbia; uandric83@gmail.com
* Correspondence: zorana@imsi.rs; Tel.: +381-63-8995-176

Received: 1 May 2019; Accepted: 14 May 2019; Published: 17 May 2019

Abstract: The acoustic analysis provides additional information on building tradition and related indoor practice that includes sound, thus deepening our understanding of architectural heritage. In this paper, the sound field of the Orthodox medieval church Lazarica (Kruševac city, Serbia) is examined. Lazarica is a representative of Morava architectural style, developed in the final period of the Serbian medieval state, when also the chanting art thrived, proving the importance of the aural environment in Serbian churches. The church plan is a combination of a traditional inscribed cross and a triconch. After the in situ measurement of acoustic impulse response using EASERA software, we built a computer model in the acoustic simulation software EASE and calibrated it accordingly. Following the parameters (reverberation time (T_{30}), early decay time (EDT) and speech transmission index (STI)), we examined the acoustic effect of the space occupancy, central dome and the iconostasis. In all the cases, no significant deviation between T_{30} and EDT parameter was observed, which indicates uniform sound energy decay. Closing the dome with a flat ceiling did not show any significant impact on T_{30}, but it lowered speech intelligibility. The height of iconostasis showed no significant influence on the acoustics of Lazarica church.

Keywords: archaeoacoustics; church acoustics; Lazarica church; architectural heritage; acoustic heritage; medieval building; reverberation time; speech intelligibility; acoustic simulation

1. Introduction

Architecture of a medieval church is a complex symbolical-metaphorical construct imbued with different references from a wide range of human experience [1]. The exclusive focus on its material reality reduces our understanding. Instead, architectural heritage requires the exploration of all the incorporated layers of meaning and their mutual relations. The multidisciplinary field of archaeoacoustics tends to introduce the balance of our comprehension of the history of a building by taking into account the aural environment influenced by various interconnected factors that create human experience in historical buildings, thus also arguing that such acoustic heritage should be considered as an intangible and inseparable layer of architectural heritage [2,3].

Christian churches have been archaeoacoustically researched for the last few decades. The comprehensive literature overview of historical church acoustics [4] considers the acoustic analysis of mostly occidental Christian churches in Europe, Asia and South America, emphasizing the need for a multidisciplinary approach that also includes perceptual and cultural aspects of the aural experience. Besides the examination of the effect of occupancy [5], speech intelligibility [6,7], and the effect of interior architectural forms and materials [8–10], it is also important to acknowledge the multidisciplinary research of acoustic vessels inbuilt in some of the historic churches [11,12], as an equally valid expression of the acoustical intentions of the builders. Archaeoacoustic research of Orthodox churches also covers a wide spectre of approaches. Among many, it includes the analysis of existing or virtually modelled church acoustics, comparison of the acoustic data from churches built in different periods [13], analysis of the live chant recorded on-site [14,15], and correlation of the development of chant and sacral architecture [16,17].

This paper contributes to the research of medieval Orthodox church heritage with its multidisciplinary approach to the investigation of mutual relatedness of architectural forms, religious practice and acoustics in the case of Serbian medieval Lazarica church. Therefore, it opens with the culture-historical context and architectural principles of building Lazarica church, leading to the acoustic requirements of the Orthodox religious indoor practice. The research methodology is based on the on-site acoustic impulse response measurement and acoustic modelling, in order to examine the acoustic influence of the presence of the faithful during the liturgy, the central dome, and the historical change in the height of the altar barrier (iconostasis). Acoustic analysis considered three parameters—reverberation time (T_{30}), early decay time (EDT) and speech transmission index (STI)—as the most relevant for two main acoustic requirements of Holy Liturgy, speech intelligibility and experience of monophonic Byzantine chant.

1.1. Culture-Historical Context of Building Lazarica Church in Kruševac, Serbia

At the end of the 14th century, the Serbian empire collapsed. After the death of the last emperor Uroš V Nemanjić in 1371, local rulers divided the land among themselves. While the Ottomans were penetrating from the south, Serbs were moving north. Among Serbian local rulers, prince Lazar Hrebeljanović who autonomously ruled the territory around the Morava rivers, most prominently tended to continue the well-established tradition of the Nemanjić dynasty [2]. However, in that restless and unstable period, new political, economic and cultural centres arose and thrived in the land of prince Lazar and his successors [4]. This inspiring zest was also reflected in church architecture, which is nowadays referred to as the Morava architectural style.

The building activity was especially induced after the reconciliation of Serbian and Byzantine church when the anathema of the Constantinople patriarch was removed and the new Serbian patriarch was elected in 1375. This was achieved thanks to, inter alia, the engagement of the monks from Mont Athos and Chilandar Monastery. The cultural relations of Serbia and Mount Athos thus grew even stronger.

Following Nemanjić's tradition of founding monasteries and church endowments, prince Lazar built the church of Ravanica monastery (1377) and then St. Stefan church in Kruševac city, also known as Lazarica [3]. The building of Lazarica (1377/81) is closely related to the work on the expansion and improvements of Kruševac city fortification [18]. Both churches—Ravanica and Lazarica—are equally important for the Morava style [19], representing the role models that achieved harmony and wholeness [18]. They marked the beginning of the most authentic building period in medieval Serbia that used the language of Byzantine constructions and shapes. It is important to note that this particular period—the end of 14th and the beginning of 15th century—was not very fruitful in Byzantine building. Due to the significant economic decay and the strengthening of the Ottoman Empire [19], the number of buildings decreased to a point of total disappearance, while in Morava Serbia construction projects flourished [20], embodying the fresh interpretation of Byzantine art [21]. The difficulty of reliably recognizing on more than one building the work of the same workshop of craftsmen leads to the conclusion that there was a strong need for the authentic art creation [20].

The development of cultural relations between Serbia and Mount Athos also affected the church architecture. The church plan in the Morava architectural style is a triconch (used in Mount Athos for monk choirs that have a significant role in the complex rites [4]) combined with an inscribed cross (traditionally used in Serbian medieval church architecture). Although applied before in Serbia, Saint Nikola Mrački church (1330), the chapel in the Rile monastery (1350) and the chapel of St. Nikola in the St. Arhangeli monastery near Prizren [7], the triconch plan became a recognizable solution for the Morava style. To designate himself as a carrier of the Nemanjić state sovereignty, it is hypothesised that prince Lazar especially looked up to the triconch church of St. Arhangeli monastery near Prizren, as an endowment of the emperor Dušan (the Mighty) Nemanjić [4]. There was also a more functional explanation as the rounded apses provided an additional chanting space on the northern and southern sides of the altar barrier, which was needed for the increased number of chanters coming to Morava Serbia from Mount Athos [8].

1.2. Architectural Design of Lazarica Church

Lazarica church was built in the Byzantine manner: alternating a row of sand stone and three rows of bricks in thick mortar couplings that emerge from the wall plane (Figure 1). It expresses a high artistic level of façade decoration, characteristic for the Morava architectural style. However, for the purpose of acoustic analysis, here we focus on the interior space organization and applied materials.

Figure 1. Lazarica church exterior view from the south-west (Photo credits: Z. Đ.).

Lazarica church has a triconch (three apses) plan combined with the compressed inscribed cross. On the intersection of cross arms rises the central dome (Figure 2). In Byzantine, and thus in Morava architecture as well, the dome had a distinctive meaning. It symbolized the heaven over the earth and complemented the image of the church as a microcosm [21] (p. 74). In Lazarica the dome is constructed on pendentives, supported by four pilasters and the attached arches (Figure 3, bottom left). The drum is circular from the inside and 8-sided from the outside. On the both ends of the transversal cross arm are chanting apses that significantly increase the space and facilitate the Holy Liturgy. The bays in naos are barrel vaulted, while the narthex is cross vaulted with a bell tower above. Since the tower is connected to the naos through one small window in the western wall of the naos, it does not affect the church acoustics.

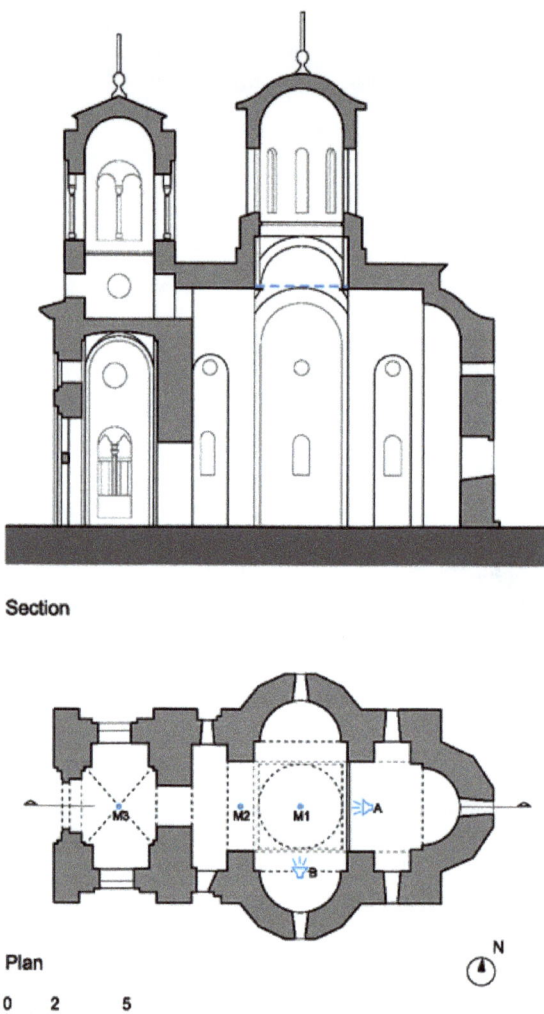

Figure 2. Plan and longitudinal section of Lazarica church with marked positions and directions of sound sources (A: Altar and B: Chanting apse) and measurement points (M1-M3). The blue dashed line in the section shows where the dome was replaced with a flat ceiling in the acoustic simulation model.

Figure 3. Lazarica church interior view from the narthex (top left), from the naos towards the altar (top right), towards the dome (bottom left), towards the southern chanting apse (bottom right) (Photo credits: Z. Đ.)

Proportional analysis of Lazarica church plan showed the combined application of two proportional methods—collapsing diagonals (based on a square) and triangulation (based on an equilateral triangle). If the length of the building brick is one module, than the side of the main proportional square is 13 modules. The dome drum circle is inscribed in this square from which all other geometric points are derived [18] (pp. 36–37).

The church interior consists of three main spaces—narthex, naos and altar—from the entrance in the west to the most sacred space on the east where the sun rises. In the altar the essence of the Orthodox Holy Liturgy takes place. Therefore, it is important to have a visual barrier which disables the faithful from approaching from the naos and to see the accomplishment of holy secrets by the priests in the altar. What we find today in Lazarica is the wooden iconostasis made in the 19th century, that is 3.7 m in width and 4.35 m in height (top of the cross is 6.5 m from the floor level, Figure 3) [18]. However, in the medieval period altar barriers were quite different. They were about 2–2.5 m high. Stone columns supported the beam, on which the curtains were hanging. Until the 14th century, the altar barrier had no icons. The central doors were beautiful gates, while the opening on the left and right had curtains. There was a short parapet wall. Comprehending that the attention of the faithful attending the liturgy is directed towards the altar barrier, over the centuries the curtains were complemented with additional rows of icons [10].

Due to the multiple architectural interventions since medieval period, in the beginning of 20th century Lazarica church was reconstructed with an intention to revive the original look from the time of prince Lazar. The main works were: taking off the façade mortar and revealing the stone-brick building play, reconstruction of the bell tower above the narthex, window details in the narthex, and rebuilding of the central dome. Nevertheless, there are no reliable data on the state of the church interior that was found. The main designer and project manager Pera Popović (1904–1908), wrote that he found the staircases leading from the narthex to the bell-tower. They do not exist anymore. There were also found segments of fresco painting from the 19th century, but it was removed during the reconstruction and the walls were mortared "straight and smooth". There is no evidence that Lazarica was fresco painted in medieval times. However, ever since Popović's reconstruction it was planned to fresco paint the church [22], but it has not been realized yet (Figure 3).

1.3. General Acoustic Requirements in Orthodox Churches in Medieaval Serbia

When we consider temple acoustics, it is important to know not only the interior shapes and finishes, but also the nature and purpose of the rites that took place there. An Orthodox temple is a place of congregation, where the Holy Liturgy is the most important act which implies communal praying, listening to the sermon, and receiving the Holy Eucharist and other sacraments. There are still in use two Liturgies written by Saint Basil the Great and Saint John Chrysostom in the second half of the 4th century. The Holy Liturgy consists of three parts. The first part is the Liturgy of Preparation or Proscomedia which is performed in silence in the northern part of the altar, where the priest prepares the bread and wine. The second part is the Liturgy of the Catechumens and the third is the Liturgy of the Faithful, both performed in a loud voice. After the Proscomedia, the church bells ringing announces the commencement of the second part of the Liturgy. The needs of the Holy Liturgy shaped the interior architecture. Thus the three parts of the Orthodox temple—narthex, naos and altar—correspond with its main acts. The narthex is for the Catechumens: those who can attend the Liturgy until a certain moment, as a part of the preparation process for receiving the Orthodox Christian Faith. The naos is the nave of the church where the faithful attend the service, standing with faces towards the altar. As the most sacred part of the temple, the altar is divided from the naos with the altar barrier, the iconostasis [23].

Byzantine chant that spread among Serbian people along with Christian faith is monophonic, with gradual melodic flow which follows rhythmically emphasized lyrics [24]. This chanting art reached its apex in the same period when the Morava architectural style developed [16]. In other words, the acoustics of a medieval Serbian church, such as Lazarica, is required to support both speech intelligibility and Byzantine chanting art. In Orthodox tradition no musical instruments are used inside a church.

2. Methodology

In accordance with ISO 3382 set of standards, monaural measurements of acoustic impulse response was carried out in the empty Lazarica church for two sound source positions (A: Altar and B: Chanting Apse) and three characteristic receiver points (M1: under the central dome, M2: in naos, and M3: in narthex), as shown on the Figure 2. As a sound source, an active loudspeaker dB Opera 110 (dBTechnologies, Valsamoggia, Italy) was used. Although the use of a directional sound source can significantly affect the obtained results of EDT and STI parameters that are analysed in this paper, the authors have decided to use such a sound source with the basic idea of establishing stronger correlation with the sound sources that emit sound energy in such spaces in real situations, which are primarily human voices. This fact should be kept in mind when it comes to the repeatability and reproducibility of measurement results. Sound field excitation was conducted from two positions that are commonly used in Orthodox religious practice, the altar and chanting apse. The excitation loudspeaker was placed at the height of 170 cm (assumed to be the average person's height) with central axe oriented towards naos.

Sine sweep signal in the range from 20 Hz to 20 kHz was used to acoustically excite the space. Acoustic impulse response measurement software EASERA (AFMG Technologies, Berlin, Germany) was used, together with Digidesign Mbox2mini audio interface and omnidirectional Behringer ECM8000 measurement microphone (Behringer, Willichm Germany) placed at assumed average height of standing listeners (170 cm).

For the purpose of sound field simulation, a detailed 3D model of the church interior was created in EASE software (AFMG Technologies, Berlin, Germany). The model was then calibrated in accordance with the results of the on-site impulse response measurements and examined for four cases: (1) empty church, (2) when faithful are present in the church, (3) when the dome is closed with a flat ceiling (Figure 2), and (4) when the iconostasis is lower, as it was in medieval times.

3. Results

3.1. Acoustic Model Calibration

The main intention of acoustic measurements conducted on-site was to investigate monophonic acoustic parameters defined by ISO 3382 set of standards, as a starting point for further analysis of basic sound field properties and comparison with values of acoustic parameters obtained by acoustic model simulation. Lazarica church interior was modelled in the EASE software and materials were assigned to all boundary surfaces. The church floor is covered with marble, which is standard material that can be found in EASE material database, but the EASE materials database does not contain a material that corresponds to Lazarica walls made of stone with a layer of mortar. Therefore, common concrete was used as the first choice for the wall material. In this respect, after comparing the first results of the simulation with the results of the on-site impulse response measurements, it was necessary to carry out calibration of the acoustic model, which was obtained by an iterative "trial and error" process. The custom values of the absorption coefficient of the church walls were defined (Table 1), which resulted in a sufficient harmonization of the results of the simulation and measurement as shown in Figure 4.

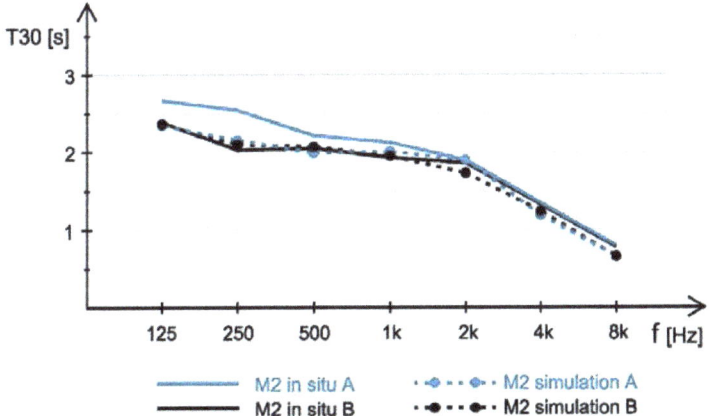

Figure 4. Measured and simulated octave values of T_{30} parameter for M2 measurement point in the case of sound excitation from the altar (A) and the chanting apse (B).

Table 1. Octave range absorption coefficient values used for wall material in the process of acoustic simulation.

Hz	125	250	500	1000	2000	4000	8000
α	0.04	0.04	0.05	0.05	0.06	0.09	0.16

3.2. Examination of the Acoustic Model

The acoustic model was used to examine the acoustic effect of (1) the space occupancy, (2) the dome and (3) the iconostasis. The first case simulates the real-life situation when faithful attend the service. The goal was to examine how the presence of people influence the church acoustic properties. An estimated 50 people are considered. The second case analysis was intended to determine the effect of the dome on the final acoustic response of the space. In this sense, a model with a flat ceiling on the place of the dome was formed (Figure 2), and its acoustic response was compared with the acoustic response of the regular model. The third case considers the altar barrier, the iconostasis, which used to be significantly lower in height in medieval period. Thus, the acoustic parameters were compared for the empty church model as it is today (with the iconostasis 4.35 m high) and as it was in medieval times (2.2 m high).

Speech transmission index (STI) is a measure of speech transmission quality, graded from 0 to 1. It is used to quantify speech intelligibility. Figure 5 visualizes the difference of the STI parameter when the church is empty (on the left) and when it is full (on the right), for the sound source positioned in the chanting apse. Table 2 shows STI values for all examined cases in Lazarica church.

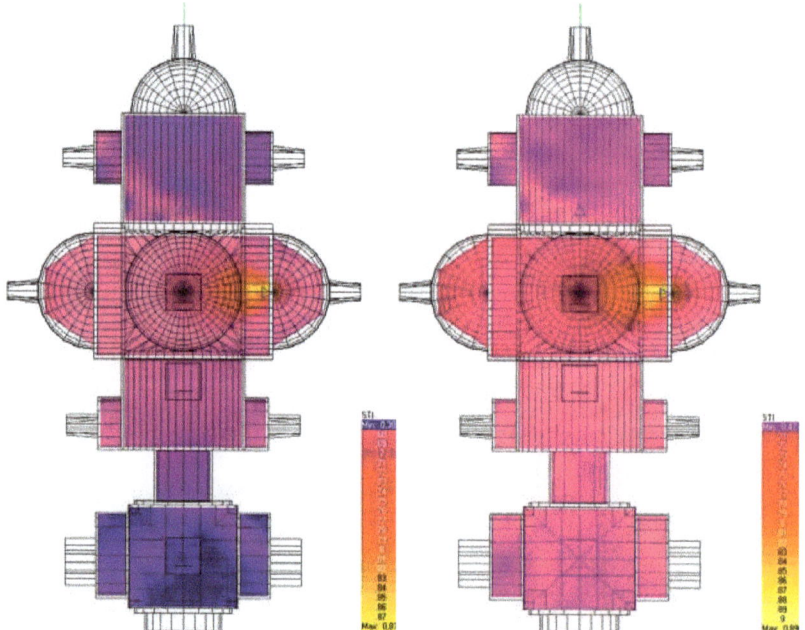

Figure 5. Mapping of speech transmission index (STI) parameter for the sound source position in the chanting apse, when the church is empty (on the left) and when the faithful are present (on the right).

Table 2. Values of speech transmission index parameter for three receiver points (M1, M2 and M3) and two sound source positions (A: Altar and B: Chanting apse) in Lazarica church.

	Speech Transmission Index (STI)					
Sound Source	A: Altar			B: Chanting Apse		
Measurement Points Case	M1	M2	M3	M1	M2	M3
Measured in situ	0.491	0.474	0.402	0.673	0.473	0.405
Empty church	0.670	0.548	0.545	0.646	0.533	0.417
Faithful present	0.709	0.627	0.657	0.685	0.583	0.554
Without the dome	0.650	0.544	0.523	0.639	0.517	0.403
Medieval iconostasis	0.651	0.557	0.537	0.663	0.547	0.418

The octave range values for T_{30} and EDT parameters measured in situ are graphically presented in Figure 6 (the exact values are shown in the Table A1, Appendix A). The similarities of T_{30} and EDT decaying curves could be observed for naos positions (M1 and M2).

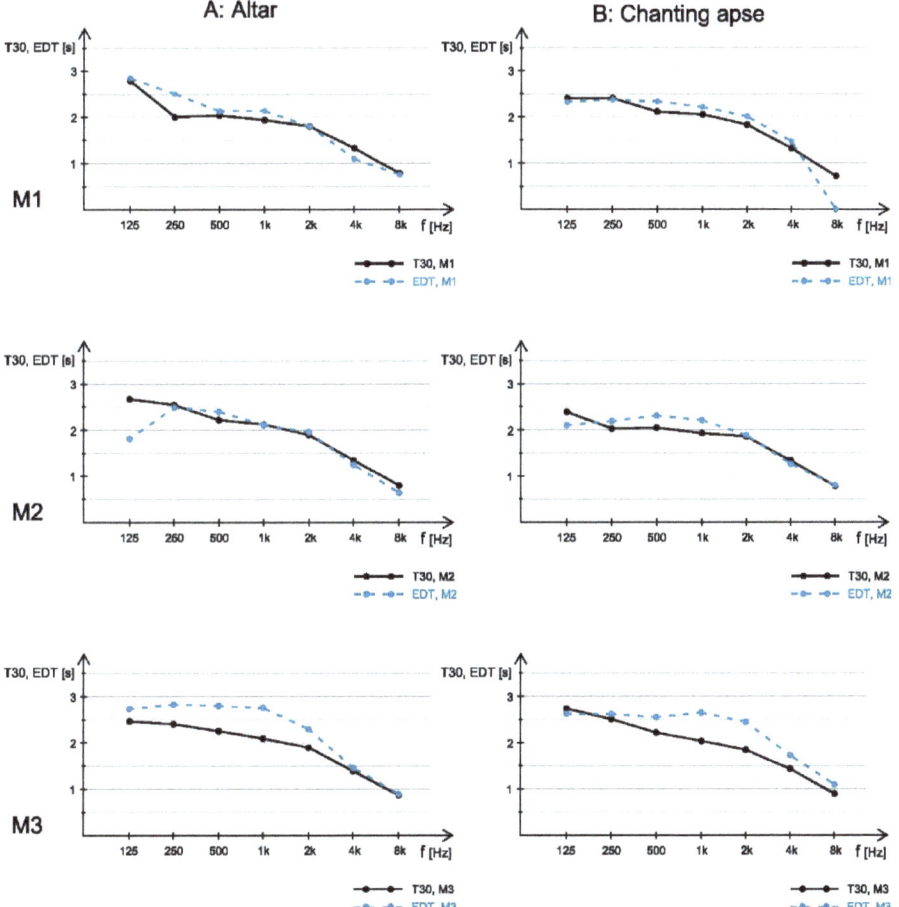

Figure 6. The octave range values for reverberation time (T_{30}) and early decay time (EDT) parameters measured in situ at M1, M2 and M3 measurement points for two sound excitation positions (altar and chanting apse).

The octave range values for T_{30} and EDT parameters for all three examined cases together with the case of the empty church obtained through the simulation process are shown in Appendix A, Tables A2–A4, respectively, for all three measurement points. As could be observed on Figure 6 there are no significant differences of T_{30} parameter values for the case without the dome, although something like this might be expected at the first glance. At the same time, the case of people present in the church derived an expected result of lowering T_{30} values (orange line on the graphs, Figure 7). The rate of this lowering is consistently around 25% at frequency of 500 Hz at all three measurement positions.

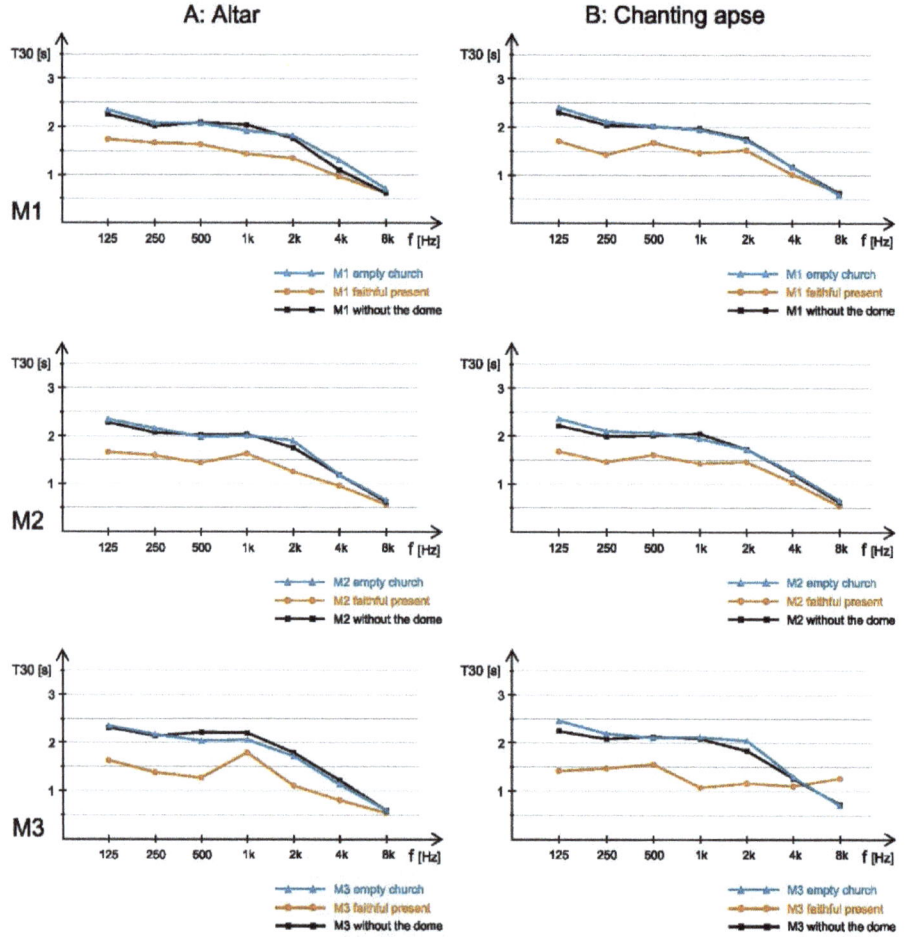

Figure 7. T_{30} octave values for three analyzed cases (case 1—blue, 2—orange, 3—black line) at M1, M2 and M3 measurement points for two sound excitation positions (altar and chanting apse).

In order to examine if the change of height of iconostasis affected Lazarica church acoustics, it is simulated in the acoustic model the lower iconostasis as it was in the time of church building (2.2 m high). Iconostasis was simulated as a single wooden plate made of 6 mm thick plywood. In the middle of the iconostasis is one arch door opening 90/230 cm. When two cases are compared—empty church with iconostasis as it is today, and empty church with iconostasis as it was in medieval times—no significant differences occur, as shown in Table 2 for the STI parameter and on Figure 8 for the T_{30} octave values.

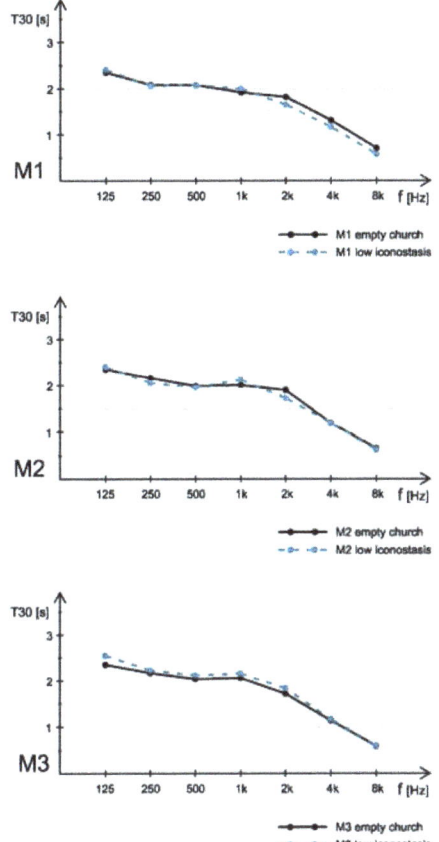

Figure 8. T_{30} octave values for cases of two different iconostasis height at M1, M2 and M3 measurement points.

4. Discussion

The basic parameter that describes church acoustics is reverberation time and its comparison with optimal reverberation time. There are several recommendations in a form of empirical relations for optimal reverberation time for two different purposes, speech and music, and only those derived for speech are of interest for the Orthodox tradition. The reason for this is that there was no instrumental music in Orthodox religious practice, and all sound excitations are exclusively made by the human voice, either as a speech or as a monophonic chanting, which is slow and monotonous, retaining in such way all the basic characteristics of speech. On the other hand, all of those recommendations were derived for churches built in the Western tradition, without any reference to the Orthodox medieval tradition. For example, Perez-Minana relations, which is especially adapted for churches, indicate that optimum reverberation time for churches of this volume (Lazarica volume 717 m^3) should be under 1s @500 Hz [25], and yet Lazarica church has reverberation time that is more than twice as long (Table A1, Appendix A). A reason for such a difference lays in different geometry and interior design materials.

Speech intelligibility is another acoustic property that is essential for churches. In situ measurements indicate that intelligibility under the dome (M1 position) highly depends of the position of the sound source. In the case of excitation from the altar, the STI parameter has a value of 0.491, and in the case of excitation from the chanting apse the same parameter has a value of 0.673.

Such a situation can be explained with specific relations between chanting apse and M1 position—a person standing under the dome is practically in front of the chanting apse, having the highest "direct to reflected" sound ratio, which directly impact speech intelligibility, while there is a physical barrier (iconostasis) between the sound source in the altar and the M1 position. At the other two positions (M2 and M3) such differences cannot be observed, which indicates that speech intelligibility does not depend on the position of the sound source.

Overall speech intelligibility inside the church can be rated as fair inside the naos (M1 and M2 position), and poor inside the narthex area (M3), as show in Table 3. As already stated, the only exception of such a description is the case when listener is standing right in front of the speaker. Although the Orthodox service requires the speech intelligibility, it also has the opposite demand to emphasize the spiritual experience of the chanting art. Therefore, besides the speech intelligibility, also the richness and specific nature of reflected sound is of a great importance.

Table 3. Intelligibility rating based on the STI value.

Intelligibility Rating	STI
Excellent	greater than 0.75
Good	0.60–0.75
Fair	0.45–0.60
Poor	0.30–0.45
Bad	less than 0.30

Measured values of the EDT and T_{30} parameters at M1 and M2 positions are very similar, as can be observed in Figure 6, which indicates at steady decay of energy inside the naos. On the other hand, in the narthex, EDT is consistently longer than the reverberation time for about 20% at the average. The naos and narthex are two different rooms linked with the door opening, which makes them acoustically coupled spaces. In the case of acoustically coupled spaces, the sound energy decay curve in the receiving room (room without the sound source - in this case narthex) can be expected to have a fairly slow early decay rate, followed by a reverberation tail which is determined by the space with longer reverberation time (in this case it is naos) [26] (pp. 445–448). Comparison of the in situ measurement results for the M2 and M3 positions showed that T_{30} is practically the same, but EDT is longer in the narthex area. One should have in mind that EDT correlates with reverberance, as audience perceive it during a music play or a speech. Also, both EDT and T_{30} parameters have different values for each measurement points, which indicates that the sound field inside the church cannot be considered as diffused and homogenous.

As for the sound field simulation conducted in EASE software, beside the case of empty church, three more cases were analysed: (1) church with people inside, (2) empty church without the dome, and (3) empty church with the lower iconostasis. The acoustic analysis of the empty church was important in order to calibrate acoustic response of the simulated model, and T_{30} parameter was used for this process. Although it was impossible to achieve perfect matching of this parameter gained through simulation with measurement results for all three positions, a satisfactory degree of T_{30} match has been achieved. At the same time, matching of the STI parameter was significantly lower, especially for the case of the excitation from the altar. It turns out that the iconostasis has much greater influence on decreasing speech intelligibility than was predicted by the simulation. One possible way to interpret such a case is that the influence of the iconostasis (as a wooden obstacle between sound source and the rest of the church) is not described precisely enough in the process of simulation. This kind of analysis is beyond the scope of this paper, but it could be of interest for some further investigation.

In the case of the presence of people in the church, the results are completely in line with the expectations. Namely, increased absorption decreased the overall reflected energy which results in consistent lowering of EDT and T_{30}. Influence of the people on church acoustics was simulated with increased absorption of the floor area, with absorption coefficients that describes human absorption.

At the same time, STI values increased, once again pointing to the fact that the presence of the people raises the level of speech intelligibility (Figure 5).

In order to investigate the effect of the dome on the overall acoustic response within the empty church, the initial model was modified by "closing" the dome and replacing it with a flat ceiling on the top of the arches that support the dome (marked with blue dashed line in the church section, Figure 2). It can be noticed that the STI parameter in the case of the model with flat ceiling has even lower values than in the initial case with the dome (although minimal, this difference exists). This implies that the dome does not have a bad influence on overall speech intelligibility, although such a thing may be expected at first glance because of the concave shape of the dome. Considering time parameters, it was expected that flat ceiling case has lower EDT and T_{30} values because of decreased overall church volume. However, although such a trend exists, it is not fully consistent. At some frequencies those parameters have greater values in the case of a flat ceiling, but their regularity is difficult to be predicted from the corresponding curve. What is possibly interesting is that differences between those two cases are the least pronounced for the M1 position, which is directly under the dome. Such result obtained by the simulation could indicate that dome has greater acoustic influence on the positions inside the church that are further from the dome. Although not definite, this conclusion can indicate some further directions for dedicated investigation of the dome influence on the overall church acoustics.

5. Concluding Remarks

In the last period of the Serbian medieval state, church building and chanting art reached their apex [16]. The goal of this paper was to examine the acoustics of Lazarica church, as a typical representative of a Morava architectural style, with a layout that combines a compressed inscribed cross and a triconch. Acoustic properties were investigated through the simulation process conducted in the EASE acoustic software. The acoustic model was calibrated upon the on-site measurement of acoustic impulse response, by defining the absorption coefficient of the wall material. The sound source positions were characteristic for the Holy Liturgy—the altar (behind the iconostasis) and the chanting apse (where is chanted and preached from). Three receiver positions were analysed: M1 under the dome, M2 on the longitudinal axe of naos, and M3 in the narthex. The results obtained by the measurements were achieved by an excitation from a directed loudspeaker, in order to simulate a "real life" situation (sound field excitation by the human voice), which should be kept in mind when it comes to repeatability and reproducibility of measurement results. Due to the Orthodox religious practice and the related organization of the church interior space, several cases were of interest for the acoustic analysis, such as the acoustic effect of (1) the central dome, (2) faithful presence, and (3) the medieval iconostasis (because it was half the height than the one standing today). Acoustic requirements in medieval Orthodox church were for the speech intelligibility, as well as the experience of monophonic chanting. Therefore, three acoustic parameters were tracked: reverberation time (T_{30}), early decay time (EDT) and speech transmission index (STI).

Regarding speech intelligibility, the acoustic analysis showed that it is increased with the presence of people, as expected, but lowered when the dome is replaced with a flat ceiling. It is highly dependable on the listener position inside the church, which might indicate on inhomogeneous sound field. The experience of monophonic chanting is correlated with T_{30} and EDT. The simulation showed that the presence of the faithful in the church causes the decrease of T_{30} for about 0.5 s, comparing to the case of the empty church (average rate of 25%). In all the examined cases, it is observed that there is no significant deviation between the T_{30} and EDT parameters, which indicates uniform sound energy decay. Closing of the dome with a flat ceiling did not show any significant impact on T_{30}. The simulation showed that the height of the iconostasis has no significant influence on the acoustics of Lazarica church.

Due to the lack of written records on medieval Serbian Orthodox church design and in what range the aural environment was considered, it is valuable to analyse their acoustics in order to better understand the complex relation between architecture, religious practice and acoustics. The examination of Lazarica church acoustics, presented in this paper, contributes to this goal.

Author Contributions: Conceptualization Z.Đ.; methodology Z.Đ. and D.N.; modelling and examination of the acoustic model U.A. and D.N.; formal analysis D.N.; writing Z.Đ. and D.N.

Funding: This research received no external funding.

Acknowledgments: The paper is a result of research on the project Theory and practice of science in society: multidisciplinary, educational and intergenerational perspectives (number 179048), financed by the Ministry of Education, Science and Technological Development of Republic of Serbia. The authors are grateful to Stefan Dimitrijević, who worked on the pilot acoustic model of Lazarica church that was presented on the conference TAKTONS 2015 in Novi Sad, Serbia.

Conflicts of Interest: The authors declare no conflict of interest.

Appendix A

Table A1. Lazarica church in situ acoustic measurements: Octave values of T30 and EDT for three measurement points (M1, M2 and M3) and two sound source positions (A: Altar and B: Chanting apse).

In situ	M1				M2				M3			
	A: Altar		B: Chanting apse		A: Altar		B: Chanting apse		A: Altar		B: Chanting apse	
Hz\parameter	EDT	T_{30}	EDT	T_{30}	EDT	T_{30}	EDT	T_{30}	EDT	T_{30}	EDT	T_{30}
125	2.50	2.78	2.32	2.40	1.82	2.67	2.10	2.39	2.74	2.47	2.63	2.74
250	2.84	2.01	2.37	2.40	2.49	2.55	2.19	2.03	2.83	2.41	2.62	2.51
500	2.13	2.04	2.33	2.11	2.40	2.22	2.31	2.05	2.80	2.26	2.55	2.22
1000	2.14	1.94	2.21	2.05	2.11	2.13	2.21	1.93	2.76	2.10	2.65	2.04
2000	1.80	1.80	2.01	1.83	1.96	1.90	1.89	1.86	2.30	1.90	2.45	1.85
4000	1.10	1.33	1.47	1.32	1.25	1.35	1.26	1.33	1.46	1.39	1.73	1.43
8000	0.76	0.79	0.01	0.72	0.64	0.80	0.80	0.78	0.89	0.87	1.09	0.89

Table A2. Lazarica church acoustic model examination: Octave values of T_{30} and EDT for receiver point under the dome (M1) and two sound source positions (A: Altar and B:Chanting apse) in three cases – when church is empty, when faithful are present, when the central dome is excluded, and when the iconostasis is 2.2 m high as in medieval times.

M1	Empty church				Faithful present				Without the dome				Low iconostasis			
	A: Altar		B: Chanting apse		A: Altar		B: Chanting apse		A: Altar		B: Chanting apse		A: Altar		B: Chanting apse	
Hz\param.	EDT	T_{30}	EDT	T_{30}	EDT	T_{30}	EDT	T_{30}	EDT	T_{30}	EDT	T_{30}	EDT	T_{30}	EDT	T_{30}
125	2.45	2.35	2.17	2.41	1.99	1.74	1.67	1.71	2.33	2.25	2.10	2.30	2.40	2.41	2.36	2.43
250	2.14	2.08	2.05	2.11	1.53	1.67	1.60	1.42	1.87	2.01	1.79	2.04	2.02	2.06	2.07	2.12
500	1.83	2.08	1.94	2.02	1.64	1.64	1.77	1.68	2.06	2.09	1.95	2.01	2.07	2.08	1.98	2.04
1000	1.94	1.92	1.80	1.95	1.28	1.44	1.47	1.46	1.87	2.04	1.79	1.97	1.62	2.00	2.06	1.97
2000	1.46	1.82	1.40	1.74	1.58	1.35	0.95	1.52	1.52	1.76	1.65	1.76	1.10	1.65	1.71	1.64
4000	1.21	1.31	0.74	1.17	0.64	0.97	0.77	1.02	1.22	1.10	0.93	1.17	0.95	1.17	0.70	1.15
8000	0.63	0.71	0.52	0.59	0.04	0.62	0.53	0.64	0.86	0.62	0.55	0.62	0.65	0.58	0.55	0.56

Table A3. Lazarica church acoustic model examination: Octave values of T_{30} and EDT for receiver point in naos (M2) and two sound source positions (A: Altar and B:Chanting apse) in three cases – when church is empty, when faithful are present, when the central dome is excluded, and when the iconostasis is 2.2 m high as in medieval times.

M2	Empty church				Faithful present				Without the dome				Low iconostasis			
	A: Altar		B: Chanting apse		A: Altar		B: Chanting apse		A: Altar		B: Chanting apse		A: Altar		B: Chanting apse	
Hz\param.	EDT	T_{30}	EDT	T_{30}	EDT	T_{30}	EDT	T_{30}	EDT	T_{30}	EDT	T_{30}	EDT	T_{30}	EDT	T_{30}
125	2.53	2.35	2.36	2.37	1.90	1.66	1.77	1.68	2.32	2.28	2.24	2.22	2.55	2.41	2.43	2.37
250	2.17	2.16	2.06	2.10	1.67	1.59	1.45	1.46	1.97	2.07	1.90	1.99	2.15	2.06	2.02	2.18
500	2.08	1.99	1.99	2.07	1.31	1.43	1.50	1.61	2.01	2.02	1.86	2.01	1.96	1.96	2.12	1.91
1000	2.20	2.01	2.04	1.95	1.53	1.63	1.43	1.43	1.88	2.04	1.83	2.05	1.69	2.13	1.74	2.02
2000	1.45	1.90	1.58	1.73	1.56	1.25	1.44	1.46	1.78	1.75	1.81	1.73	1.23	1.73	1.42	1.96
4000	1.06	1.19	0.84	1.24	1.02	0.96	0.60	1.04	1.12	1.19	1.26	1.21	0.93	1.19	1.27	1.17
8000	0.48	0.66	0.50	0.66	0.37	0.57	0.47	0.55	0.62	0.62	0.56	0.62	0.60	0.63	0.47	0.61

Table A4. Lazarica church acoustic model examination: Octave values of T_{30} and EDT for receiver point in the narthex (M3) and two sound source positions (A: Altar and B:Chanting apse) in three cases – when church is empty, when faithful are present, when the central dome is excluded, and when the iconostasis is 2.2 m high as in medieval times.

M3	Empty church				Faithful present				Without the dome				Low iconostasis			
	A: Altar		B: Chanting apse		A: Altar		B: Chanting apse		A: Altar		B: Chanting apse		A: Altar		B: Chanting apse	
Hz\param.	EDT	T_{30}	EDT	T_{30}	EDT	T_{30}	EDT	T_{30}	EDT	T_{30}	EDT	T_{30}	EDT	T_{30}	EDT	T_{30}
125	3.01	2.35	2.67	2.46	2.11	1.63	2.29	1.42	2.88	2.31	2.80	2.24	2.93	2.55	2.68	2.44
250	2.65	2.17	2.69	2.19	1.43	1.38	1.71	1.47	2.54	2.14	2.36	2.08	2.71	2.23	2.69	2.16
500	2.65	2.04	2.63	2.11	1.39	1.27	1.90	1.56	2.26	2.21	2.53	2.13	2.51	2.12	2.39	2.20
1000	2.36	2.06	2.82	2.12	1.39	1.79	1.78	1.08	2.12	2.20	2.60	2.09	2.43	2.16	2.45	2.12
2000	2.00	1.72	2.36	2.05	1.02	1.11	2.19	1.17	2.04	1.79	2.13	1.84	1.63	1.84	2.16	1.95
4000	1.22	1.14	1.72	1.30	1.22	0.81	1.21	1.11	1.22	1.22	1.67	1.27	0.83	1.17	1.45	1.25
8000	0.40	0.59	1.02	0.71	0.20	0.54	0.89	1.27	0.54	0.59	1.00	0.73	0.64	0.60	0.68	0.69

References

1. Crnčević, D. Architectural work, context, meaning: Approaches to the study of architecture of Moravian Serbia. *Hist. Rev.* **2008**, *LVII*, 93–106.
2. Đorđević, Z. Intangible tangibility: Acoustical heritage in architecture. *Struct. Integr. Life* **2016**, *16*, 59–66.
3. Suárez, R.; Alonso, A.; Sendra, J.J. Archaeoacoustics of intangible cultural heritage: The sound of the Maior Ecclesia of Cluny. *J. Cult. Herit.* **2016**, *19*, 567–572. [CrossRef]
4. Girón, S.; Álvarez-Morales, L.; Zamarreño, T. Church acoustics: A state-of-the-art review after several decades of research. *J. Sound Vib.* **2017**, *411*, 378–408. [CrossRef]
5. Alvarez-Morales, L.; Martellotta, F. A geometrical acoustic simulation of the effect of occupancy and source position in historical churches. *Appl. Acoust.* **2015**, *91*, 47–58. [CrossRef]
6. Zamarreño, T.; Girón, S.; Galindo, M. Assessing the intelligibility of speech and singing in Mudejar-Gothic churches. *Appl. Acoust.* **2008**, *69*, 242–254. [CrossRef]
7. Brezina, P. Measurement of intelligibility and clarity of the speech in romanesque churches. *J. Cult. Herit.* **2015**, *16*, 386–390. [CrossRef]
8. Alonso, A.; Suárez, R.; Sendra, J. The Acoustics of the Choir in Spanish Cathedrals. *Acoustics* **2018**, *1*, 35–46. [CrossRef]
9. Navarro, J.; Sendra, J.J.; Muñoz, S. The Western Latin church as a place for music and preaching: An acoustic assessment. *Appl. Acoust.* **2009**, *70*, 781–789. [CrossRef]
10. Martellotta, F.; Cirillo, E.; Della Crociata, S.; Gasparini, E.; Preziuso, D. Acoustical reconstruction of San Petronio Basilica in Bologna during the Baroque period: The effect of festive decorations. *J. Acoust. Soc. Am.* **2008**, *123*, 3607. [CrossRef]
11. Palazzo-Bertholon, B.; Valière, J.-C. *Archéologie du son: Les dispositifs de pots acoustiques dans les édifices anciens*; Bulletin Monumental, Suppl. 5; Picard: Paris, France, 2012; ISBN 978-2-901837-41-1.
12. Đorđević, Z.; Penezić, K.; Dimitrijević, S. Acoustic vessels as an expression of medieval music tradition in Serbian sacred architecture. *Muzikologija* **2017**, *1*, 105–132.
13. Elicio, L.; Martellotta, F. Acoustics as a cultural heritage: The case of Orthodox churches and of the "Russian church" in Bari. *J. Cult. Herit.* **2015**, *16*, 912–917. [CrossRef]
14. Antonopoulos, S.; Gerstel, S.E.J.; Kyriakakis, C.; Raptis, K.T.; Donahue, J. Soundscapes of Byzantium. *Speculum* **2017**, *92*, S321–S335. [CrossRef]
15. Gerstel, S.E.J.; Kyriakakis, C.; Raptis, K.T.; Antonopoulos, S.; Donahue, J. Soundscapes of Byzantium: The Acheiropoietos Basilica and the Cathedral of Hagia Sophia in Thessaloniki. *Hesperia J. Am. Sch. Class. Stud. Athens* **2018**, *87*, 177–213. [CrossRef]
16. Peno, V.; Obradović, M. On the chanting space and hymns that were sung in it. Searching for chanting-architectural connections in the middle ages. *Muzikologija* **2017**, *23*, 145–173. [CrossRef]
17. Mourjopoulos, J.; Papadakos, C.; Kamaris, G.; Chryssochoidis, G.; Kouroupetoglou, G. Optimal acoustic reverberation evaluation of Byzantine chanting in churches. In Proceedings of the International Computer Music Conference, Athens, Greece, 14–20 September 2014; pp. 14–20.
18. Ristić, V.; Lazarica, I.K.G. *Republički Zavod Za Zaštitu*; Spomenika Kulture Beograd: Beograd, Serbia, 1989.

19. Korać, V. *Između Vizantije i Zapada: Odabrane Studije o Arhitekturi*; Prosveta [u.a.]: Beograd, Serbia, 1987; ISBN 978-86-80371-01-6.
20. Stevović, I. Arhitektura Moravske Srbije: Lokalna graditeljska škola ili epilog vodećih tokova poznovizantijskog graditeljskog stvaranja. *Zb. Rad. Viz. Inst.* **2006**, *XLIII*, 231–253.
21. Korać, V.; Šuput, M. *Arhitektura Vizantijskog Sveta*; Zavod za Udžbenike: Beograd, Serbia, 2010; ISBN 978-86-17-16798-9.
22. Ristić, V. Restauracija Lazarice Pere, J. Popovića iz 1904–1908 godine. *Saopštenja* **1983**, *15*, 129–146.
23. Anđelković, B. Liturgija i unutrašnji poredak hrama—kanonski osnovi crkvenog graditeljstva. In *Tradicija i Savremeno Srpsko Crkveno Graditeljstvo*; Stojkov, B., Manević, Z., Eds.; Institut za arhitekturu i urbanizan Srbije: Beograd, Serbia, 1995; pp. 31–61.
24. Stefanović, D. *Old Serbian Music—Examples of 15th Century Chant*; Institute of Musicology SASA: Belgrade, Serbia, 1975.
25. Fernandez, M.; Recuero, M. Data Base Design for Acoustics: The Case of Churches. *Build. Acoust.* **2005**, *12*, 31–40. [CrossRef]
26. Kurtović, H. *Osnovi Tehničke Akustike*; Naučna knjiga: Beograd, Serbia, 1990; ISBN 978-86-23-21072-3.

 © 2019 by the authors. Licensee MDPI, Basel, Switzerland. This article is an open access article distributed under the terms and conditions of the Creative Commons Attribution (CC BY) license (http://creativecommons.org/licenses/by/4.0/).

Article

The Acoustic Environment of York Minster's Chapter House

Lidia Álvarez-Morales [1,*], Mariana Lopez [1] and Ángel Álvarez-Corbacho [2]

1. Department of Theatre, Film, Television and Interactive Media, University of York, York YO10 5GB, UK; mariana.lopez@york.ac.uk
2. Instituto Universitario de Arquitectura y Ciencias de la Construcción, Escuela Técnica Superior de Arquitectura, Universidad de Sevilla, 41012 Seville, Spain; arqangel@us.es
* Correspondence: lidia.alvarezmorales@york.ac.uk

Received: 13 December 2019; Accepted: 27 January 2020; Published: 30 January 2020

Abstract: York Minster is the largest medieval Gothic cathedral in Northern Europe, renowned for its magnificent architecture and its stained glass windows. Both acoustic measurements and simulation techniques have been used to analyse the acoustic environment of its Chapter House, which dates from the 13th-century and features an octagonal geometry with Gothic Decorated stone walls replete of geometric patterns and enormous stained glass windows, covered by a decorated wooden vault. Measured and simulated room impulse responses served to better understand how their architectural features work together to create its highly reverberant acoustic field. The authors start by analysing its acoustic characteristics in relation to its original purpose as a meeting place of the cathedral's Chapter, and end by reflecting on its modern use for a variety of cultural events, such as concerts and exhibitions. This work is part of the "Cathedral Acoustics" project, funded by the EC through the Marie-Sklodowska-Curie scheme.

Keywords: heritage acoustics; cathedral acoustics; room acoustics; York Minster; acoustic simulation

1. Introduction

Research on the acoustics of heritage sites allows us, firstly, to understand and preserve the acoustic behaviour of spaces threatened by the passage of time, or modified by human intervention, by capturing the acoustic information digitally in the form of room impulse responses (RIR). Secondly, research in the field furthers our understanding on how acoustic conditions have changed over time, opening up discussions on what such changes might have meant to our ancestors and their experience of such sites.

This paper focuses on assessing the acoustics of the Chapter House of the Cathedral and Metropolitical Church of St Peter in York, better known as the York Minster. York Minster is the largest medieval Gothic cathedral in Northern Europe, renowned for its outstanding architecture. Its Chapter House is an independent venue attached to the north transept of the cathedral, in which a great variety of events are held on a regular basis. The York Minster's Chapter House is considered a remarkable piece of architecture itself and its acoustics a major feature of its character.

This work provides new data about the acoustics of the York Minster's Chapter House, that can be used to preserve its acoustic field as well as be used as a starting point for further research on more specific aspects of the sonic environment. This research contributes to the existing literature on acoustics of cathedrals and heritage buildings, using this space as a representative example of polygonal-shaped English Chapter Houses to better understand how their proportions, verticality, rich decoration and hard finishing materials work together to create their reverberant acoustic field.

The experimental study of the acoustics of the Chapter House was conducted following the well-established methodology outlined in ISO 3382-part 1 [1], as well as additional guidelines developed for churches and cathedrals [2,3]. As in previous studies on cathedral acoustics [4–6], a thorough analysis of the acoustic behaviour of the space was conducted through the acoustic parameter values derived from the measured RIR, registered at a set of positions which reflected the wide diversity of uses of the site.

The sound reflection pattern in polygonal-shaped rooms is always a challenge because of the unwanted echoes, low frequency colouration, and non-diffusion conditions commonly found in those spaces [7]. The octagonal shape of the Chapter House, which has multiple parallel walls, is one of those shapes. Therefore, a careful inspection of the reflection pattern and the early arriving energy at each reception point has been conducted. The reflection pattern and the sound decay curves derived from the measured RIR has been analysed in previous studies of cathedrals, as their rich architectural features and complex forms sometimes give rise to particular acoustic phenomena such as coupled spaces or singular patterns of early reflections [8,9].

In addition to capturing RIR on site, the study of the Chapter House utilised computer modelling to further our understanding of the space and its use. Three-dimensional reconstruction and acoustic simulation techniques present a powerful tool to recreate the acoustic field within a space, helping us to deepen the analysis of its acoustic behaviour, while also offering the opportunity to acoustically experience sites that no longer exist through the process of auralisation [10,11]. Simulation software tools based on geometrical acoustics (GA) algorithms have been also successfully used in similar buildings before for a great variety of purposes, for instance to explore different intervention options and adapt their acoustics to a specific use [12] and to assess how their acoustics change in an occupied state [13]. In this particular case, working with the model of the Chapter House allowed the research team to analyse in detail the influence of the polygonal-shape of the room on the early reflection pattern. Such analysis was done by considering the different configurations of the space depending on its uses. Furthermore, simulations were used to give an insight on the acoustics of the Chapter House in its original state, before any restoration work or intervention was made.

Carrying the on-site acoustic measurements prior to the simulation work is crucial to the validation of the acoustic model created, first, to represent the current state of the building, and then, modified as required. However, it is important to acknowledge that the main challenges in research on acoustic simulations of heritage sites are the uncertainties on the acoustic properties of historical materials, the modelling of certain sound sources and the inherent limitations of the geometrical acoustic theory itself in which the simulation software tools are based on [14].

2. The York Minster's Chapter House

2.1. Overview of its Construction Process and Architectural Description

York Minster's history dates back to the year 627, when a wooden church dedicated to St Peter was built [15]. The construction of the medieval cathedral first began around 1225, with the construction of the transept on the foundations of the Norman cathedral, which was demolished in stages as Gothic additions were made [16]. Its constructive evolution continued throughout more than 250 years, until the north and south towers were built. The Minster was consecrated in 1472, a few years before its completion [17]. The cathedral of York is considered one of the finest medieval buildings in Europe, representing almost every stage of the Gothic style of architecture from 1230 to 1475 and having the highest proportion of medieval stained glass of any European cathedral [18]. Not surprisingly, the York Minster is one of the most-visited buildings in Britain [19].

The Chapter House has a polygonal shape, which is a peculiarity of English thirteenth-century architecture, not found in any other place in Europe (with few exceptions such as the Chapter House in Elgin cathedral (octagonal shape, 1270) in Scotland). There is evidence for 25 polygonal Chapter Houses in England, but more than half of them no longer exist [20]. There is no certainty as to the reasons behind the popularity of the polygonal form, although previous research has linked the choice of shape to acoustics [21,22], as a way to avoid the "intolerable resonance" of the rectangular chapter houses [23]. Nevertheless, Wickham [23] also stated, "the resonance of some of the polygonal houses, both of those with central pillars and of those without them, is also intolerable", and suggested that the polygonal shape emerged to serve its original purpose as a meeting place where all canons were considered equals, intended to avoid the dean sitting at the head as in the traditional rectangular Chapter Houses. Nevertheless, it seems more plausible that their shape, decoration and proportions derived from the tendency in English architecture around the year 1300 of "compiling designs for full-scale buildings or parts of buildings out of enlargements of the microarchitecture from the portals of thirteenth-century French cathedrals" [24], where the wall-framework structure, the geometric uniformity and the use of large windows had a profoundly visual value, being a prestigious architectural statement by the cathedral's canons.

Regardless of the reasons behind the octagonal shape of the York Chapter House, its architecture is remarkable, having no central pillar supporting the vaulting ceiling, which is instead supported by a spire-like timber roof structure. It is considered, along with the octagon in Ely Cathedral, one of the masterpieces of English medieval carpentry [25]. Although there is a lack of documentary proof related to the construction of the Chapter House [26], some evidence suggest that its construction process was completed around the last two decades of the 13th century [15]. Work started right after the current transepts were erected, and it was completed before the current nave was built. The space is built in Early Decorated Gothic style, so-called geometric style (1250–90) in English architecture, being designed to promote verticality and with the clear purpose of creating visual impact. The York Minster's Chapter House took the Westminster Chapter House as a source of inspiration, being the one in Westminster designed not only for the Chapter meetings but also to be used by Henry III [27], fact that also may explain its exuberance.

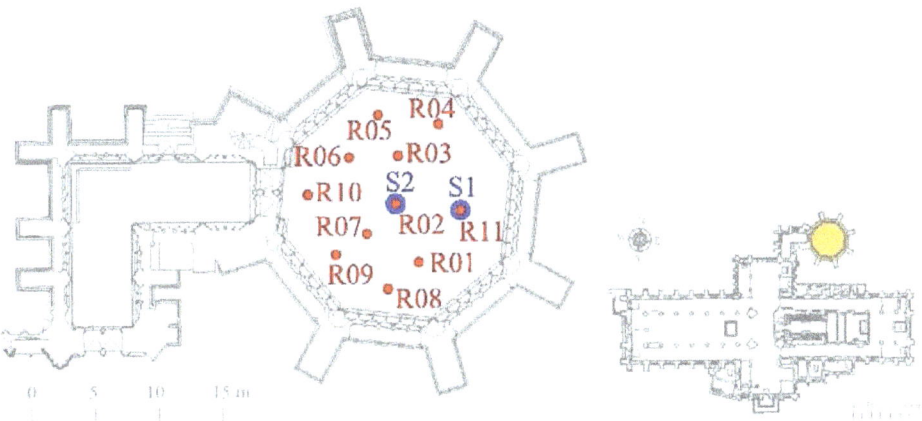

Figure 1. Floor plan of the York Minster's Chapter House with the source (S) and receiver (R) positions set for the acoustic measurements. The complete floor plan of the Minster is included for showing its location.

As can be seen in the floor plan included in Figure 1, the ground layout of the Chapter House is a quasi-regular octagon. It is annexed to the cathedral in its north transept through an L-shaped vestibule, also built in the Early Decorated style. The building is perfect in proportions, being about as high as its axis is long. For more details about its dimensions see Table 1.

Table 1. Geometrical data of the York Minster's Chapter House.

Approximate interior Volume	Max interior Height	Max interior Length	Octagon side's Length	Approximate interior floor Surface
5,280 m^3	20.3 m	19.2 m [a]	6.95 m	240 m^2

[a] Measured glass to glass.

The Chapter House is accessed by the west-side of the octagon (see Figure 1), through a rounded archway divided by a pillar with a statue of The Virgin and Child [28]. The oak doors at each side of the entrance date from the 13th century, and their original ironwork rich decoration is still visible [29].

The current polychrome Minton tiled floor (see Figure 2) is a result of the restoration of 1843-1845. There is no feasible trace of medieval floor in the Chapter House, which is thought to have been made of stone [30], although it is also possible that the decorative pattern tiles that still survive at the Westminster Chapter House once existed at York [16]. Eighteenth century views show an evenly paved floor, consisting of rectangular tiles with no visible decorative pattern [31].

Figure 2. York Minster's Chapter House interior views: looking to the West side (**left**) and looking to the East side (**right**).

Seven of the Chapter House's sides have a uniform design (the eighth side is the entrance). The lower part of the room is organised as a peripteral colonnade, in which a constant pattern of black Purbeck marble columns separates the canon stalls, 6 in each side (apparently, the six stalls on the East side were designated for the Dean and dignitaries [30]). The stalls are covered by limestone canopies which are highly decorated with a variety of forms and carves, from humorous faces to animals, which were originally coloured [30]. The upper part of the canopies forms a passage surrounding the room in front of the windows, with the exception of the doorway. Magnificent stained glasses (framed by limestone clusters of shafts or nerves) dominate the upper space. On the access side, the limestone tabernacle over the door contained coloured figures (possibly of Christ and the 12 Apostles) lost by the 17th century [32], and the blind limestone window above had a series of polychromy medieval wall paintings, removed during restoration in 1840 [33].

In the exterior, the structure of the Chapter House is reinforced by buttresses on the octagon corners. Inwardly, these buttresses become pillars on which the structure of the vault rests. The current ribbed vault, divided into sixteen segments, is a fine plaster and lath work dated from 1797, when the 14th century wood ceiling panels were replaced due to deterioration [31]. The original panels (some now stored in the Minster's crypt) were decorated with paintings of kings and bishops and the bosses were covered with silver [34], filling the spaces outside the hexadecagon. The entire ceiling was again restored in 1976 when the current decoration was applied.

2.2. A singular space for a variety of uses

As any other Chapter House, this space was constructed to provide a large and decorous place to host the regular meetings of the Chapter for the daily business of the cathedral. Forty-four seats were placed around the space as described in the previous section, a number greater than the canons in York at the time of the construction [16]. For most Chapter meetings at that time, only a minority of the stalls were occupied [26], for example, in 1310 only three canons met in the Chapter House to arrange a date for the election of a new dean [35]. However, it has a long history of being used for events other than Chapter's meetings, for example, in 1297, when the government of England was York-based, the Parliament of Edward I, king of England, met in the Chapter House in October [36]. The space is still used for the meetings of the Corporate body of the York Minster. The College of Canons (32 canons), use the space regularly. The current Chapter is made up by 9 members, but they do not always meet there. According to Harrison [29], "at the beginning of the 20th century some of the meetings were moved to the Zouche chapel, due to the difficulties to heat the space and because of the echo which every sound made in it produces". Nowadays, the configuration of the space changes depending on the meeting [37]. Such changes are not only in terms of the location of participants, but also, depending on the occasion, temporary pieces of furniture are used, mainly wooden tables and wooden or leather Victorian-style chairs [20] and the floor is sometimes covered with removable carpets. J. Archuleta, verger at the York Minster, shared some insight on the matter [38],

> "sometimes the stone seats around the space are used, perhaps one speaker at a time may stand towards the centre, and other times, there is a table placed in the centre of the Chapter House and modern chairs are used".

Contemporary uses of the Chapter House go beyond meetings, and include concerts, talks, tours and other social events held on a regular basis. The venue can also be hired to host dinners and drink receptions. L. Power, head of events and learning at the York Minster, set the capacity of the room at about 200 people, and when concerts are taking place it is a regular occurrence for the space to be fully booked [39], which mean approximately 130 spectators. Temporary stages, choir stalls, and audience seats (normally light weight upholstery portable chairs) are used as needed, and ephemeral solutions are sometimes included to meet the acoustic requirements of each use. For example, there is no permanent electroacoustic system installed in the Chapter House, but a portable system is used when needed, according with both Archuleta and Power [38,39].

Figure 3 shows the usual arrangement patterns between performers/sound sources and listeners used in the Chapter House. Figure 3a depicts the "front stage" arrangement; Figure 3b represents what is called "the amphitheatre-like arrangement" [40], and Figure 3c shows the original arrangement, in which the stone stalls are used both for performers and listeners. It is important to bear in mind that these arrangements are very flexible, and the configuration of the space is adapted to the requirements of each event. For instance, the audience area in "a" and "b" could expand to the stalls; in "c" the performers could be moved to any of the others sets of stalls depending on the location occupied by the person talking in each moment during the meeting. Configurations "a" and, occasionally, "b" are typically used for concerts and music events, and configurations "a", "b" and "c" are used for meetings and talks. Figure 3 does not include the configurations used in events like tours or dinners.

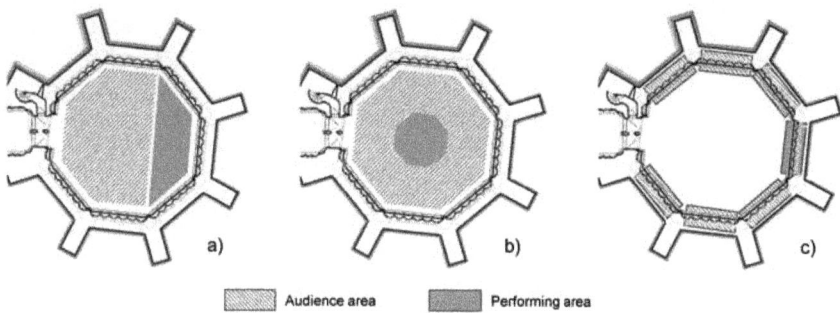

Figure 3. Examples of some typical configurations of the Chapter House: (**a**) front stage arrangement; (**b**) amphitheatre-like arrangement; and c) original surrounding arrangement.

3. Methodology of Analysis

The acoustical study of the Chapter House of the York Minster was based on acoustic measurements and simulation techniques. The room impulse responses (RIR) registered on site serve to assess the objectives acoustic characteristics of the space. Measured RIR are also used as a reference to adjust the acoustic model in order to ensure that the computational model is reliable in its representation of the soundfield, enabling conclusions to be drawn from the simulations as well as making it possible for simulated RIR to be used for auralisations that plausibly represent the acoustic environment of the space under different conditions. Simulation techniques are also utilised to assess the acoustic impact of restoration works, as well as different configurations related to the variety of uses of the Chapter House throughout history.

3.1. Acoustic Measurements in the Chapter House

The complete RIR measurement session involved the entire cathedral, including seven sound source positions selected by considering the architectural characteristics and the variety of uses of the space. Moreover, 65 receiver positions were located throughout the different audience areas. The number of source-receiver combinations characterised was selected in order to have representative data of each of the "subspaces" in the cathedral. This article focuses on the measurements and simulations linked to the Chapter House, and no acoustic information is given about the other subspaces, which will be analysed in future articles.

The acoustic measurements were carried out under unoccupied conditions taking ISO3382-1 [1] as a reference, as well as considering the specific guidelines developed for similar buildings [2,3]. The signal utilised was a 24-second-long sine-sweep emitted through a dodecahedral loudspeaker with a frequency response from 50 Hz to 10 kHz (NTi DS3 loudspeaker together with a NTi PA3 power amplifier) at a height of 1.5 m. The level and the duration of the excitation signal were set to ensure a reliable calculation of T_{30}, which means that an impulse-to-noise ratio (INR) higher than 45 dB in all octave bands would be achieved, even in those receivers located at the furthest positions from the source (see Figure 4).

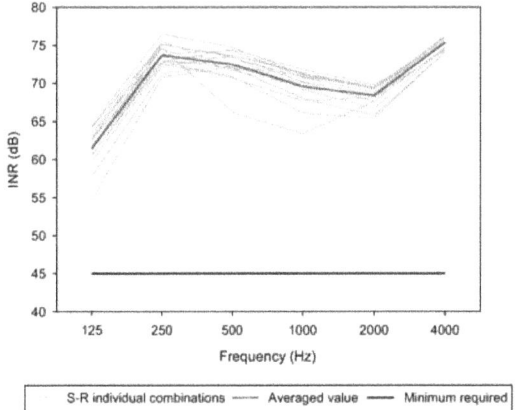

Figure 4. Impulse-to-noise ratio (INR) measured values obtained at each source-receiver combination characterised in the York Minster's Chapter House (grey lines) and the spatially averaged values (green line). The minimum value of INR required for a reliable calculation of T_{30} is also shown (red line).

The loudspeaker was placed in two different source positions inside the Chapter House: one on the symmetry axes of the space, in the East side, which is considered a common position for performers during cultural activities (Front stage configuration, Figure 3a) (S1); and another one at the centre of the octagon (Amphitheatre-like configuration, Figure 3b) (S2). Up to 11 receiver positions were distributed throughout the space from where a set of RIR were captured by using a B-format microphone (Soundfield ST450) and a binaural dummy head (Neumann KU 100). Table 2 summarises the source and receiver combinations included in the survey. The exact positions of S1 and S2 and the receiver points can be seen in Figure 1.

The digital audio workstation Pro Tools 12 and MATLAB software were used to capture and process the signals respectively. Then two commercial software tools, WinMLS2004 and IRIS 1.4, were employed for the acoustic analysis of the RIR and the calculation of the acoustic parameters.

Environmental conditions were monitored during the session, registering an average temperature of 18 °C and 60% of relative humidity.

Table 2. Summary of the source-receiver combinations included in the survey.

Source position ID	S1	S2
Num. Rec. points	10	8
Receiver points ID	R01 – R10	R01, R04 – R06, R08 – R11

3.2. Acoustic Simulation of the Space

The objective evaluation and validation of the acoustic simulation was based on the comparison of the room acoustic parameters derived from the simulation and those calculated from the RIR registered in the space. The acoustic simulation was carried out with Odeon 14.0, by using the precision algorithm. The IR length and the number of late rays were manually set to 8 ms and 150,000 rays respectively, while the early reflection settings were automatically set (8,976 of early rays with a transition order of 2). The geometrical model was imported into the acoustic software from the SketchUp modelling tool, which was used to build the geometrical model.

3.2.1. Acoustic Model

The 3D model was built from architectural plans and laser measurements taken on site, by using the modelling software tool SketchUp. The simplified model was created following the guidelines for geometrical acoustics (GA) algorithms, avoiding overly detailed surfaces and too complex forms [41]. The first significant simplification is the modelling of the highly decorated limestone canopies as a smooth surface and the simplification of the upper passage. To compensate for the limestone area lost with this simplification, the acoustic absorption coefficient assigned to this layer in the acoustic model was slightly increased. This is a common practice when modelling complex buildings, in which rich decorative patterns can make hard surfaces acoustically more absorbent [42]. The scattering coefficient was estimated taking into consideration the surface's depth variations [43].

The second simplification, is the definition of the limestone nerves that frame the stained glasses. In this regard, three different possibilities were tested: no nerves, 2D nerves and 3D nerves. Omitting the nerves implied an increment of stained glasses surface of about 120 m² at the expense of the stone surface, so, considering the differences in the acoustic properties between both materials, this option was discarded. Acoustic models including the 2D nerves and 3D nerves were tested in the simulation software. Modelling the nerves in 3D resulted in an increment of 200 m² of limestone distributed in 1,887 planes added to the model including the 2D nerves. After analysing the percentage of surface included in each configuration and considering the costs of the calculation process of each model, the option of including the nerves as 2D surfaces was chosen, and again, the absorption coefficient of the material assigned to the nerves was slightly increased to compensate for the surface lost and the scattering coefficient was estimated accordingly. Figure 5 includes a detail of one of the sides of the octagon were the simplifications described above are shown, including the limestone canopies and the modelling of the limestone nerves considered in the different options.

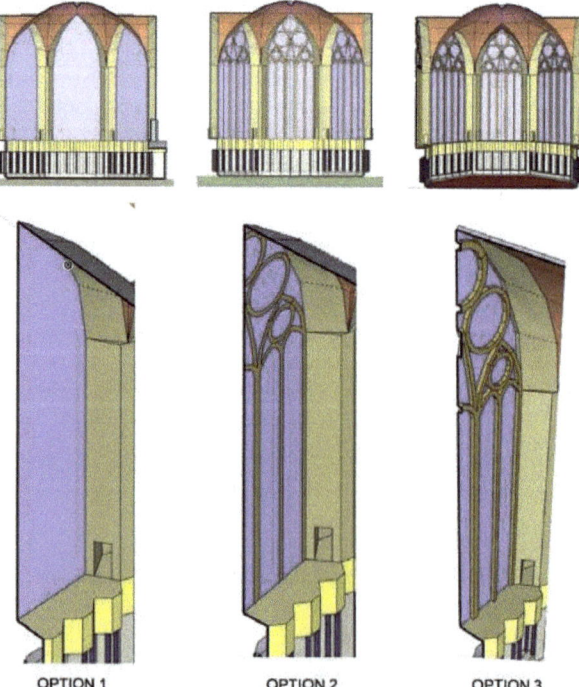

Figure 5. Acoustic model of the York Minster's Chapter House. Detail of the simplifications including the limestone canopies and the three options tested for the limestone clusters of shafts or nerves framing the stained glasses.

The final model built for the acoustic simulations, corresponding to option 2 in Figure 5, has 1,927 planes and a total surface area of 2,024.55 m².

The characteristics of the finishing materials of the indoor surfaces of the room were determined by visual inspection and by consulting bibliography [16,29–34]. The Chapter House includes ceramic tiles for the floor, Magnesian limestone for walls, marble for the columns of the stalls, lath (strips of wood) and plaster for the vaulted ceiling, and last but not least, glass and lead for the stained glasses. The initial frequency-dependent values of the absorption coefficients were assigned based on bibliography and publicly available databases [41,44,45]. The scattering properties were set to account for textures at mid frequencies, in the way that an estimation is made according to the depth of the decoration pattern of that surface as described in the Odeon manual [46].

The simulation model was adjusted through an iterative process in which on-site measurements and computer modelling results were compared [3,47]. The differences found between them are assessed in terms of the just noticeable difference (JND) values for various acoustic parameters [1,48]. The tuning process is based on the initial adjustment of the reverberation time by adapting the acoustic properties of the surfaces which present more uncertainties and which also have a significant surface area in the model, so that a relatively small change has a considerable effect. The absorption coefficients initially assigned to those surfaces are carefully changed until the simulated values of the reverberation time at all frequency bands differ by no more than 5% (1 JND) from those measured on site. Once this primary requirement is achieved, a point-by-point comparison for other parameter values (EDT, T_5, C_{80}, D_{50}) is performed considering the complete set of source-receiver (S-R) combinations. In those S-R combinations where major differences are found, a more detailed inspection of the RIR is done, analysing if there are any critical reflections added or missing in the simulation model. Having B-format or spatial RIR is crucial for this stage as it is possible to see where those reflections are coming from and implement local adjustments in the model if necessary. The model is considered adjusted once the differences between measured and simulated values are acceptable for a plausible representation of the real sound field, meaning that the measured and simulated RIR are comparable in shape, and that the measured and simulated values of the acoustic parameters differ by no more than 2 JND (no more than 1 JND in the case of T_{30}) in all the frequency bands [49]. A margin of 1 JND is desirable for all the acoustic parameters, but a margin of 2 JND is accepted taking into consideration all the approximations and limitations inherent to the measurement and simulation procedures. Furthermore, it is important to bear in mind that the calculation of the measured values through the use of different software tools may imply a variation up to 1 JND mainly due to the different algorithms they use to deal with the background noise and the arrival time detection [50].

Two different options have been tested for the adjustment of the acoustic model of the Chapter House. The first option considered the vault as the adjustment surface, since it is the finishing material that presents greater uncertainty in its acoustic behaviour. A relatively wide range of values have been published for plaster and lath works and other comparable materials [41,44,45]. The final value of the absorption coefficient of the vault resulting from the iterative adjustment procedure increased from 0.16 up to 0.24 in the 125 Hz to 4 kHz frequency bands, which is greater than the range estimated by the different authors at mid-high frequencies (0.05 to 0.10). However, this value is in perfect harmony with the results derived from previous measurement campaigns developed in similar buildings [51]. Furthermore, considering that historical plasters have been shown to be more absorptive than modern ones [52] and that the vault has a large air cavity behind, the coefficients obtained seem more than reasonable.

The second option tested was to modify the acoustic properties of the limestone to achieve the adjustment of the model. The iterative process described above was applied, showing that the values of the absorption coefficient initially assigned to the limestone at medium-high frequencies needed to be increased by 0.02 to match the measurements. On hard surfaces such as the limestone, such increment implies a relative change of almost 50%, which a priori seems large considering the polished finishing of the limestone surfaces found in the Chapter House, remarkable for the time of construction. Although the absorption coefficients of the limestone resulting from the iterative process were high for a hard surface [41], similar values have been estimated for comparable materials in historical buildings before [12], and therefore, this option was not discarded. Furthermore, in order to achieve a good match in high frequency, an additional increase of the absorption coefficient assigned to the stained glasses and the vault in the 2 kHz frequency band was required.

Table 3. Acoustic properties of the materials used in the simulation model of the York Minster's Chapter House, and the area which each of them occupies in the model.

	Absorption Coefficients						Scatt. Coef	Area (m²)
Frequency (Hz)	125	250	500	1 k	2 k	4 k	707	-
Limestone walls and seats	0.02	0.02	0.03	0.04	0.05	0.05	0.05	686.8
Stained glasses	0.25	0.20	0.14	0.10	0.05	0.05	0.15	384.0
Lath and plaster vault[1]	0.16	0.16	0.16	0.17	0.24	0.24	0.35	321.0
Tiled Floor	0.01	0.01	0.01	0.02	0.02	0.02	0.05	240.0
Limestone nerves[2]	0.03	0.04	0.05	0.05	0.065	0.065	0.15	131.0
Limestone canopies[2]	0.03	0.04	0.05	0.05	0.065	0.065	0.45	139.4
Marble columns	0.01	0.01	0.01	0.02	0.02	0.02	0.35	113.6
Wooden doors	0.14	0.10	0.06	0.08	0.10	0.10	0.20	8.8

[1] Derived from tuning process; [2] initial values raised up to compensate for the simplifications of the model.

Both options were based on reasonable assumptions, and therefore, tested in the simulation software.

After testing both options, it was decided that the first option would be applied, as it resulted in a better reproduction of the early reflection pattern in terms of their intensity which led to a better match between measured and simulated acoustic parameter values. Table 3 summarises the acoustic properties assigned to the finishing surfaces of the simulation model of the York Minster's Chapter House, and the area which each of them occupies in the model.

The final model is considered adjusted with the spatial averaged values of the acoustic parameters differing by no more than 1 JND except in the 250 Hz frequency band, where the simulated values are significantly overestimated, and in the case of EDT and T_s in the 4 kHz band, where the JND is much smaller in absolute values. Regarding the comparison point-by-point, more than 86% of the values match the criteria stated above (difference between measured and simulated values below 2 JND) including all the frequency bands except 250 Hz.

Figure 6 depicts the comparison of the spatially and spectrally averaged values (indicated by subscript "m") obtained in each source–receiver combination. Following the indications of the ISO 3382-1 [1] T_{30m}, EDT_m, T_{Sm}, D_{50m} and C_{80m} were calculated by averaging values obtained at 500 Hz and 1 kHz frequency bands and J_{LFm} was calculated by averaging values obtained from 125 Hz to 1 kHz frequency bands. The values corresponding to 1 and 2 JND have also been included in the figure to allow for the study of the margin of variation between the measured and simulated values of the acoustic parameters included in the comparison.

Additionally, when comparing results, it is important to remember that the acoustic simulation is built under the premises of the geometrical acoustic theory, and therefore, can be considered reliable for frequencies higher than four times the Schroeder's frequency [41], which, in this case, is 64 Hz, meaning that simulated results in the 125 Hz and even in 250 Hz octave bands have to be been taken with caution.

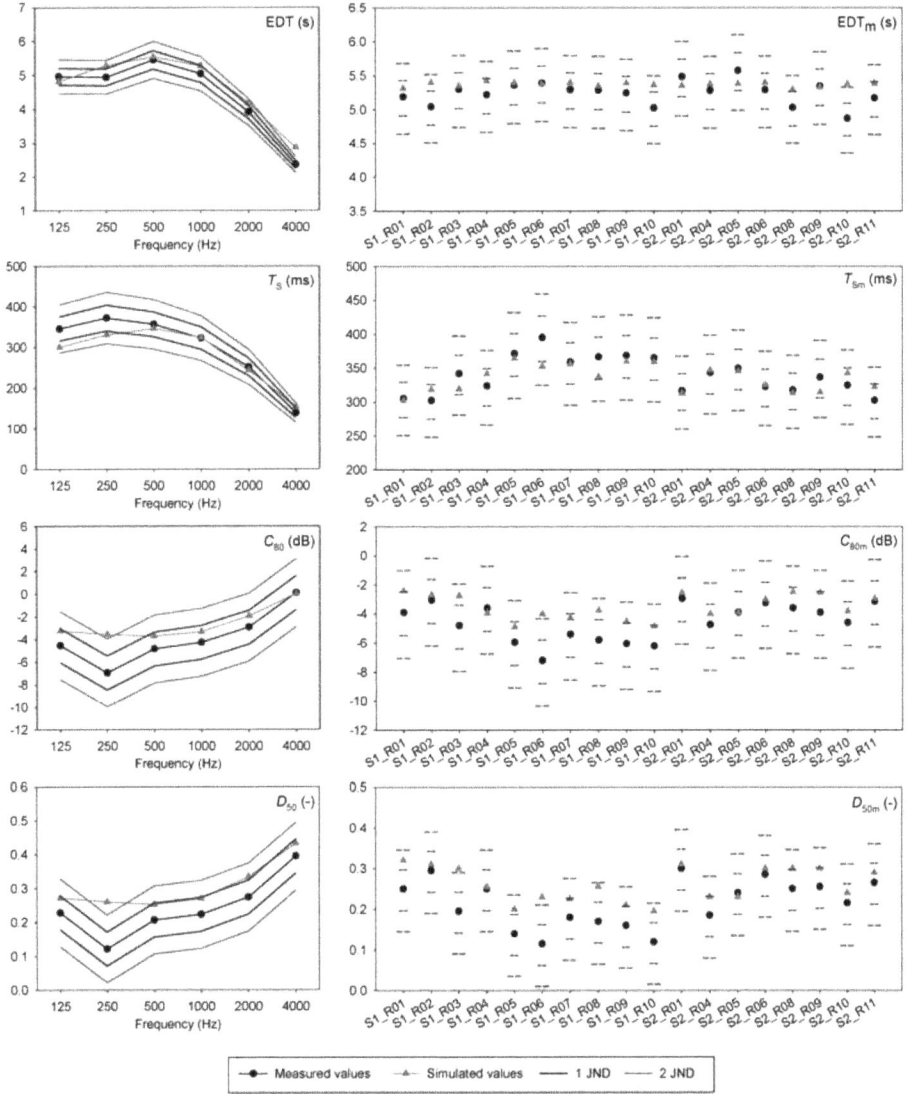

Figure 6. Measured and simulated values of several acoustic parameters (from top to bottom: EDT, T_S, C_{80}, and D_{50}): Spectral behaviour of the spatially averaged values (left column) and spectrally averaged values obtained at each source-receiver combination (right column). The values corresponding to 1 and 2 JND have been also included.

3.2.2. Acoustic Simulation Details

After the calibration process had been completed, the acoustic model was used to conduct a full analysis of the current acoustics of the Chapter House focusing on the early reflection pattern and on the behaviour of the space in the perimeter area of the room to evaluate the influence of its architectural features on sound behaviour. For this purpose, the initial range of source and receiver positions set during the measurement session was expanded, mainly to assess the acoustic behaviour of the space in the area of the limestone stalls, used in the original configuration of the Chapter meetings, shown in Figure 3c. A new source position (S3) was placed in the seat reserved for the Dean (one of the central stalls in the East wall) to represent the position of a speaker and six additional receiver positions (R12-R17) were set in various stone stalls (one in each of the sides of the octagon, excluding both the East side on which the sources are located and the West side of the entrance door) as potential listener positions (see Figure 7).

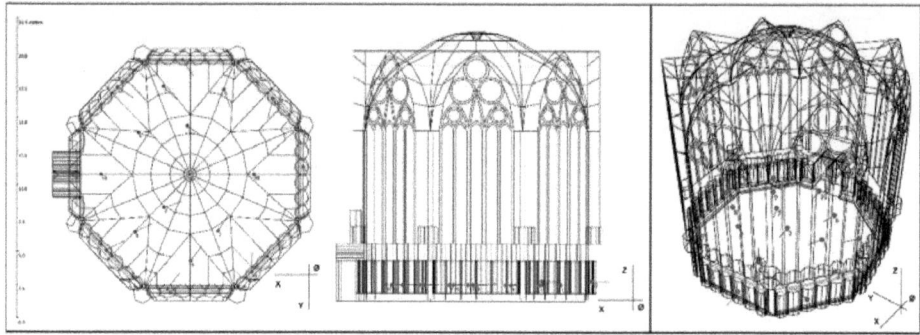

Figure 7. York Minster's Chapter House simulation model views.

The model was then modified to investigate the acoustics of the medieval Chapter House in its original state. The aim of this simulation is to offer an insight into the probable acoustical impact that the restoration works may have had on the space. It is always a challenge to try to plausibly reproduce the acoustic environment of a historic building at an earlier time, since there are usually gaps in the information needed to create the acoustic model in terms of the finishing materials used and their acoustic behaviour. In this particular case, for example, there is no trace of the original floor of the Chapter House, and it is not certain how the original ceiling and the painted walls interacted with sound, and therefore, some assumptions have to be made.

The absorption coefficient assigned to those limestone surfaces that were potentially painted (see section 2) has been reduced (0.01 at low frequency bands and 0.02 in mid-high frequency bands) taking into consideration how the paint finishing affects the acoustic characteristics of similar materials according to the literature [44].

For the 13th century wooden vault, the absorption coefficient was estimated taking as reference the published data on the acoustic behaviour of similar finishes and considering the resulting absorption coefficient after the adjustment procedure in the current construction. Previous acoustic studies on similar buildings revealed an averaged absorption coefficient for wooden vaulted ceilings close to 0.40 [53]. Such absorption coefficients were estimated from previous measurements campaigns in Italian churches [51]. Based on these results and considering the architectural characteristics of the original vault and its decoration at that time, an absorption coefficient of 0.30 at low-mid frequencies and 0.34 at high frequencies has been used in the simulations, which means an increment of approximately 10% with respect to the absorption coefficients assigned to the current vault.

Finally, and even though the current tiled floor dates from the 19th century, no change in the model has been introduced for this surface, since it is known that it was a similar hard surface although presumably with a slightly more irregular and rough finishing, due to the time of construction. Therefore, it is likely that this change did not have a major impact on the acoustics, since its absorption and scattering coefficients are unlikely to have changed significantly.

4. Results and Discussion

The results below are derived from the measured and the simulated B-format RIR, through the observation of the spectral and spatial behaviour of a series of acoustic parameters and the inspection of the reflection patterns found at each reception point. Results has been analysed and discussed in terms of perceived reverberation, clarity of sound and spatial impression, considering the different purposes this space has served through history.

4.1. Acoustics of the York Minster's Chapter House Today

The most relevant descriptor of the acoustic environment of a room is its reverberation time (T_{30}). The analysis of the decay curves of the measured RIR shows an average value of T_{30m} of 5.4 seconds. As can be seen in Table 4, similar values are obtained in the lower frequency bands, in which the greatest values of standard deviation are observed, and T_{30} values significantly decrease at the higher frequency bands, where usually the sound absorption of materials is greater and the air absorption is more pronounced. The Chapter House has a large volume considering its total surface area and it lacks acoustically absorbent finishing materials, which leads to this high level of reverberation more typically found in churches or cathedrals with considerably greater volumes [54]. Its T_{30} is extremely high for a "meeting room", since, typically, the recommended reverberation time to achieve adequate speech intelligibility in a conference or a meeting room of a comparable volume is around 1.1 seconds [55]. Its reverberation time is also above the values considered suitable for music reproduction (of around 2 s depending on the type of music), being even slightly above the limits of preferred values for organ music and medieval plainchant (set about 2–4 s) [56,57].

Table 4. Reverberation time (T_{30}) and Early decay time (EDT) values[1] measured under unoccupied conditions at the York Minster's Chapter House, together with their standard deviation (SD).

Frequency band (Hz)	125	250	500	1 k	2k	4 k
T_{30} (s)	5.00	5.17	5.56	5.24	4.19	2.82
SD_{T30}	0.13	0.13	0.06	0.05	0.03	0.02
EDT (s)	4.97	4.94	5.45	5.04	3.92	2.36
SD_{EDT}	0.53	0.49	0.30	0.17	0.10	0.06

[1] spatially averaged including all the source-receiver combinations characterised.

Looking at the Early decay time (EDT), which better assesses the subjective impression of reverberation [58], spatially averaged values are only slightly lower than those obtained for the T_{30} (Table 4). Nevertheless, SD values obtained for EDT are considerably higher than those obtained for T_{30} in all frequency bands, which indicates that the perceived reverberation depends on the relative position of the receiver in the room, meaning that the reduced source-receiver distance that certain source-receiver combinations have may emphasise the role of direct sound and the proximity to the surrounding walls may emphasise the contribution of early reflections arriving at certain receiving points. In any case, the room has an averaged EDT_m of 5.2 s, which is again above the optimal range suggested for speech (below 1.2 s) and for vocal and organ music in highly reverberant spaces, being 2.1 s and 4.2 s, respectively [57].

The analysis of the energy parameters derived from the measured RIR gives us complementary information about the balance between early and late reflections, or in other words, the clarity of sound in the space [58]. The definition, D_{50} (-), has been used to assess the clarity of the spoken word; whereas the clarity parameter, C_{80} (dB), was chosen to analyse the clarity of music. The central time, T_s (ms), which is strongly correlated with the decay time and therefore less sensitive to spatial variations [59], is useful to ascertain the clarity of sound in general terms. In Figure 8, the left column shows the spatially averaged values of the acoustic parameters measured for all the source-receiver combinations as a function of frequency. The high T_s mean values are indicators of a poor clarity of sound at low-mid frequency bands in the entire audience area. D_{50} mean values bellow 0.3 up to the 2 kHz band denote a poor clarity of the speech transmission in those bands, and the C_{80} values of less than 4 dB below 500 Hz are indicative of poor musical clarity especially at low frequencies. The error bars show the spatial dispersion in terms of the standard deviation (SD). It must be noted that they are relatively small, especially for the amphitheatre-like arrangement (S2), remaining mostly below 1 JND. Values of SD significantly exceeding this threshold are only found for the front-stage arrangement (S1) at the 500 Hz frequency band (max variation of 1.57 JND in the case of D_{50} at 500 Hz). The spectrally averaged values are also included in the figure as a function of the source-receiver distance. It can be seen that, when the source is located in the centre (S2) greater values of D_{50m} and C_{80m} are achieved, since the S-R distance remained below 7 m. As expected, there is a tendency for sound clarity indicators to decrease as the relative distance from the receiver to the sound source increases. However, it must be highlighted that the receivers closest to the side walls have higher C_{80m} and D_{50m} values than those that are at a similar relative S-R distance, but located in a more central area of the room. For instance, D_{50m} and C_{80m} values obtained at R4 when the source is located in S1 (S1-R4 dist = 6.7 m) are 1 JND above of those obtained in R3 (S1-R3 dist = 6.4 m), which means that such difference is perceptible by listeners. This fact may be due to the presence of early reflections nearer to the walls. In any case, the values of the energy parameters denote poor clarity of the sound (D_{50m} <0.3 and C_{80m} <-2dB), which indicates that how the sound energy behaves in the space is unfavourable to musical definition, and especially worrying for effective speech transmission, even at the receiver points closest to the source, for both source positions.

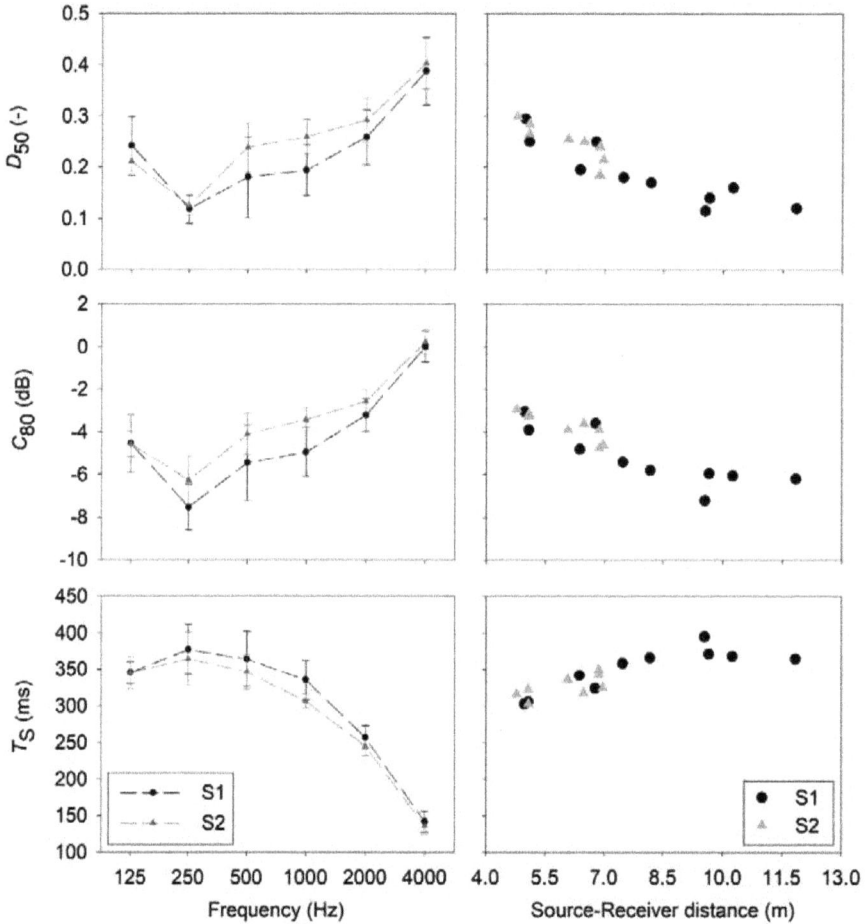

Figure 8. Spatially averaged values of the acoustic parameters as a function of frequency (left column)., and mid values as a function of source-receiver distance (right column).

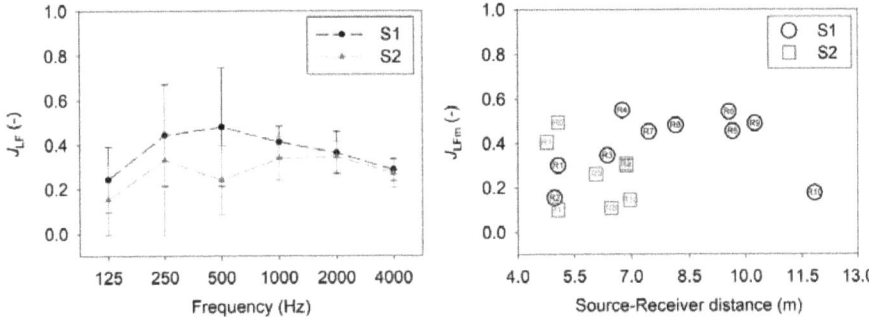

Figure 9. Spatially averaged values of the acoustic parameters as a function of frequency (left column)., and mid values as a function of source-receiver distance (right column).

Additionally, the Early Lateral Energy Fraction (J_{LF}) parameter is used to assess the spatial impression perceived by listeners in the Chapter House, since it has been demonstrated that it is related to the apparent source width (ASW) [60]. Furthermore, the J_{LF} parameter has been proven to play an important role in the way that music is experienced in this type of building together with the T_{30} and other factors depending on the music motif [57]. As shown in Figure 9, looking at the spatially averaged results, lower values of J_{LF} are found at low-mid frequencies when the source is located in the centre of the room (S2) than in S1 due to the geometry of the space, although the SD values indicate that it depends on the receiver position. This is relevant since the spatial impression is strongly correlated with the low frequency early-arriving sound [61]. Looking at the results obtained at each S-R combination, it is noted that considerably high values of J_{LFm} (around 0.5, while a typical value for a suitable spatial impression is between 0.2–0.3 [62]) are obtained in those positions located closer to the walls (R4 to R9) with S1. Nevertheless, when considering S2, only R1 and R6 reach comparable values.

Further analysis of the direction of arrival of sound reflections and their relative intensity was conducted with the aim of understanding the specific role of the polygonal shape and specific architectural elements, such as the vault, in the interaction with the sonic events that take place in the Chapter House. For this purpose, 3D sound intensity vectors derived from the measured B-format RIR were inspected by using Iris 1.4 acoustic software. A resolution of 2 ms was set, which constrains the analysis to 500 Hz and above [63]. The time interval window named as "Speech" was used for the representation in Figures 10 and 11, in which the red lines represent the direct sound (arriving at 0–2 ms), and the green and dark blue lines (arriving at 2–50 ms and 50–80 ms, respectively) represent the contribution of the early reflections. The reflections arriving after 80 ms or late reflections are coloured in sky blue.

Figure 10 shows the XY view of the 3D plots in the measured S-R combinations and allow us to observe the direction from which the early reflections, contributing to the clarity of sound, come from at each reception point depending of the position of the sound source. In the plots, the length of each ray represents their level in relation to the intensity of the direct arrival. Note that late reflections have been omitted here for the sake of clarity. It can be observed that, in the front-stage arrangement (S1), there are a considerable number of early rays arriving at those receivers that are at a greater distance from the source, with an attenuation of 5–10 dB relative to the intensity of the direct arrival which likely contribute towards a clearer perception of sound in those positions. Such rays are mainly coming from the closer lateral walls (in the case of R5, R6 and R9) and from the entrance (in the case of R10). Nevertheless, when the source is located in the centre (S2, amphitheatre-like arrangement), S-R distances are shorter and the intensity of the direct sound is higher, so the early reflections in general arrive with a greater attenuation, with the exception of the first reflection coming from the floor. It is only in those receivers closer to the walls (R5, R8 and R10) that a small number of rays approximately 5 dB lower than the direct arrival coming from the back, are found.

Focusing only on the early lateral rays we get information related to the spatial impression. With S1, a significant number of early lateral rays are observed in all the listener positions except in those receiver points located along the symmetry axe of the room (R2, R10), which is in good correlation with the high value of the J_{LFm} parameter obtained for those S-R combinations. Conversely, the intensity plots for S2 show limited lateral energy arriving at those points located in the central part of the space, which corresponds in general to lower values of the early lateral fraction parameter (Figure 9). As an example of this, the full Iris plot calculated for the receiver position R8 with both source positions is shown in Figure 11, which also provides insight into the general distribution of the energy, by including the late arrivals.

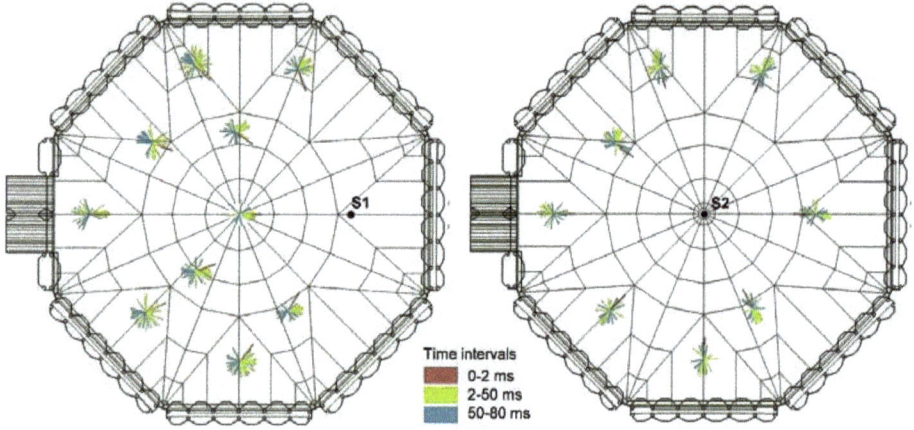

Figure 10. Representation of the early energy arriving at the receiver points (Iris plots, 2D XY plane, level range set to 40 dB with the direct sound level individually set as a reference (0dB)): front stage arrangement (left); and amphitheatre-like arrangement (right).

It can also be seen in Figure 11 that virtually no early reflections are coming from the upper part of the room (stained glasses and the vault), not even in a significant way when the receiver is located in the centre of the space immediately below the centre of the vault (R2). This lack of early reflections coming from above was expected given the vault's height (see geometry details in Table 1 and Figure 7). In general, the late energy arriving at all the receiver positions is significant and is coming from all around the space, whilst early energy is mainly coming from below and the horizontal plane, but not from above.

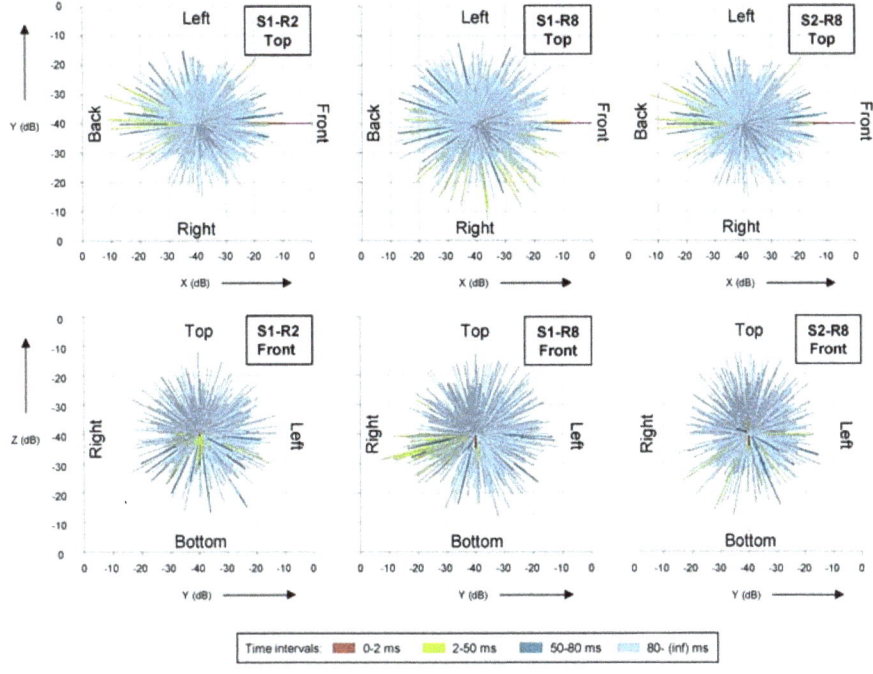

Figure 11. Representation of early and late energy arriving at selected S-R combinations (2D Iris plots, XY plane (upper row) and YZ plane (lower row), level range set to 40 dB with the direct sound level individually set as a reference (0 dB)).

Acoustic simulations were then used to assess in more detail the influence of the architectural features of the space in its function as a meeting place, including its original configuration where the speaker and/or the listeners where sitting in the limestone stalls (Figure 3c). Simulated mappings were generated to analyse the spatial distribution patterns of the acoustic parameters. The horizontal grid for the colour mappings was defined at approximate ear height (1.20 m from the floor in the centre area and 0.6 m from the sitting area of the stalls) were set. Figure 12 shows the simulated mappings of the speech transmission index (STI) and D_{50} at 1 kHz obtained for the different source positions in the current state. These parameters have been selected in this case since they are indicators of the clarity and intelligibility of the spoken word. When looking at the simulation mappings, no significant improvement in terms of speech intelligibility is observed in the stalls in comparison with the open floor area. In general, values denote "poor" speech intelligibility throughout the audience area with the three source positions analysed. Despite this, it can be seen that some STI and D_{50} values obtained in a number of receivers located in the stalls are higher than those obtained at points that are at a closer relative distance from the sound source, and this effect can be perceptible when these differences are greater than 1 JND. Furthermore, when the source is located in the stalls (S3), there is a potential lack of direct sound in those stalls located in the adjacent walls to which the sound source is located due to the geometry of the space, and also because of the marble columns, and this is detrimental to intelligibility in that section of the sitting area.

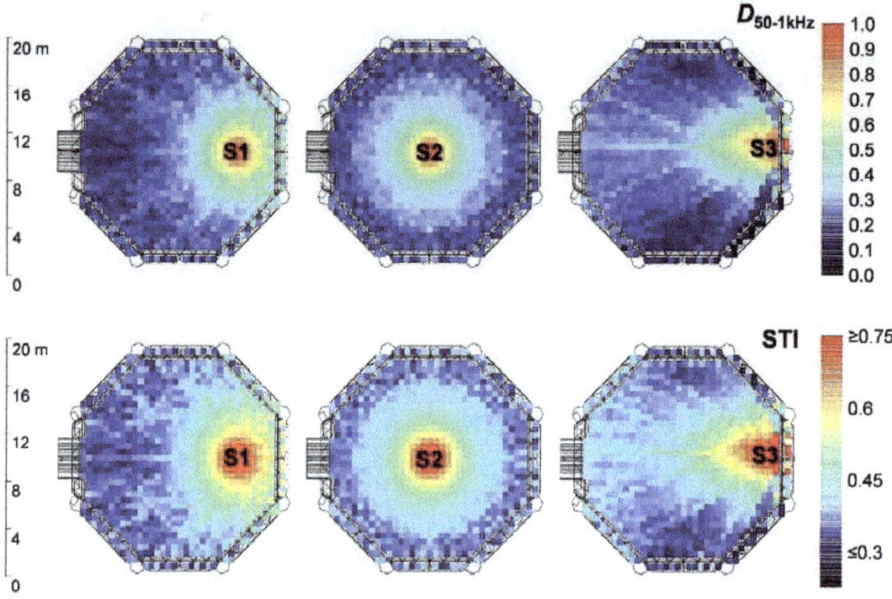

Figure 12. Simulated mappings of the D_{50} at 1 kHz (upper row) and STI (lower row) obtained for the different source positions. York Minster's Chapter House in its current state.

For more details, Figure 13 shows D_{50m} and STI values obtained at each S-R combination, including the receiver positions in the stalls. The results reveal that when the source is in the centre, higher values of D_{50m} and STI are obtained in the stalls, while with S1 and S3 no significant improvement can be seen, and even lower values are obtained. These results demonstrate that, although the early reflection pattern inspection for those receivers located in the stalls shows a group of strong early reflections of 1st and 2nd order arriving before 50 ms (even arriving before 10 ms in the case of those coming from the walls immediately behind the receivers) that reinforce the clarity of the speech in those positions, this reinforcement does not have a significant impact in terms of intelligibility or sound clarity since the reverberant field predominates at that distance from the sound source.

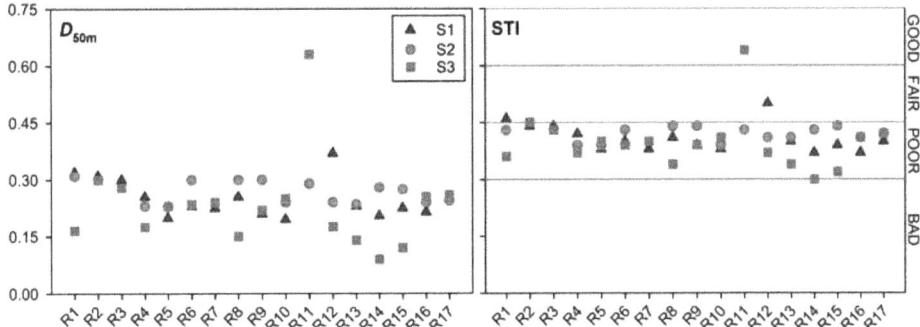

Figure 13. D_{50m} and STI results derived from the simulated RIR simulation at each source-receiver combination. York Minster's Chapter House in its current state.

4.2. Acoustics of the York Minster's Chapter House before any Restoration Work

The Chapter House's dimensions and geometrical features have remained the same since its construction date. Therefore, the visual inspection of the free path length (FPL) distribution estimated from the ray-tracing process has been used to have an initial notion of the average distance that a ray will travel between successive reflections in the model, which depends on the geometry of the room but not on the acoustic properties of its finishing materials. The FPL of the acoustic model of the Chapter House shows a high probability of the shortest path lengths (below 3 m) but also a significant concentration of the distribution around 19 m, which is clearly determined by the proportions of the room and will have a key influence in the estimation of its reverberation time [64], anticipating a high number of late-arriving energy both before and after any restoration work took place.

The acoustic simulation of the space, under the premises established in Section 3.2.2 gives us an idea of the impact that all the modifications, restorations and conservation interventions throughout its history have had on the acoustics of the space. As the most acoustically relevant modifications took place in the upper parts of the space, it is expected that they caused a significant effect on the late reflections arriving at the listeners, and consequently, in the reverberant field of the Chapter House. Table 5 shows the simulated reverberation time in the hypothesis that has been made about its original state, described in Section 3.2.2.

The results reveal that the medieval Chapter House had a reverberation time considerably shorter (around 1 s) at low and medium frequencies when compared with the reverberation time it has today. In other words, the modifications are estimated to have increased the reverberation time up to 3-4 JND in that frequency range. The replacement of the original wooden vault at the end of the 18th century with the current lath and plaster work is the most influential factor in the noticeable change of the reverberation time. However, it should be considered that a reverberation time of 4.3 s at mid frequencies was still too high for good speech intelligibility and clarity, especially considering the lack of furniture and the low level of attendance that the Chapter meetings had at that time, implying that there would not be significant additional absorption due to the presence of more people.

Table 5. A comparison of the reverberation time of the York Minster's Chapter House in its current state ($T_{30\text{-CS}}$) and the estimation in its original state ($T_{30\text{-M}}$). Simulated values and differences in terms of JND.

Frequency band (Hz)	125	250	500	1k	2k	4k
$T_{30\text{-M}}$ (s)	3.90	4.16	4.34	4.31	3.75	2.65
$T_{30\text{-M}}$ vs. $T_{30\text{-CS}}$ (s)	−1.10	−1.01	−1.22	−0.93	−0.44	−0.17
$T_{30\text{-M}}$ vs. $T_{30\text{-CS}}$ (JND)	−4.40	−3.91	−4.39	−3.55	−2.10	−1.21

5. Conclusions

The present paper explores the acoustic characteristics of the Chapter House of the York Minster. Although its acoustics is frequently mentioned in literature on the Minster [21–23], there have been no thorough attempts to analyse it in detail or to consider its sound environment as a key element in terms of the relationship between visitors and the space over time.

The first approach used to analyse the acoustic behaviour of the Chapter House is based on the acoustic measurements carried out in the unoccupied space by means of the acoustic parameters that can be derived from the captured room impulse responses. Then, acoustic simulation techniques were used to perform a more detailed analysis of the propagation of sound in the space due to its octagonal shape and its finishing materials, as well as to assess the possible acoustic impact caused by a series of restoration works. These results have been analysed and discussed paying attention to the variety of purposes and diverse uses that this space has had through history.

Despite being designed as a meeting place for the chapter, the acoustic simulation of the Chapter House in its original state, prior to any restoration work, showed a T_{30m} above 4 seconds. These results suggest that the priority in its design was to create a great visual impact, at the expense of compromising its functionality from the acoustic point of view, since its proportions and finishing materials give rise to a long tail of reverberation that would make clarity and intelligibility of speech unsuitable for meetings.

In order to assess how the architectural features of the Chapter House contribute to the acoustic behaviour of the space and to answer the frequently asked question of whether its acoustic performance is really better in the perimeter than in the centre, the early and late sound energy arriving at the listener were analysed in detail through the measured B-format RIR. Additionally, acoustic simulations were used to ascertain how it works in its original configuration as a meeting place, in which the speaker (S3) and the listeners where setting in the limestone stalls (except the reader at the lectern, possibly placed in S1 or S2 [23]).

The analysis of the sound energy distribution in the Chapter House shows that there is a significant amount of early reflections contributing to the sound clarity due to its octagonal shape. Those early reflections arriving less than 50–80 ms after the direct sound arrive from the lateral walls and stalls closer to the listeners and, to a lesser extent, from the decorated canopies. This contribution is more noticeable in those listener positions closer to the walls, and when the sound source is not located in the centre of the room (S2). The results show higher values of C_{80m} and D_{50m} at those receivers located on the peripheral area and the stalls than at those receivers that are at a similar S-R distance, but located in a more central area of the room. Although a good clarity of sound is not achieved in the peripheral positions with a greater contribution of the early reflections, such increase is above the 1 JND established for each acoustic parameter (C_{80m} and D_{50m}) in low and mid frequencies, which means that relatively more favourable acoustics for speech could be experienced by the listeners located closer to the perimeter, as mentioned by previous sources [65].

The late energy, linked to the reverberant field, is shown to be very significant throughout the entire space, which is unfavourable to speech but contributes to the perception of envelopment. The upper part of the room, including the stained glasses and the vault, play an important role here as they are at a great distance from the listener positions.

These findings suggest that, although the octagonal shape of the York Minster's Chapter House supports shorter distances between the speaker and the listeners and provides a great number of strong early reflections arriving at the listeners located close to the surrounding walls, its proportions and hard finishing materials result in a large number of late reflections that produce an excessive sound tail which clearly dominates the space and negatively affects speech intelligibility and the perceived clarity of music. Therefore, a priori, and according to the criteria established for concert halls and meeting rooms, we can conclude that the particular shape of this building does not contribute to better acoustics in terms of clarity of sound, but does allow a more balanced spatial distribution of the sound energy, which is shown by a more balanced spatial distribution of C_{80}, D_{50} and T_S values.

At present, the Chapter House has a T_{30m} of 5.4 s, which is notably higher than in its original state, mainly due to the restoration of the vault of the 18th century. The highly reverberant character of the space hinders a clear perception of the sound in the room in the 3 configurations analysed (see Figure 3), which is evidenced by the values obtained from T_S, D_{50} and STI. This condition forces speakers to adapt to the space, speaking with a leisurely pace and projecting their voices, and makes it necessary to use electroacoustic support on certain occasions. With ephemeral interventions intelligibility can be improved, by incorporating acoustically absorbent elements in the room such as carpets and soft pieces of furniture. Of course, the degree of occupation of the room during the different activities will also have an impact in the reverberation they experience, acting as acoustic absorbers in a first place; but they may be also a significant source of noise resulting in the acoustic phenomenon known as cocktail party effect, which is very pronounced in such reverberant rooms. L. Power pointed out this fact very clearly [39],

> "It doesn't take much to any other sound to start competing with you. You don't need to raise your voice very much to talk in there, but if other people are saying something even quite low, kind of conversation, it can start to build the sound very quickly".

But today, the York Minster's Chapter House is not only considered as a meeting place, it is a multipurpose room, where concerts and other cultural and social activities are held regularly.

As a concert room, it could be said that this space is too reverberant to host any vocal or instrumental music concerts, based on the low values obtained for musical clarity indicators, especially at low-medium frequencies (C_{80m} < −4 dB), both in the front-stage arrangement and the amphitheatre-like arrangement. Nevertheless, the unique reverberant atmosphere of the space, which has been described as an "uncontrolled exuberance of sound" [66], is what makes it so special, when combined with its visual majesty. Significantly, some artists have incorporated the architectural features of the space, its symmetry, and its reverberant acoustics to their performances in order to create an experience specially designed for that space, for instance changing the traditional position of the choir in front of the audience and singing around the edge, or singing a solo from the vestibule, among other things [67]. H. Daffern, singer at the *Ebor Singers* choir, shared her experience in the Chapter House [68],

> "I do love performing in the Chapter House…I like the challenge of the singing as a group and not being able to rely in what you hear and the overall sound being quite spectacular, but needs to be the right repertoire in the right space".

The findings in this article support the idea previously stated by several authors [69] that "acoustics must be considered part of the intangible cultural heritage, not only considering the sound events that take place within the space, but also strictly related to the building itself".

Furthermore, the acoustic analysis detailed in this paper has the purpose of functioning as a resource for investigations in other disciplines as well as a reference when assessing the acoustics of similar English polygonal Chapter Houses. The acoustic simulation model built for this singular space will be used to generate a set of RIR that will be used for a creative exploration of acoustical heritage [11] and in future virtual reality simulations [70].

Author Contributions: Conceptualization, L.A.M., M.L. and A.A.C.; methodology, L.A.M. and A.A.C.; software, L.A.M.; validation, L.A.M., M.L. and A.A.C.; formal analysis, L.A.M.; investigation, L.A.M., M.L. and A.A.C.; resources, L.A.M.; data curation, L.A.M.; writing—original draft preparation, L.A.M.; writing—review and editing, L.A.M., M.L. and A.A.C.; visualization, L.A.M., M.L. and A.A.C.; supervision, L.A.M. and M.L.; project administration, L.A.M.; funding acquisition, L.A.M. and M.L.

Funding: This research was funded by the European Union's Horizon 2020 research and innovation programme under the Marie Skłodowska-Curie grant agreement No. 797586.

Acknowledgments: The authors are very grateful to the Dean and the staff of the cathedral for their collaboration and their assistance during the measurements. Authors very much appreciate the additional project support from the Department of Electronic Engineering AudioLab at the University of York. Thanks to Daniel Protheroe for having made Iris software available for the analysis. Thanks to Richard A. Carter, Claudia Nader and Joe Rees-Jones for the photographs taken, and thanks to Marc Girot, Adi Keltsh and Rene Idrovo for their invaluable help during the measurement session.

Conflicts of Interest: The authors declare no conflict of interest. The funders had no role in the design of the study; in the collection, analyses, or interpretation of data; in the writing of the manuscript, or in the decision to publish the results.

References

1. ISO 3382-1. Acoustics—Measurement of Room Acoustic Parameters. Part 1: Performance Rooms; International Organization for Standardization: Geneva, Switzerland, 2009.
2. Martellotta, F.; Cirillo, E.; Carbonari, A.; Ricciardi, P. Guidelines for acoustical measurements in churches. *App. Acoust.* **2009**, *70*, 378–88.
3. Álvarez-Morales, L.; Zamarreño, T.; Girón, S.; Galindo, M. A methodology for the study of the acoustic environment of Catholic cathedrals: application to the Cathedral of Malaga. *Build. Environ.* **2014**, *72*, 102–115
4. Pedrero, A.; Ruiz, R.; Díaz-Chyla, A.; Díaz, C. Acoustical study of Toledo Cathedral according to its liturgical uses. *Appl. Acoust.* **2014**, *85*, 23–33.
5. Álvarez-Morales, L.; Girón, S.; Galindo, M.; Zamarreño, T. Acoustic environment of Andalusian cathedrals. *Build. Environ.* **2016**, *103*, 182–192.
6. Postma, B.N.; Katz, B.F. Acoustics of Notre-Dame Cathedral de Paris. Proceedings of the International Congress on Acoustics (ICA), Buenos Aires, Argentina, 5–9 September, 2016, 1–10.
7. D'Orazio, D.; Garai M. Acoustic control in octagonal geometry: case study of the Torri dell'acqua auditorium. Proceedings of the Institute of Acoustics, Dublin, Ireland, 20–22 May 2011.
8. Anderson, J.S.; Anderson, M.B. Acoustic coupling effects in St Paul's Cathedral, London. *J. Sound Vib.* **2000**, *236*, 209–225.
9. Martellotta, F. Identifying acoustical coupling by measurements and prediction-models for St. Peter's Basilica in Rome. *J. Acoust. Soc. Am.* **2009**, *126*, 1175–1186.
10. Suárez, R.; Alonso, A.; Sendra, J.J. Archaeoacoustics of intangible cultural heritage: The sound of the Maior Ecclesia of Cluny. *J. Cult. Herit.* **2016**, *19*, 567–572.
11. Murphy, D.; Shelley, S.; Foteinou, A.; Brereton, J.; Daffern, H. Acoustic Heritage and Audio Creativity: the Creative Application of Sound in the Representation, Understanding and Experience of Past Environments. *Internet Archaeol.* **2017**, *44*.
12. Alonso, A.; Suárez, R.; Sendra, J.J.; Zamarreño, T. Acoustic evaluation of the cathedral of Seville as a concert hall and proposals for improving the acoustic quality perceived by listeners. *J. Build. Perform. Simu.* **2014**, *7*, 360–78.
13. Alvarez-Morales, L.; Martellotta, F. A geometrical acoustic simulation of the effect of occupancy and source position in historical churches. *Appl. Acoust.* **2015**, *91*, 47–58.
14. Vorländer, M. Computer simulations in room acoustics: Concepts and uncertainties. *J. Acoust. Soc. Am.* **2013**, *133*, 1203.
15. Ditchfield, P.H. The Cathedrals of Great Britain. Their History and Architecture. J.M. Dent & Company: London, UK, 1902.
16. Brown, S. York Minster: an Architectural History, c. 1220-1500: "Our Magnificent Fabrick", Royal Commission on Historical Monuments: Swindon, UK, 2003.
17. The Medieval Minster: History of York. Available online: www.historyofyork.org.uk. (accessed on 12 December 2019)
18. Brown, S. Stained Glass at York Minster, Scala publishers: London, UK, 2017

19. Association of Leading Visitor Attractions. Available online: https://www.alva.org.uk/details.cfm?p=423 (accessed on 10 December 2019)
20. Bond, F. The Cathedrals of England and Wales. Jeremy Mills Publishing: Huddersfield, UK, 2007.
21. Zukowsky, J. The Polygonal Chapter House: Architecture and Society in Gothic Britain. PhD Thesis, State University of New York, Binghamton, 1977, 40–41.
22. http://www.historyfish.net/abbeys/abbeyparts/livingquarters2.html (accessed on 10 December 2019)
23. Wickham, W. A. Some notes on chapter-houses part 1 and part 2, *Transactions of the Historic Society of Lancashire and Cheshire*, **1912**, *64*, 142–247.
24. Gajewski A.; Opacic, Z. The year 1300 and the creation of a new European architecture. Brepols Publishers NV: Turnhout, Belgium, 2008.
25. Hewitt, C. English cathedral carpentry. Wayland Ltd: London, UK, 1974.
26. Morgan, C. A Life of St Katherine of Alexandria in the Chapter-House of York Minster, *J. Br. Archaeol.* Soc, **2009**, *162*, 146–178
27. Rodwell, W.; Mortimer, R. Westminster Abbey Chapter House: the history, art and architecture of 'a chapter house beyond compare', Society of Antiquaries of London: London, UK, 2010.
28. Winkles, H.; Garland, R.; Moule, T. Winkles's architectural and picturesque illustrations of the cathedral churches of England and Wales. Vol. 2. London. Publisher: Effingham Wilson, Royal Exchange, and Charles Tilt, Fleet Street. 1836.
29. Harrison, F. York Minster, 1st ed. Methuen & Co Ltd: London, UK. 1927.
30. Harrison, F. *The Painted Glass of York: An Account of the Medieval Glass of the Minster and the Parish Churches*. Society for promoting Christian knowledge, SPKC: London, UK, 1927.
31. Halfpenny, Joseph. *Gothic ornaments in the Cathedral Church of York*, J. Todd & Sons: Harrogate, UK, 1975. (Plate No 102)
32. Drake, F. Eboracum: The History and Antiquities of the City of York, from Its Origin to This Time. T. Wilson and R. Spence. 1788
33. Norton, C. Friends of York Minster Annual Report, **1996**, 67, 34–51.
34. Clutton-Brock, A. The cathedral church of York: a description of its fabric and a brief history of the archiepiscopal, G. Bell & Sons: London, UK, 1899.
35. 'Collegiate churches: York (including York Minster)', A History of the County of York, Available Online: http://www.british-history.ac.uk/vch/yorks/vol3/pp375-386 (accessed 19 November 2019).
36. Medieval York: York in political history', in A History of the County of York: the City of York, Available Online: http://www.british-history.ac.uk/vch/yorks/city-of-york/pp25-29 (accessed 18 November 2019).
37. Coldstream, N. York Chapter House. *J. Br. Archaeol. Soc.*, **1972**, *35*, 15–23
38. Jason Archuleta (University of York, York, UK), interviewed 11th October 2019
39. Lisa Power (University of York, York, UK), interviewed 22nd October 2019
40. Ballou, G. Handbook for Sound Engineers. 4th ed. Taylor & Francis: New York, NY, USA, 2013.
41. Vorländer, M. Auralization; Springer: Berlin, Germany, 2008.
42. Martellotta, F.; Álvarez-Morales, L. Virtual acoustic reconstruction of the church of Gesù in Rome: a comparison between different design options. Proceedings of Forum Acusticum 2014, Krakow, Poland, 2014.
43. Cox, T.J.; Dalenback, B.I.; D'Antonio, P.; Embrechts, J.J.; Jeon J.Y.; Mommertz, E.; Vorländer, M. A tutorial on scattering and diffusion coefficients for room acoustic surfaces. *Acta. Acust. united Ac.*, **2006**, *92*, 1–5.
44. T. J. Cox, P. D'Antonio: Acoustic absorbers and diffusers: Theory, design and application. Spon Press: Taylor & Francis Group, London and New York City, UK, 2004
45. Dalenbäck, B.I.L. *CATT-Acoustic*. Gothenburg, Sweeden, 2011.
46. ODEON Room Acoustics Software User's Manual Version 15 Published in November 2019, Available online: https://odeon.dk/download/Version15/OdeonManual.pdf (accessed on 10 December 2019).
47. Postma, B.N., Katz, B.F. Creation and calibration method of acoustical models for historic virtual reality auralizations. *Virtual Reality*. **2015**, *19*, 161–80.
48. Martellotta, F. The just noticeable difference of center time and clarity index in large reverberant spaces. *J. Acoust. Soc. Am.* **2010**, *128*, 654–663
49. Bork, I. A comparison of room simulation software-the 2nd round robin on room acoustical computer simulation. *Acta. Acust. united Ac.*, **2000**, 86, 943–56

50. Álvarez-Morales, L.; Galindo, M.; Girón, S.; Zamarreño, T.; Cibrián, R.M. Acoustic characterisation by using different room acoustics software tools: a comparative study. *Acta. Acust. united Ac.*, **2016**, *102*, 578–591
51. Cirillo, E.; Martellotta, F. Worship, acoustics, and architecture. Multi Science Publishing Company Limited: Brentwood, UK, 2006.
52. Sü Gül, Z. Acoustical Impact of Architectonics and Material Features in the Lifespan of Two Monumental Sacred Structures. *Acoustics*, **2019**, *1*, 493-516
53. Cirillo, E.; Martellotta, F. Sull'applicabilità della formula di sabine nelle chiese romaniche, Associazione Italiana di Acustica, 29° Convegno Nazionale Ferrara, Italy, 12–14 June 2002
54. Girón, S.; Álvarez-Morales, L.; Zamarreño, T. Church acoustics: A state-of-the-art review after several decades of research. *J. Sound. Vib.*, **2014**, *411*, 378–408.
55. Harris, C. M. Handbook of Noise Control, McGraw Hill: New York, NY, USA,1997
56. Barron, M. Subjective study of British concert halls. *Acustica*. **1998**, *66*, 2–14
57. Martellotta, F. Subjective study of preferred listening conditions in Italian Catholic churches. *J. Sound. Vib.* **2008**, *317*, 378–99.
58. Rossing, T. Springer handbook of acoustics. Springer: New York, NY, USA, 2007
59. Bradley, J.S. Review of objective room acoustics measures and future needs. *Appl. Acoust.* **2011**, *72*, 713–720
60. Barron, M. The subjective effects of first reflections in concert halls-The need for lateral reflections. *J. Sound. Vib.* **1971**, *15*, 475–494
61. Barron, M. Spatial impression due to early lateral reflection in concert halls: the derivation of a physical measure. *J Sound Vib.* **1981**, *77*, 211–232
62. Giménez, A.; Cibrián, R.M.; Cerdá, S.; Girón, S.; Zamarreño, T. Mismatches between objective parameters and measured perception assessment in room acoustics: A holistic approach. *Build. Environ.* **2014**, *74*, 119–131.
63. Iris software, Iris.co.nz: http://www.iris.co.nz/ (archived on 10 December 2019).
64. Šumarac-Pavlović, D.; Mijić, M. An insight into the influence of geometrical features of rooms on their acoustic response based on free path length distribution. *Acta. Acust. united Ac.*, **2007**, *93*, 1012–1026.
65. https://www.independent.co.uk/commercial/visityork/step-into-the-pulpit-at-york-minster-9077425.html (accessed on 12 December 2019).
66. Wheaton, N.S. A Journal of a Residence During Several Months in London: Including Excursions Through Various Parts of England; and a Short Tour in France and Scotland; in the Years 1823 and 1824, Available Online: https://archive.org/details/journalofresiden00whea/page/n5/mode/2up (accessed on 10 December 2019)
67. https://www.eborsingers.org/ (accessed on 10 December 2019).
68. Helena Daffern (University of York, York, UK), interviewed 22nd October 2019
69. Elicio, L.; Martellotta, F. Acoustics as a cultural heritage: The case of Orthodox churches and of the "Russian church" in Bari. *J. Cult. Herit.* **2015**, *16*, 912–917.
70. https://vpcp.chass.ncsu.edu/ (archived on 10 December 2019).

© 2020 by the authors. Licensee MDPI, Basel, Switzerland. This article is an open access article distributed under the terms and conditions of the Creative Commons Attribution (CC BY) license (http://creativecommons.org/licenses/by/4.0/).

Article

Sound Archaeology: A Study of the Acoustics of Three World Heritage Sites, Spanish Prehistoric Painted Caves, Stonehenge, and Paphos Theatre

Rupert Till

Department of Music and Drama, University of Huddersfield, Huddersfield HD1 3DH, UK; R.Till@hud.ac.uk

Received: 22 April 2019; Accepted: 23 July 2019; Published: 9 August 2019

Abstract: This paper explores the acoustics of three UNESCO World Heritage Sites: five caves in Spain that feature prehistoric paintings that are up to 40,000 years old; Stonehenge stone circle in England, which is over 4000 years old; and Paphos Theatre in Cyprus, which is 2000 years old. Issues with standard acoustic methods are discussed, and a range of different possible approaches are explored for sound archaeology studies, also known as archaeoacoustics. The context of the three sites are examined followed by an analysis of their acoustic properties. Firstly, early decay time is explored, including a comparison of these sites to contemporary concert halls. Subsequently, reverberation, clarity of speech, and bass response are examined. Results show that the caves have a wide range of different naturally occurring acoustics, including reverberation, and strong bass effects. Stonehenge has acoustics that change as the design of the site develops, with some similarities to the effects in the caves. Acoustic effects vary considerably as you move further into the centre of the stone circle, and as the stone circle develops through time; these effects would be noticeable, and are a by-product of the human building of ritual sites. At Paphos Theatre, acoustics vary from the best seats on the front rows, backwards; here, the architects have considered acoustics in the design of the building. The paper illustrates the changing acoustics of ritual sites in human cultures, showing how sound contributed to giving spaces an individual character, helping to afford a sense of contextualized ritual place.

Keywords: sound; archaeology; archaeoacoustics; acoustics; reverberation; clarity; Stonehenge; cave; theatre; EDT; music

1. Introduction

This paper asks how one best studies the acoustic ecologies of archaeological spaces? Are standard acoustics methods appropriate in such sound archaeology or archaeoacoustic research? It explores how acoustics in sites that feature ritual use change through different stages of human culture, from Palaeolithic painted caves in Northern Spain, with a context dated from 43,000 years in the past, through the monumental 5000-year-old architecture of Neolithic Stonehenge in Britain, to the public theatres of Roman culture, in Paphos on the Mediterranean island of Cyprus. All three are UNESCO World Heritage sites, and the paper explores what the changes in the acoustic effects present suggest about how human culture changes over time. It asks what problems are encountered when examining the acoustics of ancient sites. The three sites are contextualised—prehistoric caves in Spain, Stonehenge, and Paphos Theatre—followed by a metrical analysis of their acoustics, before finishing with conclusions.

To comprehensively explore changes in acoustics over time in antiquity would take a large-scale project with many more examples, which is beyond the scope of the present study. Instead, metrical results are generated through the author's study of the acoustics of these three archaeological spaces, and are compared in order to illustrate both to what extent they differ, and the range of acoustics present in different archaeological contexts and periods. This paper provides a number of novel perspectives. While previous studies of the acoustics of these and other archaeological sites have provided a range of results, this paper provides for the first time a comparison of different archaeological eras and contexts, and of acoustic metrics, such as early decay time (EDT), clarity, speech transmission, consonant intelligibility, and bass ratio. The key measure of perceived reverberation, early decay time, is discussed in some detail, and compared to results for modern concert halls, exploring results across a wide frequency range, rather than the restricted averages often referred to within industrial acoustics. The limitations imposed by approaches, such as this averaging, are discussed, and the effects are discussed of adopting an uncritical application to archaeological contexts of the standard acoustic state of the art as used in contemporary industrial research. The comparative study of three different sites illustrates the challenges for standard acoustics methods of working within this field, and provides examples of how and why one might construct a bespoke acoustics study method for these sites. Comparison is made between the caves explored, something a previous study did not consider. Acoustics of a digital model of Stonehenge at a number of receiver positions within three different phases of development of the monument are discussed. No previous study of Stonehenge has explored the acoustics of a range of receiver positions, or of the various arrangements of stones of this iconic site. Paphos Theatre has never been studied acoustically, and in general, there have been no studies comparing Greek or Roman ancient performance venues to modern concert halls. As well as adopting empirical quantitative approaches, this project includes a qualitative comparative assessment of the sites being studied, as is appropriate when studying the experience of sound within ritual sites, and working within an interdisciplinary field that addresses archaeology, sound, and music, as well as scientific acoustics.

2. Materials and Methods

2.1. Problems with Standard Acoustic Methodology

Acoustics is usually conceived of as a purely empirical scientific field. Its purpose is to come to reliable, repeatable results, focused on certainties that can be used within applied engineering, industrial, or professional contexts. The results of acoustical studies are used to adjust the architectural design of buildings and other environments, and thus a focus is on results that can provide useful solutions. This aims to produce definitive conclusions, to find answers and solutions, although even this engineering approach has to struggle against compromises between results in different positions and frequency ranges, and the impossibility of achieving perfect acoustics for all purposes. In contrast, post-processual archaeological thought accepts that archaeological data requires interpretation, and when studying human history, the further back in time one moves, the less certain results become. In an archaeological context, it is accepted that there are many possible perspectives and interpretations, and that answers depend on the context of the archaeological situation, as well as the focus of the researcher and their own interests [1–3]. Studies of archaeoacoustics, involving as they do the use of acoustics methods to inform archaeological research, have to resolve such different approaches. This situation is further problematised in music and sound archaeologies, which are often centred on even more creative and artistic perspectives [4] and the phenomenological experience of sound [5,6]. Sound is a time-based medium, and thus needs an approach that includes action and experience, which addresses embodiment and emotion as much as mechanistic rationalism [7]. Sound archaeology requires a non-representational approach (NRT), as described by Thrift [8] and Dewsbury [9], as it attempts to interpret human meanings from antiquity, where the involvement of physical activities and direct experience necessitates a fusing together of separate methodologies. In order to address these concepts, and integrate elements of experiential acoustic phenomenology, the experience of the author is deliberately discussed within this paper. Although this is perhaps unusual in acoustics publications, it is more commonplace in archaeological, musical, and sound-focused research. Indeed, field studies are common elements both of archaeology and ethnomusicology, the two subjects from which archaeoacoustics and sound archaeology have emerged. It is proposed that discussion of the author's experiences during fieldwork is appropriate content for this study.

Architectural acoustics has traditionally focused on issues related to industrial practice, such as concert hall design, traffic noise, and health and safety issues. It has as a result a range of standards, leading to a number of assumptions and set practices in standard acoustics methods. These can be problematic when using acoustics to study archaeological sites. For example, Long [10] suggests that "in a room whose dimensions are large enough that there is a sufficient density of modes, it is customary to describe the space in terms of a statistical model known as a diffuse field." (p. 327). In a diffuse field it is expected that a similar acoustic result will occur, independent of the positioning of the source of sound, or receiver, as no matter where you make the sound or hear it, reverberation is considered as consistent across the space. Such simplification is appropriate for standard acoustics purposes, in modern buildings with regular shapes.

Archaeological sites often have a complex or irregular shape, and as in any space, an experience of acoustics can vary considerably depending on listening (receiver) position, and where sound is being made (source position). Although it is customary in acoustics to assume a diffuse field, this assumption overlooks the range of sonic experiences possible for an individual within such a field. Modal effects can mean that sound near a wall is different from that in the centre, and unusually shaped rooms can have odd and varied frequency and amplitude responses. In archaeological sites, a large number of measurements may have to be taken if one wishes to comprehensively understand the acoustics of the space, for example focusing on positions of particular interest, perhaps with important archaeological features, or that seem to be a particular focus of activity.

Standard acoustics approaches often act to remove measurement of variation, aiming in many cases for an averaged result. BS EN ISO 3382-1:2009 is the British Standard for Acoustics–Measurement of Room Acoustic Parameters Part 1: Performing Spaces [11]. This describes the state of the art of engineering standards for acoustic measurement practice, which is well-known and thus not described in detail in this study, other than where sound archaeology necessitates an alternative approach. It is a useful document for commercial purposes, to enable comparable standardised results, or to describe a contemporary performance space, but can be problematic for a forensic analysis of the acoustics of an archaeological site. A cave or temple is not a performance space per se, but this standard is referred to as such sites are closer in use than to the other available options for such standards, such as a room, office, or laboratory.

The standard raises the issue of "whether single spatial averages will adequately describe the room" (p. 5), and suggests that if "the room is likely to show areas with differing reverberation times, these shall be investigated and measured separately" (p. 6). The standard suggests this option for spaces that do not feature a diffuse field, but the averaging of results is the more common practice, and no definition is provided of how different reverberation results have to be, to require individual consideration of a range of source and receiver positions. The standard requires that "a minimum of two source positions shall be used" (p. 5). It suggests (p. 20) that 6 different microphone/receiver positions are needed for a venue with 500 seats, and 10 positions are required for a venue with 2000 seats. These suggestions are for a performance space, and an archaeological site with no seats requires different consideration. The multiple source and receiver results are usually averaged; "it is necessary to average over a number of measurements at each position in order to achieve an acceptable measurement uncertainty" (p. 7). In a sound archaeology project, this study suggests that both individual and averaged results should be considered, as uncertainty is often present in any case, and certainty is not the only focus. The intention of the standard is to establish a result that describes an overall impression of a space, whereas sound archaeology may be interested in a detailed identification of the variation and distribution of different acoustic effects present, and in results in individual positions of archaeological significance.

The standard recommends a minimum distance of 1.5 m between an acoustic test signal source (such as a sine signal played through a loudspeaker) and a receiver (microphone) when carrying out an acoustic measurement (p. 3). In a sound archaeology study, one might be additionally interested in results where the source and receiver are in the same place (to mimic a performer making sound and hearing it themselves), are far away from one another, or where one or the other are moving. Proximity to a wall or floor can significantly affect the frequency response recorded, as there might be more early reflections present as a result; near a curved wall, the sound may be different than next to a straight wall. In the standard, the "distance from any microphone position to the nearest reflecting surface, including the floor, shall be at least a quarter of a wavelength, i.e., normally around 1 m" (p. 5). If one is interested in the sound next to a cave painting, sound made by scraping a prehistoric engraving into the wall, the sound made by a drum placed on the floor, or by the strange effects created by singing next to a particular surface, one might need to change this standard practice. An assessment method based on sources and receivers at set distances assumes a standard separation between a performer and an audience, a contemporary Western paradigm that is not shared in all archaeological contexts, where participation may be the norm. Although it is acceptable in acoustics to deal with a venue as though it were empty, if one wants to understand people's experience of acoustics, it may be more relevant to explore a venue's acoustics with an audience present, something that is problematic due to audience noise.

Reverberation is described with a standard metric, such as T30 or early decay time (EDT), as a set value, an invariant amount that never changes. Consideration of early reflections allows EDT to provide a better indication of perceived reverberance than T30 [12] and may be more useful as a result in some sound archaeology contexts. Reverberation varies at different frequencies and may be longer or shorter at lower bass frequencies. Acoustic studies usually state the reverberation time at 1 kHz, or as an average of results at 500 Hz and 1 kHz [11] (p. 10). In a workplace, these frequencies are significant because mid-range frequencies are most likely to damage hearing, whereas low frequencies are less problematic. The standard recommends that the range from "at least 250 to 2000 Hz" is examined, and for "engineering and precision methods, the frequency range should cover at least 125 to 4000 Hz in octave bands or 100 to 5000 Hz in one-third octave bands". At an archaeological site, one might be interested in, for example, whether reverberation at low frequencies is unusually long, or whether or not mid-range reverberation is significant; as we will see in a later section, at Stonehenge, there are frequencies of interest well below 100 Hz.

Digital acoustic modelling software is sometimes used within acoustical studies to predict acoustic effects. It can be used to analyse the original effects present where an archaeological site is no longer present or is damaged or altered. Murphy et al. [13] discusses auralisation, using digital tools to represent the acoustics of spaces using digital acoustic models. Weitze et al. [14] discusses acoustic modelling of Hagia Sofia in Istanbul. Details of a site are often simplified in such modelling to provide an approximation, but when exploring the acoustics of an archaeological site, one may be interested in the effect of the detail of the architecture. Such modelling programmes have certain deficiencies. For example, the Odeon acoustic modelling programme used in this project is excellent for high frequencies, but it is ineffective at some low frequencies and as a ray-tracing method becomes problematic when the length of the rays is close in magnitude to the wavelength of the source sound. Odeon's digital acoustic modelling software competitor CATT Acoustic, uses the same modelling method, and exhibits similar issues.

The standard acoustic measurement equipment used to carry out field tests can also be problematic. Precision measurement microphone capsules often require separate power supplies, connected by banana plugs to specialist cables designed for use in the laboratory or an office. The author used these during a study of the acoustics of caves featuring Palaeolithic paintings, with a methodology described in detail in previous project publications [4,15]. Despite taking care to look after them, due to the humidity these highly expensive but fragile pieces of equipment stopped working. During the same project, we used a dodecahedron loudspeaker. These are large sound sources designed to be omnidirectional, and to meet set IEEE standards, such as ISO 10140-5 (laboratory measurements), ISO 16283-1 (field measurements), and ISO 3382 Annex A (reverberation time measurements). Optimisation for such standards inevitably means compromises. Dodecahedron loudspeakers are designed to be omnidirectional, to radiate sound equally in all directions. This directionality is adequate for testing for compliance to standards, but because the speakers need to be small so that a number can be fitted into one housing, as the enclosure's design is optimised for directionality and in terms of meeting the ISO standard, the frequency response is not flat across a wide frequency range, often dropping off at low and high frequencies. For example, the NTi Audio DS3 dodecahedron loudspeaker, though an excellent sound source in many ways, has a frequency response that drops off above 10,000 Hz and below 100 Hz [16].

For measurement of archaeological acoustic spaces, Murphy [17] proposes the use as a source of a recording studio reference monitor loudspeaker that has a flat frequency response, accompanied by a sub-woofer speaker, in addition to a dodecahedron. This is intended to balance the accuracy of directional and frequency responses; however, this approach creates other difficulties. Murphy's setup was designed to systematically capture the acoustic character of an individual site. This method uses a microphone placed on a revolving turntable to capture impulse responses from various directions. This generates a large amount of data about a single source and receiver position. Like any method, it has advantages and disadvantages. It only records a single pair of source and receiver positions, and results could be different in other positions. Murphy goes on to describe (p. 224–225) how his approach captured an accurate representation of the centre of Scottish prehistoric monument Maes Howe, identifying that in the 125 Hz octave band, reverberation is significantly higher. Though a very useful approach, the study did not address the acoustic behaviour in the side chambers of the monument or in the entrance passage. It identifies modal resonant frequencies but does not discuss a range of comparative levels, nor does it present a comprehensive frequency response of all of the space. Murphy's method takes a considerable length of time to carry out one set of measurements, and such a large a complex set up is time consuming to move. At some archaeological sites, there may be very limited time available. In some cases, archaeologists restrict access in order to protect the site. In other cases, limited time is available where no other people (and resultant noise) are present, often due to the site being open to the public. Time may also be restricted due to the battery life of equipment where other power sources are not available, or because there is limited staff time available. An acoustic study in Malta by the author [18], for example, presented many of these issues, in part because the study explored the acoustics present in different parts of the monument.

Murphy discusses the work of Jahn et al. [19], which examined a number of UK passage grave or chamber tomb monuments similar in type to Maes Howe, such as Wayland Smithy. This earlier study swept the frequency of a sound source to identify modal frequencies with a hand-held soundmeter. It also attempted to calculate resonant frequencies of the spaces theoretically, with some limited success; although results were sometimes within 10% of the frequencies measured, they were never the same. This older method illustrates the advantages of the use of modern portable digital equipment, and techniques from the ISO standard. Jahn et al. do not provide the frequency response of the loudspeaker, and one cannot thus assess whether it accurately stimulated all frequencies; in addition, subjective opinion is used to identify the most powerful resonances present. Although various receiver positions are used, only one source position is used, and in trying to calculate theoretically the various parameters, approximations are made of the size of the spaces, with curved three-dimensional surfaces theorised as cubes. On the other hand, the study does effectively report a sense of human perception of the acoustics of a number of spaces.

The dodecahedron used in the author's initial field tests in caves in Spain was used with a subwoofer [4,15]. These speakers were large and heavy; the first cave being explored was at the top of a long climb up a muddy hill, and transporting the equipment was difficult. The dodecahedron needed a separate amplifier, and both it and the subwoofer required us to access a power supply in the depths of a cave. When we arrived at the entrance to the first site, the large size of the speakers meant that it was not possible to get the dodecahedron into the cave. Had we managed to manipulate the speakers through the entrance, transporting them down ladders deep underground would have proven similarly challenging. In other studies by the author, the size of equipment and access have also been problematic: a generator used to power equipment was noisy in one situation and affected recordings and testing; in another study, the generator broke down, meaning no further measurements could be taken; power was not available in a number of archaeological sites; moving a bulky loudspeaker set-up between measurements has proven time-consuming in archaeological sites, where available time is often restricted; and setting up and moving large equipment has proven difficult where there are fragile archaeological remains present.

Any single archaeoacoustic approach does not always work for all situations, no matter how rigorously it is created. When trying to understand the individual behaviour of the acoustics of an archaeological site, an approach is needed that is designed around specific research questions, and the individual conditions of the site, with qualitative as well as quantitative analysis where necessary. This is especially the case where one is exploring sites from the ancient past, for example, in prehistoric sites, where much is unknown or uncertain. In such cases, we must move beyond the fixity of standard methods, and design a methodology that suits the particular context and captures a range of experiences of acoustics in the space, while maintaining scientific rigour and using as much of the state of the art as possible. Sound archaeology is a young field, and approaches to methods used in research studies continue to need further exploration in order to balance these issues successfully.

2.2. The Sites

This project focuses on three archaeological sites: Caves in northern Spain, Stonehenge, and Paphos Theatre in Cyprus. The caves are part of the Cave of Altamira and Paleolithic Cave Art of Northern Spain UNESCO World Heritage Site, a large group of caves that feature Palaeolithic paintings and engravings. The caves explored are El Castillo, Las Chimeneas, La Pasiega, and La Garma in Cantabria, and Tito Bustillo in Asturias. They were selected by Dr. Roberto Ontanon (Director of the History and Archaeology Museum, Santander) to represent a range of different contexts. Stonehenge is a well-known Neolithic stone circle in Southern England, and also a World Heritage Site. Various phases of this site's development are examined. Paphos Theatre in Cyprus, or the Hellenistic-Roman theatre of Nea Paphos, was built around 300 BC, and was used until around 365 AD. It had seating for 8000 spectators and is again a World Heritage Site.

2.2.1. Caves

The caves studied feature paintings from as long as 43,000 years ago, a tradition of decoration with visual motifs that continues until the end of the Magdalenian period, 12,000 years ago. The five caves included have various features. La Garma was discovered comparatively recently in 1995, and is in pristine condition, as it was left when the entrance collapsed in prehistory. A point at the far end of the caves system from the original entrance featuring hand-shaped markings was selected for the present paper, next to a deep recess leading down to an underground river. Las Chimeneas features a small side chamber, with distinctive black paintings. La Pasiega was sampled in two positions, in a small turret-like feature filled with paintings, and at the end of a corridor that features many more paintings; overall, the cave features long narrow tunnels. These three caves are little changed and are closed to the public. El Castillo and Tito Bustillo are large publicly accessible show caves. El Castillo has a series of connected chambers and side sections, one of which was chosen for this study; this has the oldest dated cave painting, a red dot that is more than 40,000 years old [20]. Tito Bustillo has large chambers connected by corridors, and features small secluded side chambers; a position is featured here next to polychrome paintings of horses, again near a drop to an underground river. The Songs of the Caves multi-disciplinary project (SOTC) set out to explore relationships between the acoustics of the caves and positioning of paintings. Two existing publications [4,15] provide further contextual information about the caves as well as relevant archaeological references.

SOTC was inspired by the work of Reznikoff and Dauvois [21], who hypothesised that the positioning of cave art was related to the acoustics present at a particular position. Reznikoff [22] explored a number of caves in France and elsewhere, suggesting that paintings were placed in response to acoustic effects, including reverberation, echo, and low frequency resonance, which he stimulated with his low bass singing voice. His methodology was not published in detail; he did not include any kind of statistical study, providing indicative rather than conclusive or empirically evidenced results. While this is a somewhat speculative assertion, the evidence found by SOTC in Spain supports this hypothesis to some extent, establishing a statistical relationship between cave art and acoustics, and suggesting that in some of these examples, it is feasible that modal effects present at least contributed to the selection of these particular sites. In order to understand the relationship between acoustic effects and the placement of visual motifs, a wider scale comprehensive assessment of caves in Northern Spain and Southern France would be necessary, which would need to account for other notable features or causes that could have inspired decoration.

Diaz Andreu and Garcia Benito also explored the relationships between rock art and acoustics, and used a whistle and observation of what was heard in a range of source and receiver positions, to map acoustic effects present in an outdoor site featuring rock art. This approach lacked the ability to generate acoustic metrics, but identified evidence of a link between echoes and the presence of rock art in a Spanish valley [23]. More recently, Diaz Andreu led the Artsoundscapes project [24], and explored technical approaches to studying the relationships between rock art and sound [25]. In Scandinavia, Rainio et al. [26] have explored the relationships between rock art and acoustics.

SOTC captured more than 200 impulse responses in the caves, either next to Palaeolithic paintings or carvings, or in control positions with no visual motifs. Details of archaeological context were recorded in each position, and logistic regression analysis was used to examine whether there were statistical relationships between acoustical and archaeological features. The study found a statistical association: "between acoustic response and the positions of Palaeolithic visual motifs (...) in these caves. Our primary conclusion is that there is statistical, although weak, evidence for an association between acoustic responses measured within these caves and the placement of motifs. We found a statistical association between the position of motifs, particularly dots and lines, and places with low frequency resonances and moderate reverberation" [15]. SOTC indicated that acoustics were in some way linked to the caves better-known tradition of visual motif-making. The earliest known human musical instruments were also found in Palaeolithic caves [27], such as the Hohle Fels and Isturitz vulture bone flute-like pipes that date from c.40 to 30,000 BC, and a range of evidence suggests that sound played an important role in the ritual life evidenced by the cave art in this period.

A detailed account of the acoustic methodology used for collecting impulse responses has been published [4,15], and these collected responses were used for generating the acoustic metrics for caves discussed in this study. Sine sweep test signals were generated through a portable powered loudspeaker that had been characterised in an anechoic chamber, using WinMLS software on a laptop, results captured with a soundcard and small diaphragm condenser microphones. A dodecahedron was not used as it would have been too large, a smaller Bang and Olufsen Beolit speaker provided a source with an appropriate balance of portability, fidelity, and directionality. Two high quality DPA microphones captured two impulse responses at a time. These models are designed for recording studio rather than acoustics use and were used as they are more robust than traditional acoustics capsules, while being appropriately flat in frequency response. As the system was characterised in an acoustics lab before use, the measurement system could account for any irregularities. Further information is available on the project website [28]. In the present paper, one illustrative position that features interesting acoustics as well as significant cave paintings was selected from each cave. This is intended to provide a qualitative comparison between the acoustics in these caves and in other sites, in order to indicate some general trends, and illustrate the range of acoustical effects present at archaeological sites.

2.2.2. Stonehenge

Stonehenge is a Neolithic monument, much of which was built between approximately 3000 and 2200 BC. Parker Pearson [29,30] describes the site as a place of, or for, the dead, and the site contains a number of significant burials. It was used in the past for ritual activities that are related to the winter solstice, and is to this day the site of solstice celebrations in summer and winter. Darvill [31] suggests that Stonehenge was related to healing; in many cultures, rituals relating to healing, astronomy, ancestor worship, or the dead are related, and many such rituals use sound and music as a key element [32]. Research on the acoustics of Stonehenge was inspired by a pilot study by Aaron Watson [33,34], which showed that there were acoustic effects present at the present-day monument. Further study by the author [35,36] analysed the acoustics present in the final arrangement of the original complete monument, pointing out low frequency resonances, echoes, and reverberation. Field studies by the author with the acoustics researcher Bruno Fazenda [37,38], analysed the Maryhill Monument, a full-size concrete Stonehenge model in the United States, which has an approximately similar design. That research identified higher reverberation than in the remaining monument, and a powerful low frequency resonance at 47-48 Hz; similar low frequency modal effects are discussed further below. The small amount of reverberation and echo present is something that was described long ago by Thomas Hardy in his novel Tess of the D'Urbervilles [39]. Fazenda confirmed the presence of weak echoes and a small amount of reverberation in a further study [37]. It focused on ISO standard approaches, and did not discuss in detail low frequency reverberation time (although this is reported in a table), nor did it discuss the differences between results from different source positions or phases of the monument. The present study focuses on Stonehenge itself, rather than the concrete part-replica in Washington, DC, USA.

Stonehenge was examined in three different arrangements, using constructed digital models that draw upon a range of archaeological sources, in particular those of Parker Pearson [29,30] and Darvill [31]. A joint article involving both these researchers [40] laid out the complex sequence of Stonehenge's construction. The phases selected in the present study demonstrate the difference between an early version of the space with a large circle of bluestones, an arrangement of Sarsen stones, and a representation of the final state of the monument. Rather than following one of the various existing archaeological numberings of phases, in the present paper, the models presented are referred to simply as Stonehenge digital models A–C. The first phase modelled by the team, dated c.3000 BC, is not discussed. Only a single large stone, the Heel Stone, may have been present at this time, along with a ditch and bank. This phase has few acoustic effects, other than that the ditch and bank isolate the space acoustically to some extent from what is outside. Three subsequent phases of building were digitally modelled graphically and acoustically. The first is dated c.2900 BC (Stonehenge A), and sees the introduction of Bluestones, mostly spotted dolerite from Wales, which were erected in a large circle in what are now called the Aubrey holes, inside the Stonehenge circular ditch and bank (a reverse arrangement of a bank enclosing a ditch, which is what characterises a henge). This was a single wide circle of stones. Each stone reflects sound, and the combined result creates unusual effects in the centre, and varied acoustics across the space. A second digital model based on the arrangement of stones at around 2500 BC (Stonehenge B) sees the Bluestones removed, and replaced with large local Sarsen stones, including huge pairs of upright stones called Trilithons with lintels across the top, and an outer ring capped by a ring of lintels. A third and final phase is modelled, dated at 2200 BC (Stonehenge C), which sees the Bluestones returned in circular patterns amongst the Sarsen stones.

As many of Stonehenge's stones are fallen or missing today, digital modelling was used to create impulse responses representing what the space sounded like in the past. A model of Stonehenge was created by this author and colleagues [41] using point clouds of individual stones provided by English Heritage [42]. DXF (a standard digital graphical modelling format) graphical files of this new model were imported into Odeon acoustics software, which then generated impulse responses for pairs of source and receiver positions. These impulse responses were originally generated for use in convolution-based auralisation, within an interactive graphical model called the EMAP Soundgate [43], offering a phenomenological experiential approach to exploring the space. This was carried out as part of the European Music Archaeology Project (EMAP), a five year co-operation project carried out with the support of the Culture Program of the European Union. All of the sites in this paper were interactively modelled audio-visually within this app, allowing the user to see and hear what it might have been like to be in these sites in the past. The impulse responses were imported into Odeon software, which was used to generate a range of acoustic metrics describing the site quantitatively. These metrical results are discussed within this article.

Impulse responses of a number of different receiver positions were generated to gain a basic understanding of the differing acoustics present. A source position in the centre of the circle was used for each Stonehenge model. In the Stonehenge A model, a number of receiver positions were chosen running in a line from just inside the stone circle to the centre, to understand how sound changed as one moved through the space. In the Stonehenge B and C models, a receiver position was chosen next to the Heel Stone outside the stone circle, because the author had observed echoes at this position onsite at the moment. Two further receiver positions were chosen, Centre Source at the centre of the circle, and Inside Source inside the monument, towards the Heelstone but still inside the stone circle (see Figure 1).

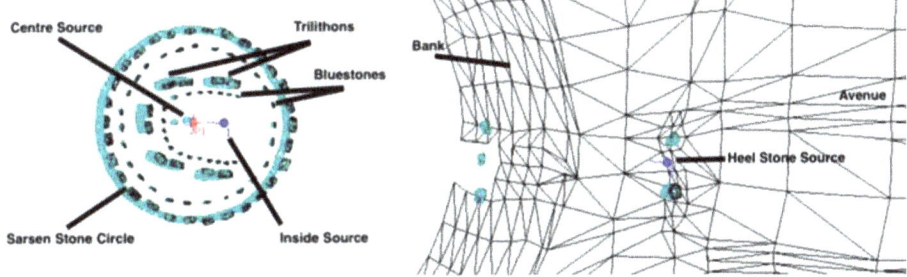

Figure 1. Source and receiver positions in the model of Stonehenge C.

2.2.3. Paphos Theatre

A monochrome archaeological digital model reconstruction of the Paphos Theatre, based on archaeological field work, was provided by the Cyprus Institute [44], and features research carried out by them with the University of Sydney [45]. Paphos was the capital of Cyprus under the Ptolemaic and then Roman administrations, the theatre was used for performance and entertainment for over six and a half centuries (c. 300 BC to the late fourth century AD). At its maximum extent during the reign of the Antonine Emperors of the second century AD, the theatre could seat over 8500 spectators. The acoustics of Greek and Roman theatres have been explored as part of the Identification, Evaluation, and Revival of the Acoustical Heritage of Ancient Theatres and Odea (ERATO) project, and in a range of publications [46–49]. The author of the present paper worked with ERATO researcher Jian Kang as part of the Acoustics and Music of British Prehistory Research Network [50], and the work of ERATO, in particular the use of raytracing-based acoustic modelling software ODEON, informed the present study. Similar to the ERATO Project, the Conservation of the Acoustical Heritage y the Revival and Identification of the Sinan's Mosques' Acoustics (CAHRISMA) project focuses on the acoustics of mosques [51].

As was the case for Stonehenge, ODEON software was used to generate a number of impulse responses and acoustic results using the digital model of Paphos Theatre. A source position on the stage was used, along with four receiver positions. Four receiver positions were chosen, in the centre of the space in front of the stage, on the third row, on the fourteenth row, and on the back row (see Figure 2). Acoustic metrics were generated for each source and receiver pair in Stonehenge and Paphos models, and for each impulse response collected in each cave space.

By the Roman period, the writings of Vitruvius [52] and the development of theatres, Odea, and other venues show that there is in Roman culture by this period an understanding that architectural design can influence the acoustics of a space. Not only do Romans appreciate acoustics, but they attempt to intentionally control them. At Stonehenge, there is no evidence that architectural design of the site is selected with the intent of controlling acoustics, but acoustics in the space change as the monument develops. Acoustic effects are noticeably present and would impact upon ritual activities in the space. In Palaeolithic caves, there is some relationship between acoustics and the paintings that are present, which suggests an appreciation of acoustics and sound in the past [15]. The acoustic ecologies of human ritual culture developed over time, from an appreciation of the acoustics of natural spaces, such as caves, to human-generated acoustics in buildings, such as Stonehenge, that were created for ritual purposes, perhaps taking advantage or notice of the acoustics present. Eventually, ritual sites feature deliberate attempts to manipulate acoustics in order to support speech and music. These sites illustrate the change of human ritual acoustic ecologies over time, as well as their variety. How the acoustics of these sites change is discussed below.

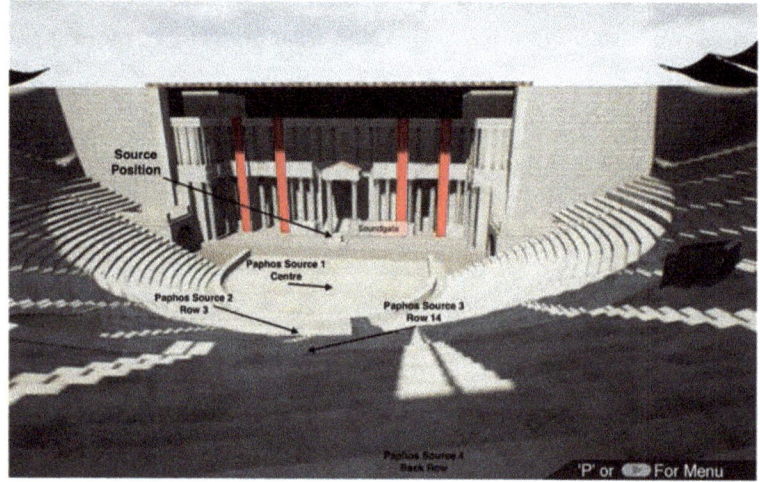

Figure 2. Source and receiver positions in the Paphos Theatre model.

2.3. Acoustic Metrics and Normal Values

Ahnert and Schmidt [53] provide a useful overview of acoustic metrics and normal values. They discuss overall parameters, including reverberation time (T30), early decay time (EDT), and bass ratio (BR); metrics related to speech, such as definition (D50), clarity (C50), articulation loss (ALcons), speech transmission index (STI or RASTI), and echo criterion for the perception of annoying reflections (EKspeech); and metrics related to music, such as the direct sound measure of the sensation of directness and nearness of the sound source (C (7)), and clarity of transparency of musical time and register (C (80)). Table 1 describes typical values of these metrics. For example, C (50) should be greater than or equal to −2 dB to avoid syllable intelligibility decreasing below 80%; this is the bottom admissible limit value for good speech intelligibility. Overall, long reverberation reduces articulation of consonants, and thus intelligibility.

In order to assess how the acoustics of ritual sites in human culture have changed over time, results for a number of these metrics in a number of sites are compared. Initially, early decay time (EDT) is discussed. Barron [54] (p. 330) compares EDT at a number of concert auditoria, and these results are used to provide contrast and a point of reference relevant to contemporary readers. As discussed above, EDT is representative of perceived reverberation [12] (p. 2), related to specific source and receiver positions. Reverb time (T30) in comparison represents the overall reverberation of the space. Comparing Barron's EDT results to those from the archaeological sites being studied is illuminating, in order to illustrate in what ways ancient and modern sites are similar and different. These results are used as a benchmark against the caves, Stonehenge, and Paphos. EDT is compared at five octaves, from 125 to 2000 Hz, as well as in two averaged results that are used in standard acoustics approaches (125–1000 Hz and 500–1000 Hz). The two cave spaces that are the closest to the concert halls are selected for comparison. The percentage difference between EDT in a selection of concert halls and in each ancient site are compared with the just noticeable difference of EDT, to provide an objective standard acoustic comparison. The variety of EDT in concert halls and caves is also illustrated. This method provides an interesting reference point and satisfactory results.

Table 1. Typical ranges [53] p. 12.

Clarity Metric	Type	Typical Range
C (7)	Directness/nearness of musical sources	>−10 to −15 dB
C (50)	Clarity (speech)	>−2 dB (>3 dB is good)
C (80)	Clarity (music)—classical music	>−1.6 dB
C (80)	Clarity (music)—romantic music	>−4.6 dB
C (80)	Clarity (music)—sacral music	>5 dB
Metric	**JND**	**Typical Range**
G	1 dB	−2 to +10 dB
EDT/T20/T30	5%	1 to 3 s
C80	1 dB	−5 to +5 dB
D50	0.05	0.3 to 0.7
Articulation Loss of Consonants (ALcons)		
ideal intelligibility		≤3%
very good intelligibility		=3% to 8%
good intelligibility		=8% to 11%
poor intelligibility		>11% to 20%
worthless intelligibility (limit value 15%)		>20%
Syllable Intelligibility		**RASTI Value**
Poor		0 to 0.3
Satisfactory		0.3 to 0.45
Good		0.45 to 0.6
Very Good		0.6 to 0.75
Excellent		0.75 to 1

Subsequently, EDT results at Paphos are compared. Results at four receiver positions are examined, as well as averaged results from these positions. EDT is then analysed in some detail at Stonehenge, the results illustrating the difficulties of standard methodology in non-standard spaces. Barron examines frequency octaves at 125, 250, 500, 1000, and 2000 Hz, whereas results for the ancient spaces are presented here an octave lower at 63 Hz and two octaves higher at 4000 and 8000 Hz. Individual octave results are explored, as are individual receiver positions, rather than a more standard approach of averaging a number of results. Frequency responses, reflection geometries, and impulse response waveforms are also examined. Finally, results at all sites are presented together, in order to address differences between EDT in different sites and eras.

A number of other metrics are then addressed. Reverberation time (T30) is discussed, showing how non-standardised distances between source and receiver positions can provide varied results. Clarity (C50) is presented at octave bands between 63 and 8000 Hz, in order to appreciate the variety at different sites and positions. Individual results for loss of consonants (ALcons), bass response (BR), and speech intelligibility (RASTI) are also presented for all sites, in order to compare a number of acoustic features. These results are then used to reveal changes in acoustics over time.

3. Results

3.1. Early Decay Time

Mostly, the caves have lower EDT times than modern concert halls, as can be seen in Table 2, but in some cases, they are remarkably similar. La Pasiega Turret has EDT that is similar to that at Royal Festival Hall, Wigmore Hall, Assembly Hall, and Free Trade Hall, and that is larger than EDT at Wembley Conference Centre.

Table 2. Difference in EDT between a position in La Pasiega cave, and a number of British concert halls.

Space	EDT 125	EDT 250	EDT 500	EDT 1000	EDT 2000	Average 125–1000 Hz	Average 500–1000 Hz
La Pasiega Turret	1.92	1.62	1.58	1.4	1.16	1.54	1.49
Royal Festival Hall, London	1.33	1.37	1.43	1.57	1.66	1.47	1.5
% difference	−31	−15	−9	12	43	−4	1
Wigmore Hall, London	1.49	1.61	1.55	1.57	1.48	1.54	1.56
% difference	−22	−1	−2	12	28	0	5
Assembly Hall Watford	1.53	1.74	1.58	1.47	1.25	1.51	1.525
% difference	−20	7	0	5	8	−1	2
Free Trade Hall, Manchester	1.47	1.47	1.65	1.77	1.62	1.60	1.71
% difference	−23	−9	4	26	40	4	15

The just noticeable difference of EDT is 5% (see Table 1), and the differences between EDT in the caves and concert halls are listed in Table 2. EDT at a number of frequencies in a number of modern venues is less than 5% different to that measured in La Pasiega, and thus not noticeably different. The average of EDT between 125 and 2000 Hz in Royal Festival Hall and Free Trade Hall, two well-known UK concert venues, is only 4% different to that measured in the La Pasiega Turret. EDT in Assembly Hall is only 1% different, and in Wigmore Hall is even more similar, only 0.26% different. If one averages EDT at only 500 and 1000 Hz, which is a standard approach in acoustics, then Royal Festival Hall is only 1% different, whereas Free Trade Hall is, in this case, 15% different, which is noticeably different. With the latter approach, Wigmore Hall is 5% different, and Assembly Hall is 2% different (Table 2). It is somewhat surprising that a cave made up of a corridor with a diameter and height of approximately 2 m, can produce reverberation which is not perceived as different to a number of modern concert halls.

Table 3 shows the equivalent results for Tito Bustillo cave. When averaged across 125 to 1000 Hz, the differences in EDT between the cave and the halls are similar to those observed at La Pasiega. Across this frequency range, EDT values in three of the halls are within 1% to 2% of that in Tito Bustillo, below the just noticeable difference. Free Trade Hall has 6% higher EDT. The difference is greater when averaged only across 500 to 1000 Hz; in this case, the differences are noticeable, between 11% to 26% different.

Table 3. Difference in EDT between a position in Tito Bustillo cave, and a number of British concert halls.

Space	EDT 125	EDT 250	EDT 500	EDT 1000	EDT 2000	Average 125–1000 Hz	Average 500–1000 Hz
Tito Bustillo	2.24	1.11	1.36	1.35	1.46	1.50	1.355
Royal Festival Hall, London	1.33	1.37	1.43	1.57	1.66	1.47	1.5
% difference	−41	23	5	16	14	−2	11
Wigmore Hall, London	1.49	1.61	1.55	1.57	1.48	1.54	1.56
% difference	−33	45	14	16	1	2	15
Assembly Hall Watford	1.53	1.74	1.58	1.47	1.25	1.51	1.525
% difference	−32	57	16	9	−14	1	13
Free Trade Hall, Manchester	1.47	1.47	1.65	1.77	1.62	1.60	1.71
% difference	−34	32	21	31	11	6	26

In Figure 3, one can see that many of the concert halls have frequency responses that are flat to different degrees. Some other halls have less flat responses, as shown in the right-hand graph. Of these, Wessex Hall and Butterworth Hall have since undergone changes that have improved their acoustics, and Royal Concert Hall now has adjustable acoustics. Before these changes, these spaces exhibited stronger low frequency EDT than at higher frequencies; this is usually a sign of modal behaviour, which means that the acoustic effects present vary in different parts of the space. For audiences, this means a different experience depending on where one sits, something that concert halls aim to minimise. Wembley Conference Centre is not designed solely as a concert hall but is used regularly as a performance space. Tito Bustillo has reverberation that is not perceptibly different to that of some of these concert halls; this time, the similarity is less surprising, as the cave is the largest studied, a cavernous open stone area. The variety present in the caves was very noticeable in situ, ranging from very dry side chambers with very little reverberation (as low as 0.11 s EDT 1000 Hz), through to large open spaces with large cathedral-like acoustic sustain (above 2.5 s EDT 1000 Hz) [15]. In La Pasiega in particular, source–receiver separation made a significant difference to perceived reverberation, and the extremely low level of background noise present (caves are very quiet) meant that reverberation seemed longer, as it could be heard for longer before being masked by environmental noise, and quite subtle effects that would be unheard outside were audible.

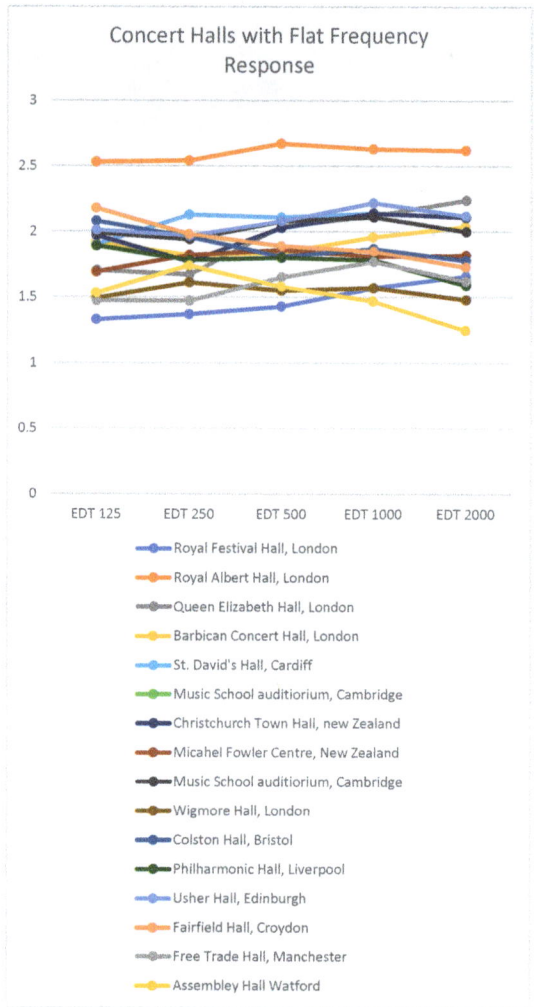

Figure 3. *Cont.*

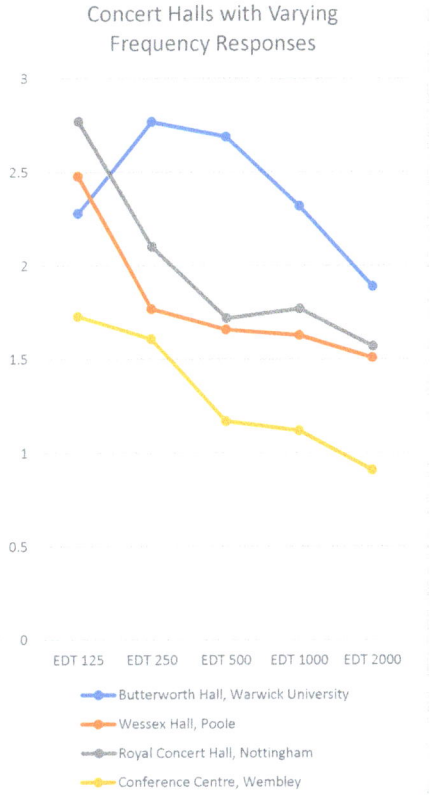

Figure 3. EDT at different frequencies in a number of British concert halls.

As can be seen in Figure 4, the caves have a range of frequency responses, but many show signs of modal resonance, evidenced by higher EDT at low frequencies. This is the case at Tito Bustillo, where EDT is noticeably different to the four concert halls discussed when one considers lower frequencies, rather than just 500 to 1000 Hz. Interestingly, the strong low frequency resonances in Tito Bustillo and La Pasiega would be missed if this study had drawn its method from the ISO standard, and only focused on 125 Hz and above. Further work is needed to explore the sound archaeology of the caves to understand the individual behaviours of the acoustics in these spaces in more detail. The existing paper by the SOTC team [15] principally studies their statistical relationships to motif positioning, further research is needed to examine the nature of the acoustic ecology of the various spaces qualitatively. The small chamber in Las Chimeneas was observed by the author when in the space to have had a noticeably powerful low frequency response. Indeed, this is the reason for its selection for use in the app discussed above [43]. Figure 4 shows high values for EDT in the 63 Hz octave, and that at higher frequencies, EDT values tail off. Las Chimeneas has some of the most dramatic images the author experienced in the caves studied (Figure 5). El Castillo has a very similar value for EDT at a low frequency of 63 Hz. El Castillo is a large cave, and the impulse response discussed here was captured in a side chamber next to a panel featuring paintings. It is likely that it is the volume of this side chamber that produces its corresponding low frequency resonance. The panel features a red dot that is over 43,000 years old and is a good example of the statistical association found by SOTC between the position of dots and lines, and low frequency resonances in particular [15] (p. 1347).

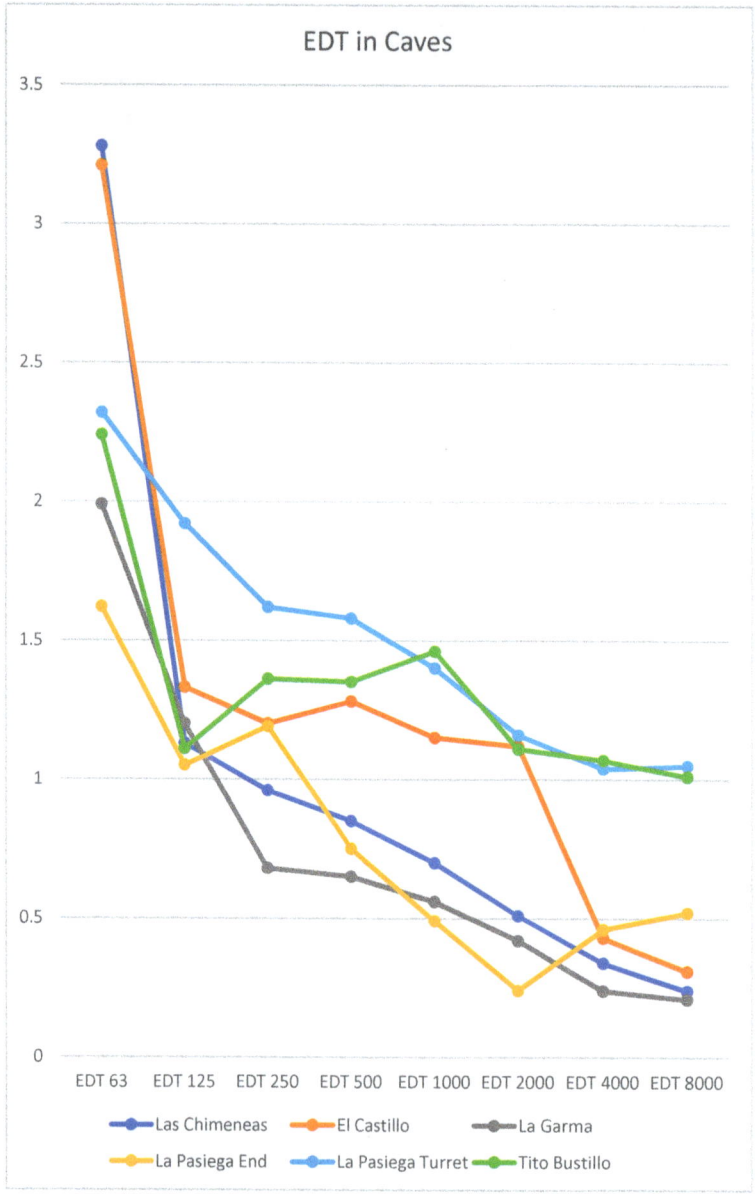

Figure 4. EDT in caves at various octave frequencies.

Figure 5. Painting of a stag from Las Chimeneas cave, image taken from the Soundgate app.

Paphos Theatre has values of EDT that are acceptably flat across a range of frequencies (Figure 6) to an extent that would make it acoustically acceptable for use as a modern concert hall. Indeed, a number of Roman amphitheatres are still in use as concert halls. Most concert halls are enclosed spaces, have a roof, but use wood or carpet for flooring. As an open roofed space, one might expect Paphos Theatre to have lower reverberation, but the stone construction and marble cladding increases reflection strength, extending reverberation time, and the stone seating contributes a stepped range of temporal reflections [47,48]. When compared across the 125 to 2000 Hz range (Figure 7), Paphos has slightly less high frequency EDT than three concert halls, but overall, it has very similar EDT values.

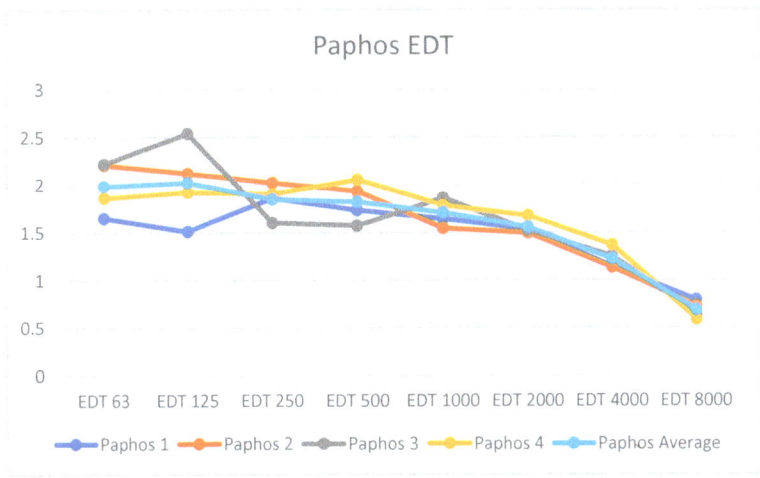

Figure 6. Early decay time (EDT) at Paphos Theatre in different source positions and at a range of frequencies.

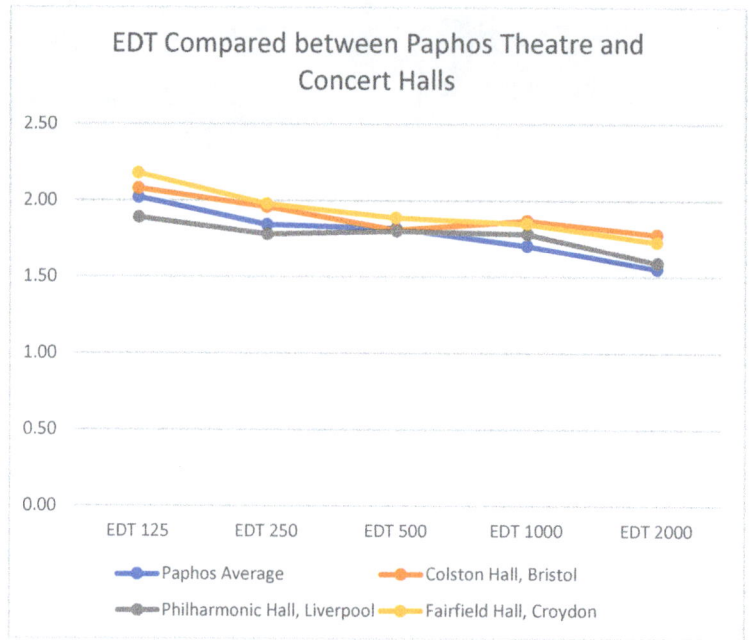

Figure 7. Comparison of EDT at Paphos Theatre with modern concert halls.

Philharmonic Hall in Liverpool is the closest match to Paphos (Table 4); the only noticeable difference between the two is in the 125 Hz octave band, Paphos has more lower frequency support, and perhaps slightly more modal behaviour. The average EDT across 125 to 2000 Hz, and the average for the standard 500–1000 Hz range is not noticeably different (2%, 1%). An audience would not notice the difference between the perceived reverberance in the two spaces according to these measures, but as we will see later, there are other differences between the acoustics of the spaces.

Table 4. Comparison of EDT at Paphos and at Philharmonic Hall in Liverpool.

	EDT 125	EDT 250	EDT 500	EDT 1000	EDT 2000	Average 125–2 kHz	Average 500–1 kHz
Paphos Average	2.02	1.85	1.82	1.71	1.55	1.79	1.76
Philharmonic Hall	1.89	1.78	1.8	1.78	1.59	1.77	1.79
% difference	−7	−4	−1	4	2	−1	2

EDT is quite different in Stonehenge (Figure 8). In Stonehenge A, a large circle of Bluestones from Wales dated to c.2900 BC [40] and with a diameter of c.87 m, there are low results for EDT both at the centre and inside the circle near the edge. The circular shape seems to cause a particular focusing of sound, with circular modes of resonance appearing. At the edge of the circle, there is a strong low frequency resonance, with an EDT of 1.31 s at 63 Hz and 1.36 s at 125 Hz, which drops off above this frequency to a very low result of 0.02 s at 2000 and 8000 Hz. This would result in a booming sound when stimulated by low frequency audio content, such as drums, thunder, or a strong wind.

Acoustics 2019, 1

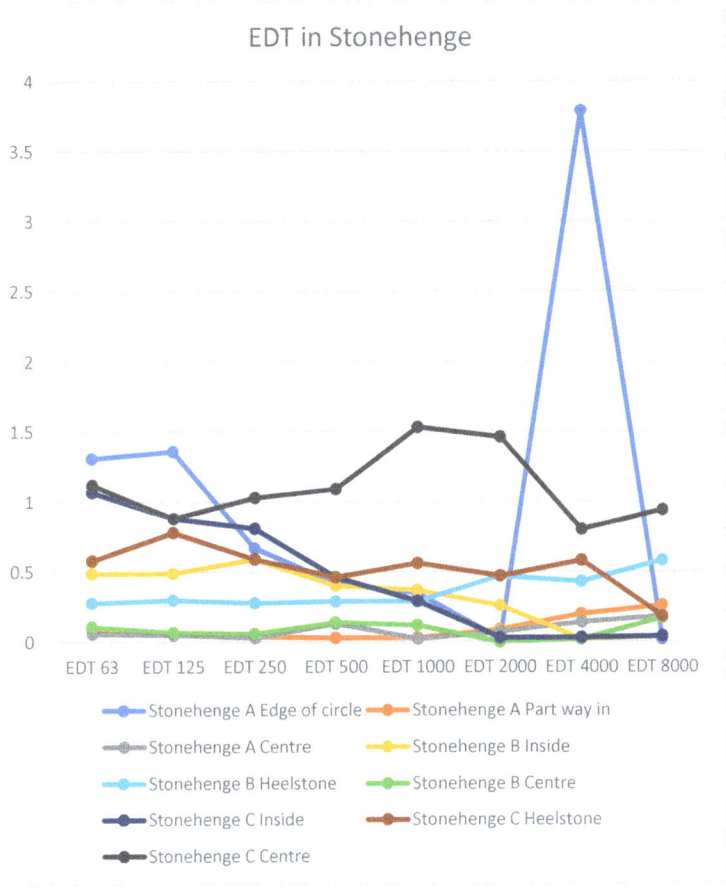

Figure 8. EDT in Stonehenge.

In the 4000 Hz octave band, there is a very high EDT result of 3.79 s. Digital modelling can cause strange results, and this figure is so extreme it seems possible this result could be a modelling error. Scattering and diffusion effects might reduce this resonance, but it seems from these results that there is a particular boost of frequencies in the 4000 Hz range. It is worth noting that 4000 Hz is where human hearing has a peak of sensitivity, and this frequency would particularly affect how one hears voices. Circular spaces can generate odd acoustic focuses and resultant modes, so this could be the result of a specific acoustic focusing. The space certainly has unusual acoustic effects that are present. This paper presents an outline of the acoustic features present, but a more in-depth investigation is warranted than is possible here to explore this and other features, and this is a suggestion for future research. Figure 9 shows the frequency response of Stonehenge A at the edge of the circle. Between 2 and 5 kHz, the space amplifies those frequencies by as much as 15 dB, whereas between 6 and 8 kHz, the level is 30 dB lower. Higher frequencies of 12 and 18 to 19 kHz are also boosted.

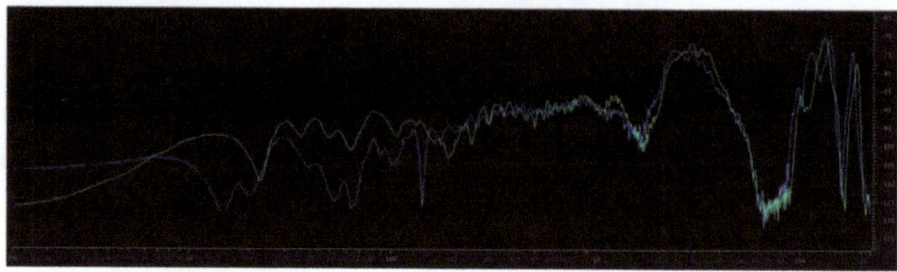

Figure 9. Stonehenge A frequency response at the edge of the circle.

Moving onwards in time to around 2500 BC, in the Stonehenge B model, the large circle of bluestones set in the Aubrey Holes is removed, and a smaller ring with a diameter of about 30 m made up of larger local Sarsen stones replaces them in the centre of the space. This arrangement is taller, with a number of stones also set inside an outer circle. It is a more complex monument, and it has less dramatic acoustic features. Low frequency resonances are present to a smaller extent, with a stronger low frequency response at the edge of the circle than at the centre (see Figure 8). The centre has very low results for EDT, and the edge of the space has EDT averaging only 0.3 to 0.4 s. This space has only subtle acoustic effects present.

The acoustics of the later (final) modelled arrangement of the site, Stonehenge C, dated c.2200 BC, has been examined in earlier publications by this author and Fazenda [35–38]. Those papers used various theoretical approaches to explore the acoustics of the site, as well as acoustic field tests at a full-size concrete replica that has a design based on Stonehenge, the Maryhill Monument in Washington State, USA. It found a T30 of 1.5 s and EDT of 1.8 s, as well as observing echoes. Two neighbouring circular/cylindrical modal resonances at approximately 47–48 Hz were identified at the Maryhill Monument by using a sine wave generator with variable frequency, as well as being calculated theoretically based on the shape of Stonehenge. In the corresponding modelled acoustics in this present study, a low level of EDT is found outside the circle, alongside little low frequency energy (see Table 5). At the centre of the circle, the EDT is over 1 s at 63 Hz, and then rises, with focused reflections from the surrounding circular stone walls generating reverberation. With an average EDT of 1.3 s (500–1000 Hz) or 1.4 s (500–2000 Hz), this suggests there would be clearly noticeable reverberation at the centre of the circle, similar results to those measured in Maryhill. A presentation of the octave band EDT does not, however, tell the whole story.

From the plotted frequency responses (Figures 10 and 11), one can see the specific resonant frequencies present. At the edge of the circle, individual low frequencies are strong, suggesting modal low frequency resonances below 200 Hz. This frequency response becomes flatter above this point, although with individual frequencies having strong resonance, for example, at around 400 Hz. A standard EDT reverberation measure based on average values at 500 and 1000 Hz would be 0.38 s, and an average of 500 to 2000 Hz is even lower at 0.26 s. EDT at 63 Hz has a value of 1.07 s when measured near the edge of the circle in the modelled acoustics, and the value would most likely be higher still at 47–48 Hz. The frequency response plot of the impulse response shows various frequencies having strong dips in EDT, perhaps caused by nodal points, which are as strong an indication of modal resonance as high values. Above 10 kHz, the EDT is very strong, another result missed by standard averages. This illustrates an issue with standard acoustic methods, in a space such as this that has odd acoustic behaviour. The space has unusual resonances at low and high frequencies, as well as less prominent but idiosyncratic mid-range responses.

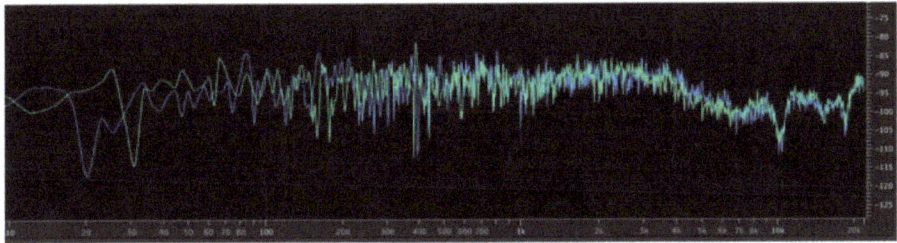

Figure 10. Frequency response of Stonehenge C digital model at the edge of the stone circle.

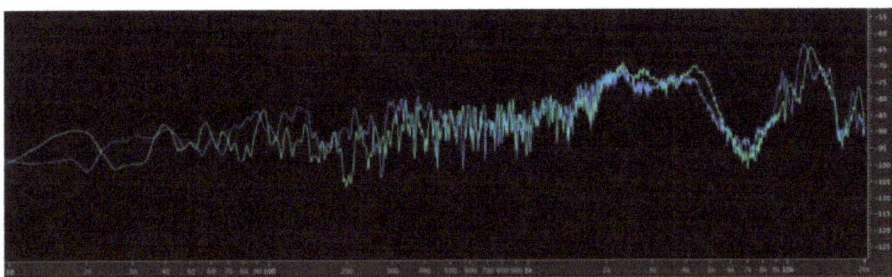

Figure 11. Frequency response of Stonehenge C digital model at the centre of the stone circle.

In both Stonehenge itself and the Maryhill Monument, the author found that speech was subjectively particularly clear, and was remarkably so at specific source and receiver positions. When walking around the space, suddenly one would hear a voice from another part of the circle with great clarity, despite coming from some distance. It is suggested that this effect is due to circular focusing, which may take place in elliptic, circular, parabolic, or hyperbolic patterns (Figure 12) [55], depending on the source position.

Figure 13 overlays an image of Stonehenge model C over these focusing effects. Elliptic focusing would mean that the voice of someone standing just inside the circle, between the arms of the horseshoe shape of Bluestones, might find their voice focused between the stones of the largest stone trilithon, just past the altar stone. Archaeologists now suggest that the horseshoe shape illustrated here was actually an complete oval shape, and the source position illustrated would be just inside this oval, which conveniently marks out where one might stand in the space to achieve elliptic focusing effects. The central position in Stonehenge could produce a particularly strong focusing of sound for someone standing there. Making sound just in front of the largest trilithon would produce parabolic focusing, adding support to a voice or other sound source. The focusing could also work in reverse, and sound from far off may be easier to hear. A source just inside the Bluestone circle, the second circle inside the outer sarsen circle but outside the horseshoe shape of Sarsen trilithon pairs, may have created hyperbolic focusing, which could make a sound made just inside the stone circle, or when hidden behind the largest trilithon, appear to come from outside the stone circle. When one explores these focusing effects on the larger Stonehenge A model, this position of focus outside the circle is at the position of the Heel Stone. The two circles of stones would produce combined focusing effects, and with a number of people present in the circle, acoustic focusing effects could have been complex, unpredictable and confusing, subtle but certainly noticeable, and possibly considered by Neolithic people to be supernatural in a monument to ancestors within an animist culture.

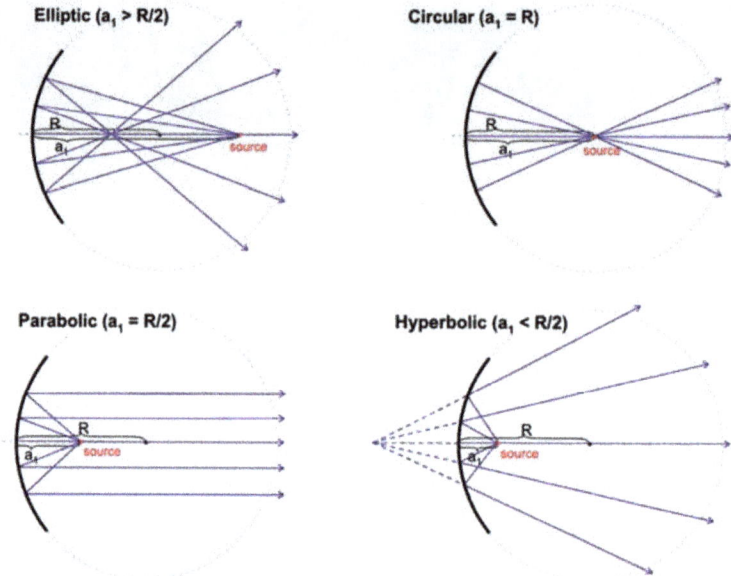

Figure 12. Focusing effects in a circle.

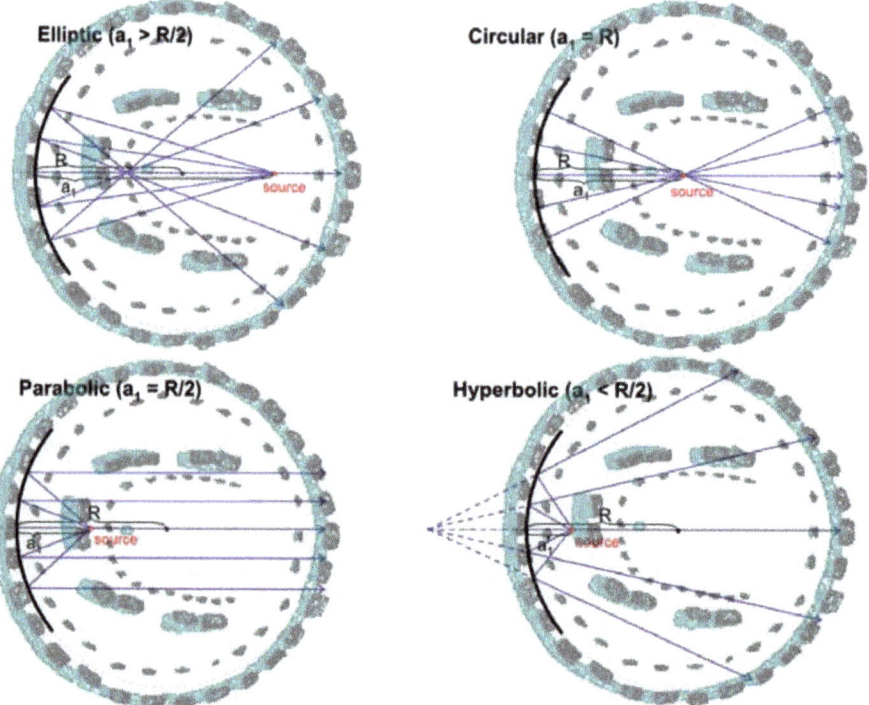

Figure 13. Possible circular acoustic focusing effects at Stonehenge.

Figure 13 also illustrates how sound bounces around the space, creating echoes. These echoes can be seen in the representations of impulse responses in the space that can be seen in Figure 14. Figure 15 and Table 5 illustrate EDT in various spaces. Overall, Stonehenge has acoustics that are irregular, and are very different to those at Paphos. They are closer in nature to some of the caves discussed, in that both have strong low frequency effects, as well as a non-linear acoustic response. The caves have varied EDT results, some of the which are closer to those in Paphos.

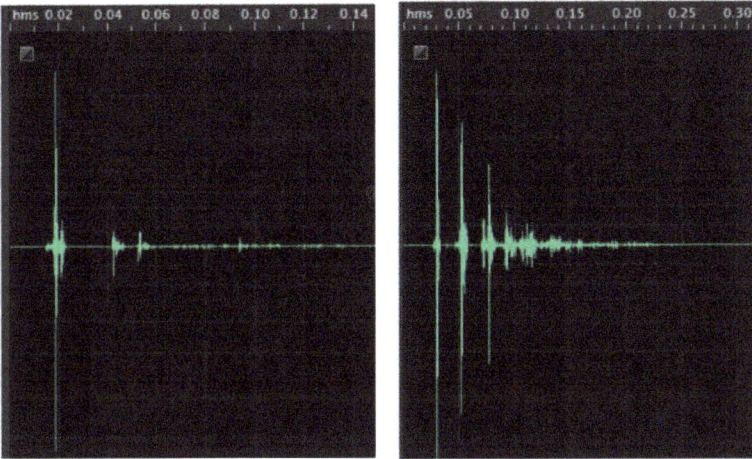

Figure 14. Waveform of Stonehenge C digital model at the edge (**left**) and centre (**right**) of the stone circle.

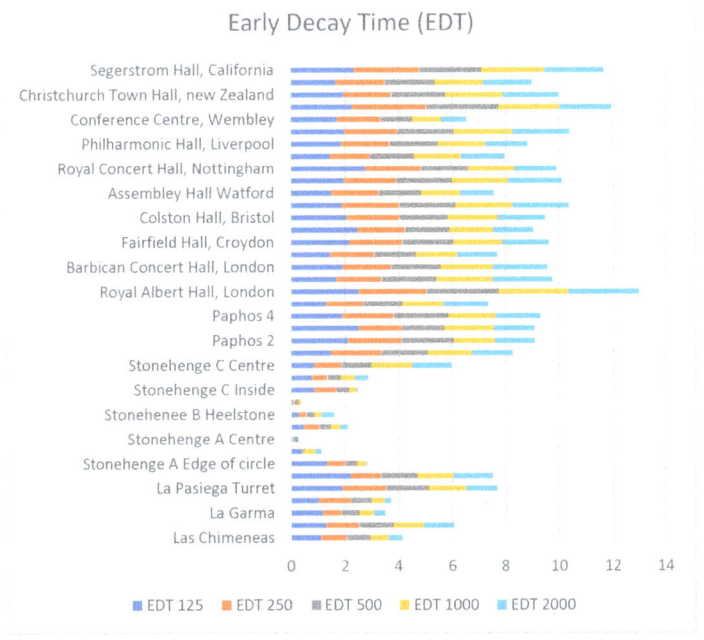

Figure 15. EDT in various sites.

Table 5. EDT in various spaces.

Space	EDT 125	EDT 250	EDT 500	EDT 1000	EDT 2000
Las Chimeneas	1.13	0.96	0.85	0.7	0.51
El Castillo	1.33	1.2	1.28	1.15	1.12
La Garma	1.2	0.68	0.65	0.56	0.42
La Pasiega End	1.05	1.19	0.75	0.49	0.24
La Pasiega Turret	1.92	1.62	1.58	1.4	1.16
Tito Bustillo	2.24	1.11	1.36	1.35	1.46
Stonehenge A Edge of circle	1.36	0.67	0.42	0.35	0.02
Stonehenge A Part way in	0.43	0.05	0.03	0.43	0.2
Stonehenge A Centre	0.05	0.03	0.13	0.02	0.07
Stonehenge B Inside	0.49	0.59	0.4	0.37	0.26
Stonehenge B Heelstone	0.3	0.28	0.29	0.29	0.47
Stonehenge B Centre	0.07	0.06	0.14	0.12	-
Stonehenge C Inside	0.88	0.81	0.46	0.29	0.03
Stonehenge C Heelstone	0.78	0.59	0.46	0.56	0.47
Stonehenge C Centre	0.88	1.03	1.09	1.53	1.46
Paphos 1	1.51	1.86	1.73	1.64	1.53
Paphos 2	2.12	2.02	1.93	1.54	1.49
Paphos 3	2.54	1.6	1.57	1.86	1.52
Paphos 4	1.92	1.9	2.05	1.78	1.67
Royal Festival Hall, London	1.33	1.37	1.43	1.57	1.66
Royal Albert Hall, London	2.53	2.54	2.67	2.63	2.62
Queen Elizabeth Hall, London	1.7	1.67	2.04	2.12	2.24
Barbican Concert Hall, London	1.92	1.82	1.84	1.96	2.04
Wigmore Hall, London	1.49	1.61	1.55	1.57	1.48
Fairfield Hall, Croydon	2.18	1.98	1.89	1.85	1.73
Wessex Hall, Poole	2.48	1.77	1.66	1.63	1.51
Colston Hall, Bristol	2.08	1.96	1.81	1.87	1.78
St. David's Hall, Cardiff	1.9	2.13	2.11	2.14	2.11
Assembly Hall Watford	1.53	1.74	1.58	1.47	1.25
Music School auditorium, Cambridge	1.99	1.94	2.08	2.11	2
Royal Concert Hall, Nottingham	2.77	2.1	1.72	1.77	1.57
Free Trade Hall, Manchester	1.47	1.47	1.65	1.77	1.62
Philharmonic Hall, Liverpool	1.89	1.78	1.8	1.78	1.59
Usher Hall, Edinburgh	2.01	1.97	2.08	2.22	2.12
Conference Centre, Wembley	1.73	1.61	1.17	1.12	0.91
Butterworth Hall, Warwick University	2.28	2.77	2.69	2.32	1.89
Christchurch Town Hall, New Zealand	1.97	1.77	2.03	2.14	2.11
Michael Fowler Centre, New Zealand	1.69	1.82	1.86	1.81	1.82
Segerstrom Hall, California	2.36	2.44	2.3	2.38	2.2

3.2. Reverberation, Clarity, and Bass Response

T30 is a measure of reverberation that relates to the overall reverberation of the space, rather than the relationship between source and receiver positions. In the caves, the results for T30 are mostly similar to those for EDT, other than in La Pasiega End (Table 6). In this space, T30 is very long, 18.4 s at 125 Hz, reducing gradually at higher frequencies. The space is very long and narrow, a corridor or tunnel. While in the space, the research team heard little reverberation when the source and receiver were close together, whereas there was long complex reverberation when the source and receiver were further apart. At Stonehenge A, T30 was higher than EDT. At Stonehenge B, T30 was a little higher inside the circle, while at Stonehenge C, T30 was a little longer than EDT, but results were less extreme. At Paphos, results for T30 were comparable to EDT results, but with less high frequency drop off.

Table 6. T30 Reverberation in La Pasiega Cave.

	T30 63	T30 125	T30 250	T30 500	T30 1000	T30 2000	T30 4000	T30 8000
La Pasiega End	7.2	18.4	13.7	8.8	7.2	4.3	2.5	0.9

Speech clarity C (50) should be greater than or equal to −2 dB to avoid syllable intelligibility decreasing below 80%, this is the bottom admissible limit value for good speech intelligibility; greater than −3 dB is a good value. In Table 7, results for C (50) that indicate poor speech clarity are marked in red; typical ranges are all taken from British Standards [11] (p. 12).

Table 7. C (50) Clarity of Speech results.

	C (50) 63	C (50) 125	C (50) 250	C (50) 500	C (50) 1000	C (50) 2000	C (50) 4000	C (50) 8000
Las Chimeneas	−8.3	−3.7	0	2	2	4.9	8.2	12
El Castillo	−22.2	−26.4	−32.2	−30.2	−31.8	−37.3	−37.5	−39.5
La Garma	−2.9	−1.8	3.6	3.3	4.2	7.3	11.1	12.3
La Pasiega End	−9.5	−7.7	−6.7	−4.2	−4.6	−2.3	−1.7	−5.5
La Pasiega Turret	−15.3	−16.9	−22.5	−26.7	−32	−35.7	−35.3	−36.4
Tito Bustillo	−13.7	−18.9	−19.4	−24.4	−28.5	−34.5	−32.4	−32.6
Stonehenge A 1m in circle	1.6	0.2	5	7.8	11	26.4	34.9	35.4
Stonehenge A Inside	24.8	19.1	21.4	17.3	18.5	26	29.5	36.3
Stonehenge A Centre	25.1	27.4	26.3	25.3	24.6	32.5	35.2	39.2
Stonehenge B Inside	5.1	6.2	5.7	8.5	10.7	17.1	18.9	21.4
Stonehenge B Heelstone	10.5	8.2	9.5	11.3	10.6	10.5	11.3	17.2
Stonehenge B Centre	24.5	26.9	28.2	29.6	29.3	-	41.5	44.6
Stonehenge C Inside	−0.2	0.4	2.7	6	10.8	19	21.3	21.8
Stonehenge C Heelstone	−2.8	−2.4	−1.4	0.3	−0.4	4.8	4.5	12.9
Stonehenge C Centre	−10.5	−12	−13.7	−16.9	−14.9	−7.7	−7.8	−7.9
Paphos 1	−26.5	−19.3	−16.4	−14.2	−10.2	−3.2	−1.5	4.4
Paphos 2	−8.9	−7.7	−5.6	−2.3	1	1.1	2.3	2.6
Paphos 3	−3.7	−5.3	−3.4	−4.6	−6.7	−4	−0.7	3.1
Paphos 4	−11.7	−6.6	−6.8	−6.9	−4.8	−2.7	−0.5	2.1

In the caves, there are very mixed C (50) results, with Las Chimeneas and La Garma having very high values, while in the other caves clarity is poor. Other measures of intelligibility produce differing results. ALcons measures loss of consonants due to acoustic effects. These results were averaged across the frequency spectrum, and all the caves have results between 3 and 8, which indicates very good intelligibility. RASTI values for the caves indicate that speech intelligibility would be very good or excellent, according to the typical values in Table 1.

At Stonehenge, C (50) speech intelligibility is very good (Table 7) except in the final stage of development at the centre of the circle. At this position, C (50) results suggest that at each octave speech clarity is poor. However, this is a measurement in which the source and receiver position are both near the centre, and as we have seen, focusing could produce very different results at specific source and receiver positions. ALcons shows little loss of consonants, and RASTI shows mostly excellent results with some very good speech intelligibility (Table 8). At Paphos, the measurements represent a source on the stage, and receivers placed in the centre, in the third row, near the front, in the middle of the seating in row 14, and right at the back. C (50) results suggest that the first position is not good in terms of clarity. Indeed, this is a position for performers, rather than the audience. All positions have poor clarity in low frequencies, but position 2, the third row, has the best speech clarity, with good results at 1000 and 2000 Hz, while on row 14 and at the back, results suggest intelligibility is not good. From 4000 Hz and above, all positions have good speech clarity. ALcons and RASTI results for intelligibility are all good.

Table 8. ALcons, RASTI, and bass ratio results.

	ALcons	RASTI	BR (SPL)
Las Chimeneas	5.01	0.68	2.2
El Castillo	4.51	0.68	11
La Garma	3.74	0.73	5.3
La Pasiega End	3.88	0.77	2
La Pasiega Turret	5.88	0.62	3.2
Tito Bustillo	6.98	0.64	5.5
Stonehenge A 1m in circle	2.07	0.91	−1.5
Stonehenge A Inside	1.22	0.97	−2.8
Stonehenge A Centre	1.91	1	−7.8
Stonehenge B Inside	1.74	0.87	−4.6
Stonehenge B Heelstone	1.91	0.88	0.9
Stonehenge B Centre	1.14	1	−2.9
Stonehenge C Inside	2.03	0.86	−5.3
Stonehenge C Heelstone	5.33	0.67	−2.2
Stonehenge C Centre	4.52	0.71	−2.8
Paphos 1	9.65	0.5	1.8
Paphos 2	9.05	0.55	−2.1
Paphos 3	10.02	0.52	−0.6
Paphos 4	10.66	0.5	0.3

Speech intelligibility in the spaces varies. In Las Chimeneas and La Garma caves, voices with low frequency content, men with particularly low voices for example, would have an effect added to their voices, which would blur consonants, and yet allow for speech to be intelligible as sibilant higher frequencies, for example, were unaffected. At the end of La Pasiega, low frequencies would be most affected, but all frequencies would be modulated. In El Castillo, the La Pasiega Turret, and Tito Bustillo, men with higher voices and women would find that their voices would be most affected, as they contain greater high frequency content. In these caves, speech would be more intelligible at some frequencies than others, but an effect would be added to all voices. At Stonehenge, speech intelligibility is good. In the final phase of development, someone standing in the centre would find the lower range of their own speech somewhat unintelligible; others listening to it from another position would still be able to hear it, underlining again the significance of this specific position. At Paphos, speech intelligibility results tell us that the third row would be a far preferable seating position to the 14th row, or back row; one would still understand speech at the back, but it would be less clear. All of these results would be affected by people filling the space, as people act as high frequency absorbers, further increasing clarity.

Musical clarity C (80) is very low in El Castillo and La Pasiega Turret, as well as in the centre of Stonehenge C, and at position Paphos 1, the stage position. Directness of musical source C (7) results are typical in Las Chimeneas and La Garma, and very low in other caves. Musical sources directness is typical in all Stonehenge models at the inside receiver positions for low frequencies, but music with higher frequencies would be highly direct. At the Heelstone positions of models B and C, musical sources do not seem near or direct, especially at low frequencies. In model A, musical sounds would be the most direct, whereas in model B, results are typical, and in model C, directness is very low. Once more, this illustrates the unusual acoustic nature of the central of Stonehenge. Low frequency musical instruments, such as large drums, could have had an interesting effect at Stonehenge.

Bass ratio, BR(SPL), describes reverberation time at low frequencies. For music, the modern standard desirable BR is 1.0 to 1.3; for speech, it should be 0.9 to 1.0 [53]. In the caves, all results are higher than what one might desire for speech. Other metrical results at low frequency octaves also suggest that bass effects are present in some positions. BR measures the ratio between reverberation energy in the 125 and 250 Hz octaves and in the range between 500 and 1000 Hz. This gives no information about the 63 Hz octave range; it is a measurement designed to assess the warmth of a concert hall acoustic, and most concert halls have results close to 1. In all of the caves, the results are considerably higher than one would expect in a concert hall, and there is far more bass than is normal or desirable. At Stonehenge, all but one of the results are negative, and the ratios are much higher than one would wish. Again, the bass response is unusual and high. At the Heel Stone receiver position of Stonehenge B, the bass response is more acceptable, but this is quite some distance from the source. In the Paphos model, the bass ratio is very high, and at the back it is very low; it is again negative in some positions. There may be a particular position in the space where the bass ratio is perfect, but it seems that the bass response is irregular and uncontrolled at Paphos.

4. Conclusions

The study of the acoustics of these sites involved working with digital models, and as mentioned above, these were later integrated into an audio-visual museum exhibit and app for PC, Mac, tablets, and smartphones, the EMAP Soundgate [43], which provided an easily appreciable understanding for a diverse audience. The impulse responses discussed in this paper are all auralised, integrated into that app, and one can explore and compare the acoustics intersubjectively in a kind of experimental phenomenology. Further research will integrate a greater range of impulse responses into the app to more accurately represent the acoustic ecologies of the sites. As these sites do not have diffuse reverberant fields, and because the variations in acoustics in the sites are interesting, a large number of impulse responses are needed to adequately represent the acoustic effects present. Although the mixed methodology in this paper should be relevant to a range of readers, the technical terminology means that such use of apps assists with dissemination to a wide audience. This paper explored a very small number of impulse responses, and further research is needed to understand and appreciate the acoustic effects present in each of these sites, and what we might learn from them, including an in-depth individual assessment of each of the three types of site.

Further work might include: An exploration of lateral fraction and envelopment metrics; exploration of low frequency resonances; mapping of patterns of modal resonance; exploring echoes; further identification and exploration of positions of particular sonic interest; and gaining a better understanding of why some of these effects are present. Many other archaeological sites have had no study of their acoustic properties, and a more complete survey of a range of sites will help to provide a better understanding of how acoustic ecologies have changed over time. Further experimental work will explore the use of virtual reality headsets in order to immersively experience the auralised acoustics of ancient sites. While this study provides examples of how acoustic effects present can differ in archaeological sites, such as Palaeolithic caves, Neolithic stone circles, and classical Hellenistic antiquity, much further work is required to comprehensively survey how human experiences of acoustics changed across the millennia. Comparison of the acoustics in a wider range of painted French and Spanish Palaeolithic painted caves would help to better characterise their acoustic ecology. Study of a larger number of source and receiver positions in the various phases of Stonehenge would help us to understand the effect of acoustics on ritual activities related to this famous stone circle. Exploration of the different phases of development of Paphos Theatre, as well as the effect of different surface materials, would assist in knowing more about audiences' experiences in the space.

This research illustrated the importance of constructing a specific set of methods suited to sound archaeology rather than uncritically adopting standard methods from contemporary building or industrial acoustics, which often has quite different research problems and questions. In this case, the methods adopted differ from standard approaches in that they: include comparison of different spaces; use a range of different source and receiver positions; explore a wider frequency range, including 63 and 8000 Hz octaves; evaluate individual octave results for metrics such as EDT, T30, and C (50); explore metrics related to musical use, such as C (80), C (7), and BR, as well as speech-related metrics, such as RASTI, C (50), and ALcons; explore low frequency effects in detail; compare results to those for contemporary performance venues, such as concert halls; use a consumer B&O Beolit loudspeaker as a source instead of, for example, a dodecahedron, because of its portability; and use recording studio/live professional audio DPA microphones as receivers, rather than acoustic testing capsules and power conditioners, due to their robustness.

The acoustic metrics examined produced a number of findings. The caves and Stonehenge have varied EDT and T30 reverberation, with modal responses at low frequencies in particular, meaning that there are acoustic effects present in particular positions, and specific frequency ranges. Speech clarity metrics are different depending on whether you use C (50), ALcons, or RASTI values. In some cases, speech is very clear, in others, effects are added to voices or speech is incomprehensible. In the caves, speech is clear when judged using ALcons or RASTI, but at different octaves identified by C (50), clarity of speech can be either extremely high or low, indicating speech is understandable, but may be changed by support or transformation of one or other frequency range. At Stonehenge, speech transmission is good, but the voices of people standing at the centre of the monument are particularly transformed. Stonehenge A has a particular boost of frequencies between 3 and 6 kHz, which would aid the hearing of human voices, and has a further boost of very high frequencies, affecting the perception of sibilants or other high frequency consonants. At Paphos, speech is transmitted clearly in general, but clarity is better when sitting near the front of the seating, with sound becoming less good as one moves towards the back of the venue; as one might expect, the best seats in the house are at the front. Overall, the musical clarity results similarly imply acoustics effects are present, rather than the acoustic supporting or detracting from musical performance.

The bass response metric shows that the caves have strong low frequency support, Stonehenge has strong and unusual low frequency effects, and that at Paphos, the bass response is variable and uncontrolled. Low frequencies would not be common in antiquity, and it is notable that all three sites feature low frequency effects, something that may have been a particular marker of special experiences. Low frequencies are often enhanced in these spaces, sometimes in an extreme manner. Sounds are transformed as one enters or moves through a space, with effects added that contribute to a perception of its special or sacred nature. Each cave is inhabited by an individual sonic character that plays a part in how one perceives the space, providing a contextualisation, a sense of place, and of identity. In a prehistoric culture dominated by animism, in which all of nature was alive with spirits, the acoustics of each cave helped to bring the space to life, providing intersubjective agency that transformed the experience of anyone present.

Acoustics at Stonehenge change during its developing phases of construction, and change as one moves towards its centre, suggesting the site is more dynamic than might be expected of a contemporary home for the dead. The centre of the final phase of development of the site provides a particular focus of sonic affect. Focusing effects in the space are complex and unpredictable in the outdoor, natural, and windswept environment, and the changes in sound present, depending as they do on the specific position where sound is made and heard, would have contributed a sense that this was a place that was alive, as well as making individual positions in the space of particular importance, or adding a sacred sonic effect to the person standing there making sound or listening. The human architectural choices of Stonehenge's construction create an interaction between space, place, and acoustic ecology, allowing human agency to creatively participate in the generation of the voice of a home for the ancestors. Echoes, low frequency effects, sound focusing, amplification of particular frequencies, and reverberation provide acoustic effects that change and shift in different circumstances, changing voices and the sound of other ritual activities in a variety of ways. Specific positions, occupied by ritual specialists, afford human culture the opportunity to structure communication with sacred otherworldly powers. Paphos Theatre has a more refined sound, comparable to that of a concert hall, human designers overtly working to create an acoustic ecology controlled by them, and that supports the secular-if-ritualised performances occurring in the building. Such human rituals develop social relationships, with the most wealthy or powerful occupying prime positions, visible to those seated behind, and lower down the social stratigraphy.

The deeper parts of caves provided a very particular and powerful acoustic for humans in Palaeolithic times. In an animist cosmology, the lack of background and environmental noise differentiated caves from outdoor spaces. With no experience of stone buildings, these were alien natural spaces that featured variable reverberation, low frequency effects, and transformation of sounds made by human speech and movement. These were natural formations, humans entered into them, becoming enveloped by these other worlds' acoustics, going into an environment over which they had no control, leaving as a record of their presence visual motifs charged with spiritual meanings, and engaging with altered states and what they probably regarded as powerful supernatural forces. Stonehenge is human-built, representing or a home for ancestral spirits or a place of healing, with acoustics that its builders could observe changing as they redesigned the site in a number of phases of development.

Over time, the acoustics of the site become more sophisticated, much as ritual culture similarly developed, the landscape being imbued with sonic character by human agency. Humanity began to exert control over its environment, clearing forests and moving from a hunter-gatherer lifestyle towards agriculture, beginning to construct an architectural tradition to monumentalise these cultural changes, with acoustics an artefact and evidence of that process. At Stonehenge, acoustic ecology is human-made, but a by-product of wider cultural factors. At Paphos, design choices are made to overtly manipulate the acoustics present, in order to facilitate the support of speech and music. Humans began by this point to attempt to control their acoustic ecologies, to consciously design their surrounding soundscapes. Over time, acoustics become more predictable, as ritual acoustics move from a Palaeolithic variety of selected natural spaces in which the acoustics present are appreciated by early humans; to the construction of Neolithic monuments, which resulted in human-made sites with a range of unusual acoustic effects present; and finally, a literate Hellenistic culture, which consciously manipulated acoustics designed to flatter speech and music. It is perhaps these changes in acoustic cultures that are a more significant development than the individual variations in the acoustic metrics present.

The earliest (cave) sites have natural sonic effects, which seem to have played some part in attracting Palaeolithic humans to carry out ritual acts within them. These effects are replicated in the human built stone circle at Stonehenge, and over time, the development of the architecture of the site made these effects more focused on specific positions, affecting low frequencies while allowing higher frequencies associated with speech to retain clarity. Although we cannot know to what extent acoustic concerns were in the minds of those designing the arrangement of stones, the change in acoustic ecology would have been noticeable and may well have become associated with the rearrangement of the site in different phases; specific acoustic effects may well have become associated with particular architectural shapes. The reverberation and echoes present would have diminished or disappeared when stones were removed, and reappeared when new stones were erected. Such acoustic changes and effects may have been attributed to supernatural forces. One possible interpretation could be, for example, that they were thought of as the voices of spiritual forces associated with individual or groups of stones.

It should be stressed that there is no evidence that the design of Stonehenge was based on acoustics, but the acoustic effects present in each stage are prominent and different enough to suggest that acoustics may well have begun to be associated with specific architecture, much as concert hall architects draw upon the designs of existing spaces with good acoustics, when planning new buildings. It is at least possible that Stonehenge's designers believed that animist or ancestral spirits were pleased by the circular arrangements of the stones when they were erected, when they heard the voices of these spirits seem to appear through acoustic effects. This may even have provided confirmation that these spirits had successfully inhabited the monument, validating the design choices made. This might have perhaps transferred a tradition of spiritual power from sites with similar acoustics, such as at the Welsh source of the bluestones, or through a vestigial memory of acoustics in sacred caves. With no written sources for confirmation, such suggested interpretations remain inevitably speculative. We do know that by the Roman period these processes of human architecture had developed enough to allow for the conscious design of performance venues with supportive acoustics, as is evidenced in the theatre at Paphos. Caves present a naturally existing acoustic, a place occupied by spirits, which humans could visit to attempt to interact with these acoustically manifesting spiritual powers; Stonehenge has a human-made acoustic in which a complex ritual culture collaborates with audible ancestral spirits in order to exert human control over natural forces; Paphos illustrates a sophisticated rational culture, in which a human logic has established how acoustic effects can be manipulated, and interact with social stratification, free from the influence of gods.

Changes in cosmology are reflected and represented in the acoustic ecologies present in these sites, from a strong relationship as part of nature in hunter-gatherer prehistoric cultures in caves, through the monumental efforts of early farming communities to entreat spiritual powers to control the land, to the larger societies and complex performative social interaction of Rome. The acoustic ecology of these archaeological sites contributed to their ritual significance and afforded particular effects that made the activities within seem alive, or even larger than life, something different to the natural spaces outside, places where meanings were manipulated and transformed. These spaces have different acoustic characters, but each has noticeable acoustic effects present, which would have been integrated into any activities in the space, contributing to the generation of meaning, and playing an important part in creating a sense of set and setting. Over time, those sets and settings became increasingly controlled and manipulated by their users, enhancing the experiences of those within, turning space into place, position into context. This process of change at these three types of sites illustrates the emergence of architectural acoustics, and outlines its significance to human culture.

Funding: This research was funded by the UK Arts and Humanities Research Council and Engineering and Physical Sciences Research Council, (grant number AH/K00607X/1) as part of the Science and Heritage Programme; and by the Culture Program of the European Union (EACEA reference number 536370).

Acknowledgments: Access to the caves was only possible because of the support and commitment of the Gobierno de Cantabria and Gobierno Del Principado de Asturias, the regional government bodies in the area. Research on Stonehenge was assisted by the kind advice and support of English Heritage and Historic England, as well as Mike Parker Pearson and Ben Chan. Paphos Theatre research was with the support of the Cyprus Institute and University of Sydney.

Conflicts of Interest: The authors declare no conflict of interest. The funders had no role in the design of the study; in the collection, analyses, or interpretation of data; in the writing of the manuscript, or in the decision to publish the results.

References

1. Hodder, I. *Symbolic and Structural Archaeology*; Cambridge University Press: Cambridge, UK, 1982.
2. Hodder, I.; Hutson, S. *Reading the Past: Current Approaches to Interpretation in Archaeology*, 3rd ed.; Cambridge University Press: Cambridge, UK, 2003.
3. Renfrew, C.; Bahn, P. *Archaeology: Theories, Method and Practice*, 4th ed.; Thames & Hudson: London, UK, 2004.
4. Till, R. Sound Archaeology: Terminology, Palaeolithic Cave Art and the Soundscape. *World Archaeol.* **2014**, *46*, 292–304. [CrossRef]
5. Solomos, M. A Phenomenological Experience of Sound: Notes on Francisco Lopz. *Contemp. Music Rev.* **2019**, *38*, 1–2. [CrossRef]
6. LaBelle, B. *Acoustic Territories: Sound Culture and Everyday Life*; Continuum: New York, NY, USA, 2010.
7. Ansell-Pearson, K.; Schrift, A.D. *The New Century: Bergsonism, Phenomenology and Responses to Modern Science*; Taylor and Francis: Abingdon, UK, 2014.
8. Thrift, N. *Non-Representational Theory: Space, Politics, Affect*; Taylor and Francis: Abingdon, UK, 2008.
9. Dewsbury, J. Witnessing space: Knowledge without contemplation. *Environ. Plan.* **2003**, *35*, 1907–1932. [CrossRef]
10. Long, M. *Architectural Acoustics*, 2nd ed.; Academic Press: Waltham, MA, USA, 2014.
11. British Standards Institute. *Acoustics: Measurement of Room Acoustic Parameters. Part 1: Performance Spaces (ISO 3382-1:2009)*; International Organization for Standardization: Brussels, Belgium, 2009.
12. Bradley, J.S. Review of objective room acoustics measures and future needs. *Appl. Acoust.* **2010**, *72*, 713–720. [CrossRef]

13. Murphy, D.; Shelley, S.; Foteinou, A.; Brereton, J.; Daffern, H. Acoustic Heritage and Audio Creativity: The Creative Application of Sound in the Representation, Understanding and Experience of Past Environments. *Internet Archaeol.* **2017**, *44*. [CrossRef]
14. Weitze, C.A.; Rindel, J.H.; Christensen, C.L.; Gade, A.C. The Acoustical History of Hagia Sophia revived through Computer Simulation. In Proceedings of the Forum Acusticum, Sevilla, Spain, 16–20 September 2002.
15. Fazenda, B.; Scarre, C.; Till, R.; Pasalodos, R.J.; Guerra, M.R.; Tejedor, C.; Peredo, R.O.; Watson, A.; Wyatt, S.; Benito, C.G.; et al. Cave Acoustics in Prehistory: Exploring the Association of Palaeolithic Visual Motifs and Acoustic Response. *J. Acoust. Soc. Am.* **2017**, *142*, 1332–1349. [CrossRef] [PubMed]
16. NTi-Audio Online, Dodecahedron Speaker Set Product Data. Available online: https://www.nti-audio.com/Portals/0/data/en/NTi-Audio-Dodecahedron-Speaker-Set-Product-Data.pdf (accessed on 2 March 2019).
17. Murphy, D. Archaeological Acoustic Space Measurement for Convolution Reverberation and Auralization Applications. In Proceedings of the 9th International Conference on Digital Audio Effects (DAFx-06), Montreal, Canada, 18–20 September 2006.
18. Till, R. An Archaeoacoustic Study of the Ħal Saflieni Hypogeum in Malta. *Antiq. J.* **2017**, *91*, 74–89. [CrossRef]
19. Jahn, R.; Devereux, P.; Ibison, M. Acoustical resonances of Assorted Ancient Structures. *J. Acoust. Soc. Am.* **1996**, *99*, 649–658. [CrossRef]
20. Pike, A.W.G.; Hoffman, D.L.; García Diez, M.; Pettitt, P.B.; Alcolea González, J.; Balbín Behrmann, R.; de González, S.C.; de las Heras, C.; Lasheras, J.A.; Montes, R.; et al. U-series dating of Palaeolithic art in 11 caves in Spain. *Science* **2012**, *336*, 1409–1413. [CrossRef]
21. Reznikoff, I.; Dauvois, M. The sound dimension of painted caves (original in French). *B. Soc. Prehist. Fr.* **1988**, *85*, 238–246. [CrossRef]
22. Reznikoff, I. The evidence of the use of sound resonance from Palaeolithic to Mediaeval times. In *Archaeoacoustics*; Scarre, C., Lawson, G., Eds.; McDonald Institute Monographs: Cambridge, UK, 2006; pp. 77–84.
23. Diaz-Andreu, M.; Garcia-Bnito, C. The sound of rock art: The acoustics of the rock art of Southern Andalusia (Spain). *Oxf. J. Archaeol.* **2014**, *33*, 1–18. [CrossRef]
24. Available online: http://www.archeoacustica.net/home-page/ (accessed on 1 April 2019).
25. Diaz-Andreu, M.; Mattioli, T. Archaeoacoustics of Rock Art: Quantitative Approaches to the Acoustics and Soundscape of Rock Art. In *CAA 2015, Keep the Revolution Going: Proceedings of the 43rd Annual Conference on Computer Applications and Quantitative Methods in Archaeology*; Campana, S., Scopigno, R., Carpentiero, G., Cirillo, M., Eds.; Archaeopress: London, UK, 2015.
26. Rainio, R.; Lahelma, A.; Aikas, K.; Okkonen, J. Acoustic Measurements and Digital Image Processing Suggest a Link Between Sound Rituals and Sacred Sites in Northern Finland. *J. Archaeol. Method Theory* **2018**, *25*, 453–474. [CrossRef]
27. Conard, N.; Malina, M.; Munzel, S.C. New Flutes Document the Earliest Musical tradition in Southwestern Germany. *Nature* **2009**, *460*, 737–740. [CrossRef] [PubMed]
28. Songs of the Caves. Available online: http://SongsoftheCaves.Wordpress.com (accessed on 2 April 2019).
29. Parker, P.M. *Stonehenge: Exploring the Greatest Stone Age Mystery*; Simon and Schuster: London, UK, 2012.
30. Willis, C.; Marshall, P.; McKinley, J.; Pitts, M.; Pollard, J.; Richards, C.; Richards, J.; Thomas, J.; Waldron, T.; Welham, K.; et al. The Dead of Stonehenge. *Antiquity* **2016**, *90*, 337–356. [CrossRef]
31. Darvill, T. *Stonehenge: The Biography of a Landscape*; History Press: Stroud, UK, 2006.
32. Gilbert, R. *Music and Trance: A Theory of the Relations Between Music and Possession*; University of Chicago Press: Chicago, IL, USA, 1985.
33. Watson, A.; Keating, D. Architecture and Sound: An acoustic analysis of megalithic monuments in prehistoric Britain. *Antiquity* **1999**, *73*, 325–336. [CrossRef]
34. Watson, A. (Un)intentional sound? Acoustics and Neolithic monuments. In *Archaeoacoustics*; Scarre, C., Lawson, G., Eds.; McDonald Institute for Archaeological Research: Cambridge, UK, 2006; pp. 11–22.
35. Till, R. Songs of The Stones: The Acoustics of Stonehenge. In *BAR 504 2009: The Sounds of Stonehenge, Centre for the History of Music in Britain, the Empire and the Commonwealth*; CHOMBEC Working Papers No. 1; Banfield, S., Ed.; Archaeopress: Oxford, UK, 2009; pp. 17–42.
36. Till, R. Songs of the Stones: An Investigation into the Musical History and Culture of Stonehenge. *IASPM J.* **2011**, *1*, 1–18. [CrossRef]
37. Fazenda, B. The Acoustics of Stonehenge. *Acoust. Bull.* **2013**, *38*, 32–37.

38. Fazenda, B.; Drumm, I. Recreating the Sound of Stonehenge. *Acta Acust. United Acust.* **2013**, *99*, 110–118. [CrossRef]
39. Hardy, T. *Tess of the d'Urbervilles: A Pure Woman Faithfully Presented*; Harper and Brothers: New York, NY, USA, 1892.
40. Darvill, T.; Marshall, P.; Parker, P.M.; Wainwright, G. Stonehenge Remodelled. *Antiquity* **2012**, *86*, 1021–1040. [CrossRef]
41. Unver, E.; Taylor, A. Virtual Stonehenge Reconstruction. In *Progress in Cultural Heritage Preservation, Proceedings of the 4th International Conference, EuroMed 2012, Lemessos, Cyprus, 29 October–12 November 2012*; Lecture Notes in Computer Science Subseries: Information Systems and Applications, Incl. Internet/Web, and HCI, 7616 (XXV); Springer: Berlin, Germany, 2012; pp. 449–460.
42. Bryan, P.G.; Clowes, M. CAA96, Computer Applications and Quantitative Methods in Archaeology, BAR International Series 845. In *Stonehenge—Mapping the Stones*; Lockyear, K., Sly, T.J.T., Mihăilescu-Bîrliba, V., Eds.; Archaeopress: Oxford, UK, 2012; pp. 41–48.
43. EMAP Soundgate App. Available online: http://www.emaproject.eu/content/soundgate-app.html (accessed on 28 March 2019).
44. 3D Model of the Antonine Final Alteration of the Paphos Theatre (Phase 5). Available online: http://www.omnia.ie/index.php?navigation_function=2&navigation_item=%2F2020720%2FDR_5d8cc774e23c67c16ffba8653d4ee24d&repid=1 (accessed on 2 April 2019).
45. Paphos Theatre Archaeological Project. Available online: http://www.paphostheatre.org (accessed on 4 April 2019).
46. Kang, J.; Chourmouzuadou, K. Acoustic Evolution of Greek and Roman Theatres. *Appl. Acoust.* **2008**, *69*, 514–529.
47. Lynge Christensen, C.; Rindel, J.H. A new scattering method that combines roughness and diffraction effects. In Proceedings of the Forum Acusticum, Budapest, Hungary, 29 August–2 September 2005.
48. Gade, A.C.; Lisa, M.; Lynge Christensen, C.; Rindel, J.H. Roman Theatre acoustics: Comparison of acoustic measurement results from the Aspendos Theatre, Turkey. In Proceedings of the 17th ICA, Kyoto, Japan, 4–9 April 2004.
49. Farnetani, A.; Prodi, N.; Fausti, P.; Pompol, R. Acoustical measurements in ancient Roman theatres. *J. Acoust. Soc. Am.* **2004**, *115*, 2477. [CrossRef]
50. Acoustics and Music of British Prehistory Research Network. Available online: http://AMBPNetwork.wordpress.com (accessed on 28 March 2019).
51. Karabiber, Z. The conservation of acoustical heritage. In Proceedings of the First International Workshop on 3D Virtual Heritage, Geneva, Switzerland, 26–27 September 2002; pp. 286–290.
52. Maconie, R. Musical acoustics in the Age of Vitruvius. *Musical Times* **2005**, *146*, 75–82. [CrossRef]
53. Ahnert, W.; Schmidt, W. Fundamentals to Perform Acoustical Measurements. Available online: http://renkusheinz-sound.ru/easera/EASERAAppendixUSPV.pdf (accessed on 19 April 2019).
54. Barron, M. Interpretation of Early Decay Times in Concert Auditoria. *Acta Acust. United Acust.* **1995**, *81*, 320–331.
55. Ottley, M. Designing for Speech in a Circular Room. In Proceedings of the ACOUSTICS, Sydney, Australia, 6–9 November 2018; Available online: https://au.marshallday.com/media/2802/speech-in-circular-rooms-aas2018-mott.pdf (accessed on 3 March 2019).

© 2019 by the author. Licensee MDPI, Basel, Switzerland. This article is an open access article distributed under the terms and conditions of the Creative Commons Attribution (CC BY) license (http://creativecommons.org/licenses/by/4.0/).

Article

The Contribution of the Stage Design to the Acoustics of Ancient Greek Theatres

Nikos Barkas

Department of Architecture, School of Engineering, Democritus University of Thrace, 671 00 Xanthe, Greece; nbarkas@arch.duth.gr

Received: 15 November 2018; Accepted: 19 March 2019; Published: 23 March 2019

Abstract: The famous acoustics of ancient Greek theatres rely on a successful combination of appropriate location and architectural design. The theatres of the ancient world effectively combine two contradictory requirements: large audience capacity and excellent aural and visual comfort. Despite serious alterations resulting from either Roman modifications or accumulated damage, most of these theatres are still theatrically and acoustically functional. Acoustic research has proven that ancient theatres are applications of a successful combination of the basic parameters governing the acoustic design of open-air venues: elimination of external noise, harmonious arrangement of the audience around the performing space, geometric functions among the various parts of the theatre, reinforcement of the direct sound through positive sound reflections, and suppression of the delayed sound reflections or reverberation. Specifically, regarding the acoustic contribution of the stage building, it is important to clarify the consecutive modifications of the *skene* in the various types of theatres, given the fact that stage buildings were almost destroyed in most ancient Greek theatres. This paper attempts to demonstrate the positive role of the scenery in contemporary performances of ancient drama to improve the acoustic comfort using data from a sample of twenty (20) ancient theatres in Greece.

Keywords: ancient Greek theatre; Classical Era; scenery; acoustic design

---

1. Introduction

In ancient drama the term used for scenery is opsis, as Aristotle mentions in the Poetics. Opsis has to do with a wide variety of elements relating to the skene (stage building), such as their decoration and infrastructure, the positions of the actors, or the props used in performances. Although Aristotle's text was written at a time when the Theatre of Dionysus in Athens was undergoing its final construction phase during the Hellenistic Era (334/330 BC, at the same time as its reconstruction in stone under Lycurgus), it gives us quasi direct insights into the theatrical world of the Classical Era. All the other references to the conditions of those performances (the accounts by Pausanias, Plutarch and Strabo, the descriptions by Julius Pollux, Vitruvius' analyses and Horace's comments, as well as the depictions on vases) are indirect information from later periods. In reality, the texts that have survived from the dramatic output of the 5th and 4th centuries BC constitute the only authentic records of the performance requirements of Classical drama [1]. As a result, the present study draws data on the form and evolution of the skene from the dramas of the classical period, combined with archaeological, architectural and acoustic findings.

Nevertheless, the precise position of the opsis behind the orchestra, the length, width and height of the scenery, and the possible installation of a raised proskenion are all complex, thorny issues which have turned Classical theatre into a highly controversial subject [2] (p. 376), [3] (p. 27), [4] (p. 259), [5] (p. 142).

During the last few years, there has been a bulk of research on ancient theatre acoustics. The most common method used for studying the acoustic behaviour of ancient theatres combines on-site measurements and computer simulations. The software most commonly used in such computer simulations is CATT and ODEON, although use is also made of EASE, RAYNOISE and OTL-Terrain. The data from the acoustic measurements are used to weigh up or evaluate the results of the simulations. Various hypothetical scenarios are studied (such as the influence of the audience, the ancillary use of loud speakers, skene installations and scenery) [6–19]. This is often followed by a detailed examination of the effects of natural frequencies and diffraction on individual architectural elements (particularly resonators) [6,20,21]. A rather restricted amount of research focused specifically on the acoustic contribution of the stage by comparing theatres with different type of stage ruins [22], by assessing the impact of noise barriers and background scenery wall for optimum performance conditions [23], or by proposing and evaluating the impact of different scenic elements aiming to establish guidelines for stage design [6,24,25].

However, with regard to the available software, concerns have been recorded about its suitability in terms of its capacity to take into account or assess the importance of the special characteristics of the ancient theatres, or the effects of outdoor noise in the modern environment [26,27].

To a lesser extent, use is made of scale simulations which are used to investigate and assess the acoustic properties of building materials, as well as additional acoustic phenomena in ancient theatres, such as diffraction [28,29]. In connection with this, correlations are drawn between various functional parts of ancient Roman theatres (e.g., the wall of the skene, the perimetric passageway around the amphitheatre, the slope of the cavea, the addition of scenery or an ancillary microphone installation) [30]. Some researchers also investigate to a limited extent the subjective criteria of the acoustic quality of theatres by using questionnaires distributed to the spectators of performances or listeners of recordings. Finally, listening tests are also carried out on sound data produced by software programs in varying sound absorption conditions [31–35].

A limited amount of research has been carried out on the effects of environmental noise during the contemporary use of ancient theatres, the nature and the characteristics of nuisance sources and the monitoring of noise levels in the surrounding environments of ancient theatrical monuments. The findings of this research highlighted the impact of environmental noise on contemporary performances and emphasised the acoustic contribution of the scenery for the re-use of ancient theatres [26,36,37]. By coincidence, all of these studies examined the theatre of Philippi and its soundscape. Chourmouziadou and Kang (2011) estimated that traffic noise could be avoided by the application of a scenery which could reduce SPL by 5dB and improve the performance conditions due to early reflection distribution. Additionally, they proposed that a barrier near the source could restrict noise by reducing direct sound [26]. Barkas (2004) and Barkas-Vardaxis (2011) have calculated that the positive contribution of the scenery reflector is min +3 to +5 dB, and the reduction of the environmental noise due to the barrier of the scene is min −2 to −4 dB, namely a total of min +6.5 dB in the acoustic comfort [36,37]. According to all authors, such applications, with adaptations to the specific local conditions, could be used for other archaeological sites with similar results.

Given the above, the present paper attempts to demonstrate the acoustic contribution of the scenery in terms of improving the acoustic comfort in contemporary performances of ancient drama. Data from a wider, long-lasting research on a sample of twenty (20) ancient theatres in Greece were used, focused on the positive role of the scenery.

However, to study the acoustic contribution of the skene, it is necessary to define the form of the scenery in the original conditions, meaning during the classical Era.

2. The Evolution of Skene during the Classical Era

In the early 5th century, when the cult of Dionysus was formally established in Athens, a sanctuary to Dionysus was laid out at the foot of the Acropolis, and next to it was created a circular threshing-floor for devotional performances. Originally, the spectators gathered on temporary tiered wooden benches (ikria) on the south slope. However, following the collapse of the ikria (in theatre games held between 499 and 496 BC) with large loss of life, the Athenian state created a safe and permanent theatre in the sanctuary of Dionysus, a development that marked the birth of the Athenian or Attic Theatre of the Classical Era. The theatre of Dionysus is considered as the basic model for all the other theatres of the Antiquity, in Greece, in Sicily and Italy, in Asia Minor, and in all the Mediterranean area. Furthermore, this was the place where all ancient dramas were performed and the model of the theatrical space for all Athenian poets [38].

The functional parts of this theatre building in the early Classical Era were as follows:

- the orchestra/threshing-floor (with a radius of approximately 12 m) on a fill in the lowest flat opening;
- the thymele, an altar to Dionysus on a stepped base in the orchestra;
- the pranes, a supporting wall approximately 2.5 m high behind the fill which separated the orchestra from the sanctuary and concealed a roughly constructed theatre storeroom (the actors' dressing-rooms);
- the parodoi, side-entrances affording access for the public, the chorus and/or the actors between the retaining walls of the cavea and the natural ground of the sanctuary;
- the koilon or cavea, with tiers of seats for the spectators, either made of wood or hewn out of the rock, arranged concentrically around the orchestra and laid out in an arc of 240° on the natural slope of a hill [39] (pp. 30–35), [40] (pp. 40–44).

If we place the dramas surviving from the Classical Era (44 tragedies and comedies) in chronological order, we find that the references to the drama space begin with allusions and gradually evolve into clear, detailed descriptions [38,41–44]. Aeschylus' early work, therefore, required an indeterminate setting, with an altar or mound, while the thymele ensured that actors could stand out clearly above the chorus or as a counterpoint to the appearance of a chariot (Persians, Seven Against Thebes, The Suppliants).

Figure 1 presents the relative connection among the orchestra/threshing-floor, the pranes and the old temple according to Doerpfeld [2]. The height of the old temple and the possible presence of a small storeyroom (a dressing room for the actors) in a lower level behind the pranes are given by estimation [39].

Figure 1. Representation of Dionysus theatre from the koilon before the burning of Athens (480 BC).

The first substantial breakthrough in the representation of theatre space appears in the prologue of Aeschylus' trilogy, when the watchman appears on the roof of the palace, sees the light signalling the fall of Troy and runs to announce the good news inside the building, while Clytemnestra silently emerges from it in order to offer sacrifices. These movements indicate a stage with a number of conventional possibilities, as the action acquires physical boundaries and the performance space material form, distinct areas and a clear setting. This composite representational function marks the beginning of an evolutionary process in stage design which broadened the representational possibilities of the conventional space, increased the functional requirements of stage infrastructures and created new acoustic conditions [1].

As excavations in the area of the cavea at the Theatre of Dionysus have revealed, successive layers on the south slope (laid between 460 and 440 BC) created a steeper incline in the cavea and increased the theatre's capacity. A new orchestra (with a radius of approximately 9.9 m) was formed within the orchestra/threshing-floor and was moved closer to the auditorium, leaving a free space behind it facing downhill, in which a wooden stage framework with scenery was installed (across the central axis of the cavea). This served to focus the dramatic interest behind the orchestra, separated the visible façade of the skene from its invisible rear and provided a tall support for the roofing [2] (p. 130), [3] (pp. 27–31), [45] (p. 15), [46] (p. 66), [47] (p. 10).

Figure 2 presents a roughly crafted stage (theatrical storeyroom) behind the new orchestra (in the free space of the old orchestra/threshing-floor according to Allen [3]. The dimensions of the stage façade, and the height of the new temple are given by estimation [44]. The walls of the stage are wooden, the roof is covered with tiles. The position of the koilon and the walls of the side-entrances (analemata) according to Doerpfeld [2]. The exact configuration of the wooden benches, on the given tilting of the koilon, remains unclear.

Figure 2. Representation of Dionysus theatre from the koilon at the time of "Oresteia" (460 BC).

During Athens' great construction boom (between 449 and 429 BC, under the rule of Pericles) great changes took place in the precincts of the Theatre of Dionysus (such as the construction of the Odeion in the upper eastern section of the hollow in which it lay, and the construction of the stone retaining walls of the outer terraces and the staircases leading from the horizontal diazoma). In the area of the skene, the paraskenia were created-new theatre storerooms which projected from either side of the stage set-and formed (in the shape of a Π) a proskenion at ground-floor level [38] (p. 50), [41] (p. 68), [48] (p. 55).

The extension of the skene (across the main axis of the theatre), and the formation of numerous openings in it, constituted a second development concerning the functional possibilities of the skene infrastructure during the Classical Era. The use of one and (necessarily) only one doorway for movement of the actors and the ekkyklema (a kind of removable raft) between the front and rear sections of the skene also appears for the first time in the Oresteia and more specifically in The Libation

Bearers, during the episode in which a series of murders takes place inside the palace. In later years, the existence of a single opening in the scenery is pointed out deliberately in moments of drama where the doorway is supposedly closed or blocked for some reason (until about 427/425 BC) in Alcestis, Hippolytus, Andromache and Oedipus the Tyrant. Evidently, during quite a long period (about thirty years) after the introduction of scenery, the lengthening of the skene led to the establishment of a new convention through the incorporation of the paraskenia into the stage buildings and the use of their openings as auxiliary doorways for the movements (not mentioned in the texts) of the prop masters. Finally, the paraskenia, as parts of the skene infrastructure, were incorporated into the stage design (after 430 BC), permitting the representation of lateral buildings with independent doorways next to the central stage building (Andromache, The Acharnians, Oedipus the Tyrant, Herakles, Clouds, Peace, The Trojan Women, Lysistrata, Philoctetes, The Bacchae, Frogs, Ekklesiazousai) [1].

Figure 3 presents the stage set with the lateral buildings (paraskenia) and the proscenium between the orchestra and the stage according to Allen [3]. The floor plan dimensions of Π according Allen's theory of the two orchestras [3]. The height and the roofing of the lateral buildings are given by estimation [1]. The walls of the stage are wooden, the roof is covered with tiles. The position of the koilon and the walls of the side-entrances (analemata) according to Doerpfeld [2]. The exact configuration of the wooden benches, on the given tilting of the koilon, remains unclear.

Figure 3. Representation of Dionysus theatre from the koilon (period 430–415 BC).

One of the most notable architectural developments in the skene infrastructure (during the period of the Peace of Nicias, 421–415 BC) was the construction of the skenotheke (behind the skene), a building buttressed by the so-called "Foundation T", the stone base at the level of the orchestra that facilitated the rolling-out of the ekkyklema and supported the weight of a small wooden propylon in the centre of the stage. On either side of this base two symmetrical, matching rows of conical stone apertures were constructed (similar to those found in the Classical ruins of the theatres at Pergamon, Oropos and Corinth), which housed the supporting beams of an extensive, two-storeyed wooden skene [2] (p. 150), [3] (p. 15), [47], (p. 15). In the dramatic output of the following period, this secure stage installation made it possible to stage certain artistic performances of a symbolic and abstract nature, as can be seen in the prologues of Sophocles and Euripides (Ion, Herakles, Iphigenia in Tauris, Helen, The Phoenician Women, Orestes), where there are numerous verbal descriptions of the structure representing a propylon before the central doorway of the stage set, on a distinctly higher level in the area of the proskenion that was 2–3 steps above the orchestra (Ion, Philoctetes, Wasps, Lysistrata, Ekklesiazousai) [1]. In other words, it appears that in a slow and conservative process that lasted for about fifty years and involved a series of changes and gradual modifications, the temporary skene structures eventually evolved into a wooden installation with a relatively high propylon before the central doorway and another two lateral doorways in the facades of the paraskenia, the episkenion in the superstructure and the theologeion on the roof [49].

This third major development concerns the upward extension of the skene. The plays that have survived from the decades following the Oresteia contain a multitude of divinities or supernatural creatures (Ajax, Medea, Andromache, Hippolytus, Herakles, Euripides' The Suppliants) that appear (with or without the aid of a stage machine) on the span roof of the stage building. This high position-which later came to be called the theologeion—served for the (almost exclusive) appearances of divinities that formed a typical ending to dramas (Iphigenia in Tauris, Helen, Euripides' Electra, Philoctetes, Iphigenia in Aulis).

Later, however, literary references of this kind reveal the use of yet another distinct level—which later came to be known as the logeion—on the first storey of the stage building. In other words, it appears that the gradual separation of the actors from the chorus and their confinement to the proskenion led to the dramatic action taking place, either sequentially or simultaneously, on the three successive levels of the stage building (Wasps, Peace, Lysistrata, The Phoenician Women, Orestes, Ekklesiazousai, Rhesus) [1,49]. This form of stage design ultimately crystallised into the stone stage building that was incorporated into the definitive reconstruction of theatres that took place in the Hellenistic Area (338–326 BC) [2] (p. 28), [46] (p. 70).

Figure 4 presents the two-storeyed wooden skene with the slightly elevated propylon in the central doorway of the ground floor and the episkenion in the superstructure according to Flickinger [46]. The front part of the paraskenia was designed by combining the theories of Fiechter [45] and Doerpfeld [2]. The floor plan dimensions of stage were set according to Doerpfeld's theory for the two rows of foundation holes in front of the skenotheke building [2]. The height of the paraskenia, the storeys, and the roof are given by estimation. The walls of the stage are wooden, the roof is covered with tiles. The basement of the propylon in made with pieces of marble. The position of the koilon and the walls of the side-entrances (analemata) according to Doerpfeld [2]. The exact configuration of the wooden benches, on the given tilting of the koilon, remains unclear [45].

Figure 4. Representation of the theatre of Dionysus from the koilon (after 415 BC).

3. The Acoustic Design of Ancient Greek Theatres

Ancient Greek theatres, unlike Roman ones, made empirical use of the slope of a hill in order to create tiers of seating for the spectators around the performance area. Ancient theatres were not all aligned in a specific direction (e.g., towards the south, as Vitruvius claims), nor were they governed by certain rules dictating their precise location in the urban plan (e.g., in the agora, on a sacred site, within a complex of public buildings or in an organised sanctuary) [50]. The central axis of an ancient theatre can be aligned in any direction, depending on the prevailing winds and the sources of external noise (across the axis of a valley, along the axis of a ravine, in the hollow of a bay or on an artificial embankment), so that, with the aid of flexible adjustments and creative on-the-spot solutions, the structure can fulfil the theatrical, architectural and acoustic requirements of any historical era [4,40].

The stage building was not a primary element in theatre and acoustic design. The use, initially, of a temporary wooden framework gradually led to the establishment of a variety of theatrical conventions that served as a backdrop for the action on stage and facilitate the focus of the dramatic interest behind the orchestra. The later stone structures of the Hellenistic skene (4th and 3rd BC), apart from serving other expediencies, also provided an even more effective reflective surface compared with the original wooden structures of the permanent scenography of the Classical Era. However, when the skene evolved into a two-storey or three-storey building, like those in Late Hellenistic and Graeco-Roman theatres (3rd and 2nd BC), there was a danger that delayed sound reflections might be created by the extensive façade of the skene and the orchestra. In order to deal with this problem and control reverberation, the facades of stone stage buildings acquired a typical form of decoration (with hollows, relief elements and groups of sculptures) which were designed to diffuse the sound. Later, in order to further minimise the acoustic side-effects created by the construction of a raised proskenion, like those found in Graeco-Roman and Roman theatres, once again temporary forms of scenery (the sound-absorbing surfaces of wooden screens or cloth decorations) were adopted in the numerous openings of the skene and the hyposkenion [39,40,51].

The simultaneous functioning of two strong reflectors (the orchestra and the stage set) that was typical in Classical theatres inevitably posed the problem of homogenising the discrete sound reflections and incorporating them into the direct sound. The combined use of these reflectors obliged the actors to limit their movements on the proskenion area, in front of the stage set. The interpretations of archaeological finds and the geometric relationship between the orchestra, the skene and the parodoi lead to the conclusion that the proskenion was a narrow, ground-level space confined between the tangent of the orchestra, the projections of the paraskenia and the skene [2,48]. This space formed a "Haas zone", a useful acoustic zone in front of the scenery, outside where the combined sound reflections from the orchestra and the façade of the stage building began to be noticeable as discrete entities, meaning reverberation increased and the speech of the performers became less intelligible [52]. Acoustic experiments have demonstrated the superiority of ground-level or low proskenion platforms over higher platforms and have confirmed the archaeological and theatrical theories regarding the ground-level proskenion of the Theatre of Dionysus (up to the late 5th cent.), where a formal propylon gradually developed—a miniature version of those in public buildings of the Classical Era—resting on a low base of limited length (abutting the so-called "foundation T") in front of the central doorway of the temporary wooden stage set [4,40]. On the other hand, the gradual raising of the proskenion platform and its extension (initially by one third of the radius at the theatre on Delos, then by one half of the radius at various Graeco-Roman theatres and finally around the whole of the orchestra semicircle in Roman theatres), in conjunction with the placing of mobile seats in the free semicircle of the orchestra (in Graeco-Roman and Roman theatres), cancelled out the positive effects of the orchestra's reflector and increased the amount of sound absorbed by the bodies of the spectators. This development led to a decrease in the amount of sound energy that was produced and degraded the intelligibility of the theatrical message, a situation which (in tandem with other social and cultural changes) resulted in a necessary reduction in the capacity of Roman theatres [39,40,50,51].

The paraskenia, as lateral projections at the aisles of the stage set, directed the useful, central beams of reflected sound from the stage set to the auditorium. At the same time, they helped to protect the sides of the skene from combined reflections between the retaining walls (analemmata) and the skene, channelling the delayed, lateral beams of reflected sound towards the parodoi, which served to defuse the reflected sound (either directly by absorbing it or indirectly by diffusing it in the stone structures of the retaining walls). The parodoi gradually declined in use in Graeco-Roman theatres and gave way to the vomitoria (vaulted arcades faced with rubble masonry to help deal with the delayed lateral sound reflections), while movement towards the tiers of seating was assisted by staircases on the outer edges of the cavea [4,53].

Finally, when Roman theatres were relocated in noisy urban centres, it became necessary to drastically increase the size of the scaenae frons (the stage façade) directly opposite the amphitheatre, so that a solid, uniform barrier on the perimeter of the theatre building could increase the sound protection of the theatre space [4] (p. 260), [39] (p. 41), [52] (p. 75).

Beginning with the on-site acoustic research of Fr. Canac in the theatres at Orange and Vaison (and later in ancient theatres in Italy and Greece) during the 1950's, successive acoustic studies have shown that ancient theatres display the basic principles that apply in the design of open-air venues, and also that, despite the damage and modifications that these theatres have suffered throughout the centuries, they continue to possess their own peculiar acoustic properties [7,52–54]:

- quiet acoustic environment (elimination of external noise and parasitic disturbances);
- harmonious arrangement of the spectators around the performance space within the limits of the human vocal and acoustic range;
- smooth transmission of direct sound, which is reinforced through early, positive sound reflections (from solid sound-reflective surfaces near the actors and small sound-diffusing surfaces near the spectators);
- low reverberation and increased intelligibility of speech due to the reduction of harmful delayed sound reflections and elimination of echo (small diffusive surfaces near the actors and elimination of solid sound-reflective surfaces near the spectators) [37];

As was mentioned earlier, the independent or combined acoustic contribution of the individual building parts of ancient Greek theatres includes;

- the siting of theatres on the lee slopes of hills and the steep, stepped arrangement of the cavea (at angles of between 21° and 30°, with steeper inclines in the upper diazoma) favour open-air sound propagation conditions;
- the reflector of the orchestra reinforces the direct sound—mainly in the lower diazoma—with almost no time delay (<5 ms);
- a stage set placed in a suitable position behind the orchestra (a reflective zone extending approximately 3.2 m above the level of the orchestra) amplifies the voices of the actors with small time delays (40~85 ms), particularly in the higher tiers;
- the combined effect of the orchestra and the scenery (with time delays of 45~110 ms) keeps the length of the reverberation time down to one that is suitable for speech thanks to the sound absorption of the unwanted sound reflections by the bodies of the spectators and the atmosphere [40].

The incorporation of the positive sound reflections into the direct sound makes up for the energy losses of the sound propagation over a great distance (between +2.5 and +5.5 dB, depending on the actor's position) and ensures an even level of acoustic comfort for the spectators (+/−3 dB over the height of each diazoma and +/−1.5 dB between the central and outermost tiers of seats in the auditorium). However, the most suitable and necessary condition for ensuring that the beneficial effects of the available sound reflectors are obtained is for the actors to gather in the intermediate proskenion area (in a "Haas zone" 3–3.5 m deep). We are dealing here with a complex geometric function (which Fr. Canac called "l'équation canonique" of the ancient theatre) which tends to produce the optimum result, provided that the difference in height between the proskenion and the orchestra is reduced to a minimum. The gathering of the actors in the proskenion area enabled the monologue and dialogue parts of dramas (spoken by individual voices of limited acoustic power) to be heard in the highest, outermost and most distant parts of the auditorium [4] (p. 34), [36] (pp. 382–386), [52] (pp. 103–127).

In addition, the suitable projection of the paraskenia at the ends of the stage (the ratio of the depth of the paraskenia to the length of the façade as quantities derived from the radius of the orchestra) reinforced the direct sound in the low outer tiers of seats. Later, this development led to the abolition of the outer tiers of seats in the upper section of the auditorium, as in Hellenistic theatres, and at the same time facilitated the concentration of delayed lateral sound reflections (with time delays of >200 ms) in certain sections of the low, lateral areas of the auditorium. In other words, these small (though crucial) projections of the paraskenia gave ancient open-air theatres all the advantages of closed spaces for speaking conditions, without the disadvantages at the sides [52], (pp. 78–80).

The appearances of actors in the theologeion caused a considerable reduction in the loudness of the actors' voices (−3 to −3.5 dB in the central seats and approximately −4.5 to −5.5 dB in the upper and outer tiers), compared with the voices of actors standing on the ground-floor level of the skene. This acoustic disadvantage (i.e., the weak voice of a divine or supernatural presence in the theologeion), which required the assistance of the other co-functioning signs of the theatrical code [55] to compensate for it, was remedied by the formation of the logeion in the episkenion. Thanks to a reflective zone approximately 2 m high behind the actors, the sound losses were limited to around −2 dB in the auditorium as a whole (approximately −1.5 to −3.5 dB in the lower or central tiers), compared with the sound levels produced by actors on the ground-floor level of the skene [4] (p. 34), [49].

It is clear, then, that the acoustic comfort in ancient Greek theatres was due to a good proportion of overall useful sound energy compared with the initial intensity of the words spoken by the actors. There was a natural (passive) amplification of the actors' voices that compensated for the energy losses due to the sound propagation in the open air, a phenomenon that was particularly evident in the upper tiers of the auditorium. To implement such a design in large-capacity open-air theatres without electro-acoustic systems, architects and engineers used a variety of solutions, depending on the positions of surfaces in relation to the actors and the chorus, such as open-air auditoriums (sound absorption by the atmosphere), large smooth surfaces in the orchestra and the scenery (low sound-absorptive capacity), and the use of small irregular elements in the retaining walls, the passages running around the top of the cavea and the high walls of the stage facades of Graeco-Roman and Roman theatres (sound diffusion) [4,51,53].

4. Acoustic Problems in the Case of 20 Greek Ancient Theatres

The famous acoustics of the ancient Greek theatre rely on the amplified acoustic response of the space, which is related to the replacement of the energy losses, thanks to early, strong-though of a limited number-sound reflections, in the specific performing occasion when the theatrical message is delivered by vocal trained and experienced actors providing clear distinction of the successive parts of the linguistic chain [36,49,52]. In the diagram "time-sound intensity", the emergence of the message is the visible part of the sound energy that is not distorted by background noise. The acoustic emergence depends on objective criteria such as: spectral density, ratios direct/total intensity, early decay of sound, and reverberation time, which are all connected to the basic subjective criteria of a space acoustic quality, such as colorization and intimacy, clarity, and finally, speech intelligibility [40,56].

In open air performances, the acoustic environment is dominated by urban noises and unexpected reactions caused by the audience (whispers, coughs and movements). The background noise covers a portion of the useful signal producing a kind of sound mask either permanently or occasionally. During the theatrical communication, the masking of the message is a complex psycho-acoustic process related to the visual comfort or the hearing angle of each spectator. It has been established that the parasitic signals of the continuous spectrum may eliminate the intelligibility of speech, even in low levels intensity circumstances (20 dB lower than the intensity of the useful signal). The values of the sound emergence (namely the acoustic comfort AC) in theatre spaces are evaluated according to the following behaviour: excellent (>25 dB), good (20–25 dB), acceptable (15–20 dB), non-acceptable (<15 dB) [56,57].

Long-lasting research, aiming at monitoring the current status of the ancient theatres in Greece (modifications, destructions, protection works, and environmental noise levels) and evaluating their acoustic quality for contemporary operation conditions includes a sample of twenty (20) ancient Greek theatres: Amphiaraion at Oropos, Argos, Delphi, Dilos, Dion, Athenian theatre of Dionysus Elefthereus, Dodoni, Epidauros, Eretria, Larisa, Mantineia, Maroneia, Megalopolis, Messini, Orchomenos at Veotia, Philippi, Thasos, Thira, Thorikos, Zea at Peiraeus. [27,37]. Figure 5 presents the distribution of the theatres of the sample.

Table 1. The sample of twenty (20) ancient Greek theatres.

Index	Theatre	Contemporary Location/Type of Noise	Noise Sources	Min SPL
1	Amph. Oropos	natural/occasional	traffic 37/birds 41	32
2	Argos	urban/permanent	urban activities 43/traffic 46	37
3	Delphi	semi urban/occasional	traffic 46/tourist activities 50	34
4	Dilos	natural/occasional	wind 40	27
5	Dion	natural/occasional	birds 36/agricultural activities 39	31
6	Ath. Dionysus	urban/permanent	traffic 45/tourist activities 54	38
7	Dodoni	natural/occasional	wind 34/restoration activities 36	27
8	Epidauros	natural/occasional	wind 36/tourist activities 46	27
9	Eretria	semi urban/permanent	urban activities 41/traffic 47	39
10	Larisa	urban/permanent	urban activities 45, traffic 62	43
11	Mantineia	natural/occasional	restoration act. 41/airplane noise 48	32
12	Maroneia	natural/occasional	wind 45/agricultural activities 68	37
13	Megalopolis	natural/occasional	restoration activities 35/wind 41	32
14	Messini	natural/occasional	wind 40/agricultural activities 46	35
15	Orchomenos V	semi urban/permanent	religious activities 46/traffic 57	38
16	Philippi	semi urban/permanent	wind 44/traffic 52	40
17	Thasos	natural/occasional	wind 38/coastal activity 44	36
18	Thira	natural/occasional	wind 38/airplane noise 58	29
19	Thorikos	natural/occasional	birds 37/wind 48	34
20	Zea Piraeus	urban/permanent	traffic 52/airplane noise 67	44

Research data are briefly presented in Table 1. Column "Theatre" shows the exact location of each theatre, and the column "Index" has the serial number for every theatre in the Figure 5. "Contemporary Location" refers to the type of environment (urban, semi-urban or natural). "Type of Noise" refers to whether the noise sources are occasional (e.g., an airplane crossing in the theatre of Thira, or seasonal agricultural works in the theatre of Dion) or permanent (e.g., distant traffic noise in the theatre of Dionysus or recreational activities in the case of the theatre in Larisa). The column "Noise sources" has the different type of noises with the recorded sound levels (maxL in dB for civil/agricultural/tourist/recreational activities and natural environment, Leq,h in dB[A] for ring road/highway traffic and airplane noise). Finally, in the last column the background noise in minimum SPL appears. The aforementioned noise levels are in global values. The frequency values for every noise source case will be discussed in a future paper.

Figure 5. The distribution of the theatres of the sample (the index number for each theatre in the Table 1).

Figure 6 presents a concise recording of the sound intensities measured in the environment of every theatre. For every research period and in each theatre, 12 measurements were conducted in the following positions: 2 at the remains of the stage, 1 at each parodos, 2 at the orchestra and 3 at the middle gradient of each part of the amphitheatre (1 at the central axis and 2 at the sides).

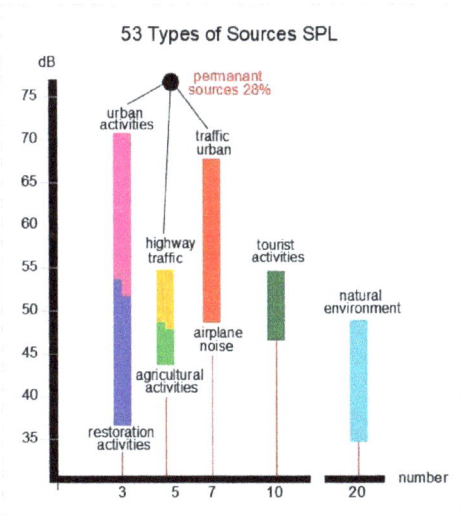

Figure 6. Measured intensities of various noise sources.

In order to evaluate the sample for contemporary performances or for potential future re-use, we have laid out a numerical model to calculate the acoustic comfort (AC), as the effective signal rising, based on the following assumptions [27,37]:

- the actor is at the back of the orchestra (not in the centre, but at the intersection of the potential scenery with the main axis of the koilon) 1.7 m above the level of the orchestra, while the audience is 1,1m above the gradients
- Lo, the initial intensity (the human voice of an experienced actor) is 82 dB [A]/1 m (normal intensity) or 87 dB [A]/1 m (strong intensity), in spherical wave conditions (with no electrical reinforcement). The above-mentioned sound intensities do not correspond to the relative levels of an ISO, but have been already measured in ancient drama performances with trained and experienced Greek actors,
- Ld, the corresponding decrease of the direct sound due to distance in a ray starting from the actor and corresponding to the 75% of the existing seats, either in lateral or central positions,
- Ro/sc, the natural (passive) loudspeaker amplification of the theatre space (+3 dB) thanks to reflections coming exclusively from the orchestra (for minimum predicted values), or (+6.5 dB) thanks to the reflector of the orchestra, plus the reflector of the scenery, plus the combination of all the positive reflections near the actor (for maximum predicted values)
- Nbn is the background noise in every theatre, plus the noise amplification (+5 dB) due to the presence of the spectators during the performance
- the predicted values accounted in accordance to the formula: AC = Lo − Ld + {R} − (Nbn + 5)

As shown in Figure 7 the ancient Greek theatres of the sample are classified as follows:

- for normal human voice intensity and in cases that active reflections come only from the orchestra (black index and bullets)
- for strong human voice intensity and in cases that active reflections come from the orchestra, plus the scenography background, plus the combination of all the active positive reflector near the actor (blue index and bullets).

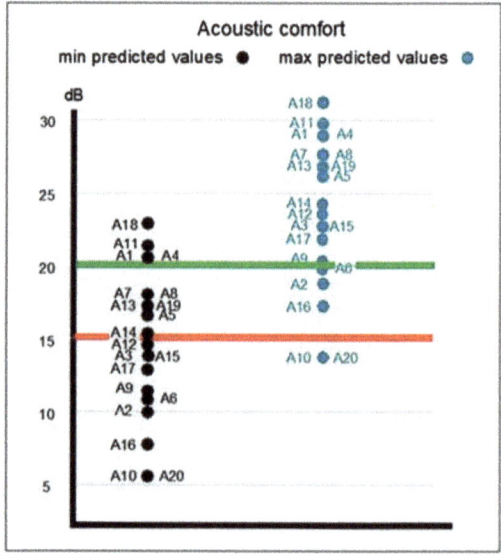

Figure 7. The acoustic comfort for min/max predicted values.

The major issue in all Greek ancient theatres is the condition of the stage. As Figure 8 shows regarding the theatres of our sample:

- in 2 cases there was never a stage building or an infrastructure for mobile stage (scenea ductilis) was used
- in 7 cases, only the remains of a low stage building (hypo-scenium) from the Roman period exists
- in 11 cases, only the ruins of a foundation from the Hellenistic period still remain.

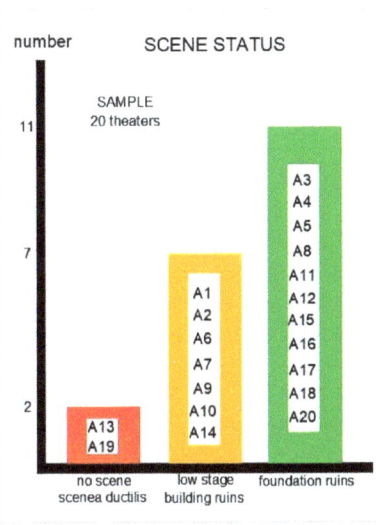

Figure 8. The current status of the scenes in the theatres of the sample.

Generalised destruction of the stage buildings is the most important problem in all cases. This fact, as discussed in previous paper, makes unavoidable the presence of a movable, low and lean scenery (a mobile stage background in the correct position and of a suitable size to be used during a performance), which could contribute mainly as an active sound reflector and secondly as a small noise barrier, as it has been proposed by other researchers, too [9,11–13,22–25].

Adding a scenery at the place of the ruins of the stage could ensure good or even satisfactory conditions in 90% of the cases, but most important, it could improve the acoustic comfort (acceptable conditions) in 45% of the cases. Specifically,

- only two (2) theatres seem to remain in non-acceptable conditions (AC < 15 dB): Larisa, Zea at Piraeus (urban environment, traffic and activities),
- only three (3) theatres would have acceptable acoustic conditions (15 < AC < 20 dB): Argos, Athenian of Dionysus Elefthereus, Philippi (urban environment with rural activities or traffic),
- the other theatres (15) would have good (AC > 20 dB) or excellent conditions (AC > 25 dB) [27].

The above estimation is rather conservative, because the usual absence of a scenery in contemporary performances obliges the actors to move towards the cavea in an attempt to compensate sound loss. As a result, the main sound reflector of the orchestra is also cancelled. The exact position, the size and the materials to be used for such a scenery depend on the specific conditions (theatre type, cavea size, environmental noise conditions).

5. Conclusions

It is difficult to compile an exact record of the types of scenery and skene installations used in the Classical Era either because they have been completely destroyed or because they have been incorporated into later reconstructions. Various studies using modern software programs have attempted to make an acoustic evaluation of ancient theatres, based on the model of an "ideal" fully-formed architectural structure [6]. The whole question would be merely of academic interest were it not for the fact that, since the beginning of the 20th century, there has been a popular tendency for ancient drama to "return" to its natural home. However, there are main contradictions for the current re-use of ancient Greek theatres:

- the performance of Classical dramas takes place in the ruins of later construction phases or Roman modifications;
- the existing remains of the stage buildings constitute palimpsests of different construction phases, and restorations of ruins often include elements of Roman conversions of the theatres into arenas;
- the restoration works do not always take into account the "immaterial" acoustics in ancient Greek theatres.

The contemporary and proper use of ancient Greek theatres requires organisers of theatrical performances to understand and fulfil the architectural and scenographic requirements relating to the acoustic design and the demands of performing ancient drama. The present paper has attempted to highlight the interdisciplinary aspect of the contemporary use of ancient Greek theatres. The extent of the problems concerning open-air productions are connected with the anthropometric (phonetic and acoustic) factors involved in a theatrical performance and directly depend on the contribution of the scenery to the natural (passive) amplification of sound in the theatre.

First of all, the re-use of ancient theatres in Greece should begin with the necessary establishment of a quietness criterion similar to those used internationally to achieve the quietness required in open-air cultural venues. Furthermore, in theatres where contemporary use is both feasible and desirable, it is essential to provide a removable, plain and aesthetically neutral background in the right position and of a suitable size [27,36,58], to counterbalance the absence of the stage reflector during the classical Era.

The present paper demonstrated that a scenery could improve the acoustic comfort in most of the theatres. The addition of a temporary, removable scenery during performances can provide a crucial, auxiliary reflector (plus a sound barrier), of specific characteristics for each theatre, without restricting the artistic freedom and without jeopardising the protection of ancient theatrical monuments.

Funding: Since 2009 the research has been funded by the former Insurance Fund for Engineers (TSMEDE).

Acknowledgments: Gianis Tompakidis designed the 3D theatre model. Since 2004, several groups of students from the Department of Architecture of DUTh Greece have collected data during their undergraduate research: Thomy Nikaki-Dimitris Sakoulis; Heliana Andoniadou-Nikos Vardaxis-Anna Moyses; Alexandra Tsatsaki; Mary Katsafadou-Christos Lagoudas; Melina Vyzika.

Conflicts of Interest: The author declares no conflict of interest.

References

1. Barkas, N. The Opsis in Classical Drama. In *Proceedings 2nd Conference "History of Building Construction"*; Democritus University of Thrace: Xanthi, Greece, 2017. (In Greek)
2. Doerpfeld, W.; Reisch, E. *Das Griechische Theater*; Barth und von Hirst: Athens, Greece, 1896.
3. Allen, J. *The Greek Theatre of the Fifth Century B.C.*; Semi Centennial Publications of the University of California: Los Angeles, CA, USA, 1918.
4. Izenour, G. *Theater Design*; Mc Graw-Hill: New York, NY, USA, 1977.
5. Gogos, S. *The Ancient Theatre of Dionysus*; Militos Emerging Technologies & Services: Athens, Greece, 2006. (In Greek)
6. Angellakis, K.; Barkas, N. Contemporary methods in studying the acoustics of ancient Greek and roman theatres. In *Proceedings of 6th Conference HELINA "ACOUSTICS 2012"*; Hellenic Institute of Acoustics: Corfu, Greece, 2012. (In Greek)

7. Vasilantonopoulos, S.; Mourjopoulos, J. Simulation and analysis of the acoustics of ancient open-air theatres. In *Proceedings 2th Conference HELINA "ACOUSTICS 2004"*; Hellenic Institute of Acoustics: Thessaloniki, Greece, 2004; pp. 367–374. (In Greek)
8. Gade, A.; Angelakis, K. Acoustics of ancient Greek and Roman theatres in use today. In Proceedings of the 4th Joint Meeting of ASA and ASJ, Honolulu, HI, USA, 29 November 2006.
9. Angelakis, K.; Rindel, J.H.; Gade, A. Theatre of the Sanctuary of Asklepios and the Theatre of Ancient Epidaurus: Objective measurements, computer simulations and listening tests. In *Proceedings International Conference "The Acoustics of Ancient Theatres"*; HELINA and EAA: Patras, Greece, 2011.
10. Goussios, C.; Tsinikas, N.; Chourmouziadou, K.; Kalliris, G. The Roman Odeion of Nicopolis: Observations of architectural elements affecting its acoustics. Measurements and calculations of acoustic indices. In *Proceedings International Conference "The Acoustics of Ancient Theatres"*; HELINA and EAA: Patras, Greece, 2011. (In Greek)
11. Nianounakis, T.H. The Acoustics of the Ancient Theatre of Hephaistia. In *Proceedings International Conference "The Acoustics of Ancient Theatres"*; HELINA and EAA: Patras, Greece, 2011. (In Greek)
12. Bo, E.; Kostara Konstantinou, E.; Lepore, F.; Shtrepi, L.; Puglisi, G.E.; Astolfi, A.; Barkas, N. Acoustic Characterization of the Ancient Theatre of Tyndaris: Evaluation and proposals for its Reuse. In Proceedings of the 23th International Conference ICVR, Athens, Greece, 21–23 July 2016.
13. Rindel, J.H.; Frederiksen, R.; Vikatou, O. The Acoustics of the Pi-shared Greek Theatre of Kalydon. In *Proceedings International Conference "Euronoise 2018"*; HELINA and EAA: Crete, Greece, 2018.
14. Vassilantonopoulos, S.; Mourjopoulos, J. The acoustics of roofed Ancient odeia: The case of Herodes Atticus Odeion. *Acta Acust. United Acust.* **2009**, *95*, 291–299. [CrossRef]
15. Vassilantonopoulos, S.; Hatziantoniou, P.; Tatlas, N.; Zakynthinos, T.; Skarlatos, D.; Mourjopoulos, J. Measurements and analysis of the acoustics of the ancient theatre of Epidaurus. In *Proceedings International Conference "The Acoustics of Ancient Theatres"*; HELINA and EAA: Patras, Greece, 2011.
16. Iannace, G.; Trematerra, A.; Masullo, M. The large theatre of Pompeii: Acoustic evolution. *Build. Acoust.* **2013**, *20*, 215–227. [CrossRef]
17. Hak, C.; Hoekstra, N.; Nicolai, B.; Wenmaekers, R. Project ancient acoustics part 1 of 4: A method for accurate impulse response measurements in large open-air theatres. In Proceedings of the ICSV, Athens, Greece, 10–14 July 2016.
18. Hoekstra, N.; Nicolai, B.; Peeters, B.; Hak, C.; Wenmaekers, R. Project ancient acoustics part 2 of 4: Largescale acoustical measurements in the Odeon of Herodes Atticus and the theatres of Epidaurus and Argos. In Proceedings of the ICSV, Athens, Greece, 10–14 July 2016.
19. Van Loenen, C.; van der Wilt, M.; Diakoumis, A.; Hooymans, K.; Wenmaekers, R.; Hak, C. Project ancient acoustics part 3 of 4: Influence of geometrical and material assumptions on ray-based acoustic simulations of two ancient theatres. In Proceedings of the ICSV, Athens, Greece, 10–14 July 2016.
20. Karampatzakis, P.; Zafranas, V.; Polychronopoulos, S. A study on Aristoxenus acoustic urns. In *Proceedings International Conference "The Acoustics of Ancient Theatres"*; HELINA and EAA: Patras, Greece, 2011.
21. Polychronopoulos, S.; Skarlatos, D.; Kougias, D. The use of resonators in ancient Greek Theatres. In *Proceedings International Conference "The Acoustics of Ancient Theatres"*; HELINA and EAA: Patras, Greece, 2011.
22. Wenmaekers, R.; Nicolai, B.; Hoekstra, N.; Hak, C. Project Ancient Acoustics part 4 of 4: Stage acoustics measured in the Odeon of Herodes Atticus and the theatre of Argos. In Proceedings of the 23th Congress on Sound and Vibration, Athens, Greece, 10–14 July 2016.
23. Chourmouziadou, K. Open-Air Theatres vs Contemporary Noise Sources. In *Proceedings 4th Conference HELINA "ACOUSTICS 2008"*; Hellenic Institute of Acoustics: Xanthe, Greece, 2008; pp. 27–34.
24. Bo, E.; Astolfi, A.; Pellegrino, A.; Pelegrin-Garcia, D.; Puglisi, G.; Shtrepi, L.; Rychtarikova, M. The modern use of ancient theatres related to acoustic and lighting requirements: Stage design guidelines for the Greek theatre of Syracuse. *Energy Build.* **2015**, *95*, 106–115. [CrossRef]
25. Bo, E.; Astolfi, A.; Shtrepi, L.; Rychtarikova, M.; Pelegrin-Garcia, D.; Gen-Ta, G. The acoustic influence of the scenery on the audience sound perception: The case of the ancient theatre of Syracuse. In Proceedings of the Forum Acusticum, Krakov, Poland, 7–12 September 2014; pp. 7–12.

26. Chourmouziadou, K.; Kang, J. Architectural & Scenery Design Implementation for the Improvement of the Soundscape of the Ancient Theatres. In Proceedings of the International Conference "The Acoustics of Ancient Theatres", HELINA and EAA, Patras, Greece, 18–21 September 2011.
27. Barkas, N. Ancient Greek Theaters: Current Operation vs Contemporary Noise Environment. In Proceedings of the International Conference "Euronoise 2018", HELINA and EAA, Crete, Greece, 27–31 May 2018.
28. Farnetani, A.; Prodi, N.; Fausti, P. Validation of a numerical code for edge diffraction by means of acoustical measurements on a scale model of an ancient theatre. In Proceedings of the International Conference "The Acoustics of Ancient Theatres", HELINA and EAA, Patras, Greece, 18–21 September 2011.
29. Economou, P.; Charalampous, P. The significance of sound diffraction effects in simulating acoustics in ancient theatres. *Acta Acust. United Acust.* **2013**, *99*, 48–57. [CrossRef]
30. Prodi, N.; Farnetani, A.; Pompoli, R.; Fausti, P. Acoustics and architecture in ancient open-air theatres. In Proceedings of the International Conference "The Acoustics of Ancient Theatres", HELINA and EAA, Patras, Greece, 18–21 September 2011.
31. Mota, M. Meter matters: Embodied rhythms at stage as a challenge to the acoustics of ancient theatres. In Proceedings of the International Conference "The Acoustics of Ancient Theatres", HELINA and EAA, Patras, Greece, 18–21 September 2011.
32. Blauert, J. Cognitive aspects of listening in performance spaces. In Proceedings of the International Conference "The Acoustics of Ancient Theatres", HELINA and EAA, Patras, Greece, 18–21 September 2011.
33. Blesser, B. An analysis of the aural experience of ancient spaces. In Proceedings of the International Conference "The Acoustics of Ancient Theatres", HELINA and EAA, Patras, Greece, 18–21 September 2011.
34. Cocchi, C. Theatre Design: Science or opportunity. In Proceedings of the International Conference "The Acoustics of Ancient Theatres", HELINA and EAA, Patras, Greece, 18–21 September 2011.
35. Vovolis, T.H. Mask, Actor, Theatron and Landscape in classical Greek theatre. In Proceedings of the International Conference "The Acoustics of Ancient Theatres", HELINA and EAA, Patras, Greece, 18–21 September 2011.
36. Barkas, N. Acoustic Comfort in the Contemporary Use of Ancient Theatres. In *Proceedings of 1st Conference "Appropriate Interventions for Safeguarding of Monuments and Historical Buildings"*; Technical Chamber of Greece: Thessaloniki, Greece, 2004; Volume 1, pp. 376–390. (In Greek)
37. Barkas, N.; Vardaxis, N. Current Operations of Ancient Greek Theatres: The problem of environmental noise. In Proceedings of the International Conference "The Acoustics of Ancient Theatres", HELINA and EAA, Patras, Greece, 18–21 September 2011.
38. Balldry, H. *The Tragic Theater in Ancient Greece*; Edition Kardamitsa: Athens, Greece, 1981. (In Greek)
39. Athanasopoulos, C. *Problems in Contemporary Theatrical Developments*; Edition Sideris: Athens, Greece, 1976.
40. Barkas, N. The Acoustic Parameter in Ancient Greek Theater Design. In *MONUMENT & Environment'*; Technical Chamber of Greece: Thessaloniki, Greece, 1994; Volume 2, pp. 39–56. (In Greek)
41. Blume, H. *Introduction in Ancient Theater*; MIET: Athens, Greece, 1986. (In Greek)
42. Kitto, H. *Ancient Greek Tragedy*; Edition Papadimas: Athens, Greece, 1975. (In Greek)
43. Dover, K. *The Aristophanes' Comedy*; Edition MIET: Athens, Greece, 1981. (In Greek)
44. Chourmouziadis, N. *Terms and Transformations in Ancient Greek Tragedy*; Edition Gnosi: Athens, Greece, 1984. (In Greek)
45. Fiechter, E. *Das Dionysos Theater in Athens*; W. Kohlhammer: Stuttgarts, Germany, 1935; Volume 1.
46. Flickinger, R. *The Greek Theatre and its Drama*; Chicago University Press: Chicago, IL, USA, 1936.
47. Arnott, P. *Greek Scenic Conventions in the 5th Century B.C.*; Clarendon Press: Oxford, UK, 1962.
48. Allen, J. *The Key to the Reconstruction of the 5ht Century Theatre*; California University Press: Berkeley, CA, USA, 1918.
49. Barkas, N. The Acoustic Function of the Superstructure of the Skene in Ancient Greek Theatre. In Proceedings of the 3th Conference HELINA "ACOUSTICS 2006", Hellenic Institute of Acoustics, Heraklion, Crete, 25 July 2007; pp. 191–198. (In Greek)
50. *Vitruvii: De Architectura*; edition Plethron: Athens, Greece, 1997. (In Greek)
51. Canac, F. *Nouvelles Recherches sur les Theatres en Plein Air*; CRSIM: Marseille, France, 1957.
52. Canac, F. *L' Acoustique des Théâtres Antiques*; CNRS: Paris, France, 1967.
53. Lehmann, R. *L' Acoustique des Théâtres Antiques, Grecs et Romaine*; Université de Maine: Le Mans, France, 1983.

54. Polack, J.D. The acoustics of antique theatres: Canac's life work revisited. In Proceedings of the International Conference "The Acoustics of Ancient Theatres", HELINA and EAA, Patras, Greece, 18–21 September 2011.
55. Fischer-Lichte, E. *Semiotic des Theaters*; Gunter Narr: Tübingen, Germany, 1983.
56. Liepp, E. *Qualités Acoustiques des Lieux d'Ecoute*; CNRS: Paris, France, 1981.
57. Lehmann, R. *Eléments de Physio et de Psycho Acoustique*; Dunod: Paris, France, 1969.
58. Kostara-Konstantinou, E.; Lepore, F.; Astolfi, A.; Barkas, N. Scenery design through a geometrical model: An application in the theatre of ancient Tyndaris. In *Proceedings of the 8th Conference HELINA "ACOUSTICS 2016"*; Hellenic Institute of Acoustics: Athens, Greece, 2016; pp. 245–252.

© 2019 by the author. Licensee MDPI, Basel, Switzerland. This article is an open access article distributed under the terms and conditions of the Creative Commons Attribution (CC BY) license (http://creativecommons.org/licenses/by/4.0/).

Article
Acoustic Simulation of Julius Caesar's Battlefield Speeches

Braxton Boren

Audio Technology Program, Department of Performing Arts, College of Arts & Sciences, American University, Washington, DC 20016, USA; boren@american.edu; Tel.: +1-202-885-1482

Received: 14 September 2018; Accepted: 10 October 2018; Published: 14 October 2018

Abstract: History contains many accounts of speeches given by civic and military leaders before large crowds prior to the invention of electronic amplification. Historians have debated the historical accuracy of these accounts, often making some reference to acoustics, either supporting or refuting the accounts, but without any numerical justification. The field of digital humanities, and more specifically archaeoacoustics, seeks to use computational techniques to provide empirical data to improve historical analysis. Julius Caesar recalled giving speeches to 14,000 men after the battle of Dyrrachium and another to 22,000 men before the battle of Pharsalus during the Roman Civil War. Caesar's background and education are discussed, including his training in rhetoric and oratory, which would have affected his articulation and effective sound pressure level while addressing his troops. Based on subjective reports about Caesar's oratorical abilities, his effective Sound Pressure Level (SPL) is assumed to be 80 dBA, about 6 dB above the average loud speaking voice but lower than that of the loudest trained actors and singers. Simulations show that for reasonable background noise conditions Caesar could have been heard intelligibly by 14,000 soldiers in a quiet, controlled environment as in the speech at Dyrrachium. In contrast, even granting generous acoustic and geometric conditions, Caesar could not have been heard by more than about 700 soldiers while his army was on the march before the battle of Pharsalus.

Keywords: acoustics; history; Julius Caesar; digital humanities; archaeoacoustics; acoustic simulation; historical speeches; general's harangue; military history

1. Introduction

Throughout most of human history, electronic amplification of the spoken word was unavailable; therefore, all human gatherings were effectively limited in size by the acoustic range over which the person speaking could be heard intelligibly. Many of the largest reported crowds in history are records of generals delivering a speech or "harangue" to an army [1], since the army's numbers were counted and thus represent a relatively rare instance of a counted crowd in ancient history (although there is certainly a large range of error in all—ancient and modern—crowd-counting methods [2,3]). These accounts were generally accepted by modern historians (e.g., [4,5]) until a landmark paper by Mogens Herman Hansen in 1993, which cast doubt upon the entire historicity of a general's speech to the army [6]. Hansen made many historical and textual arguments that need not be reproduced here in detail, but among these, he made acoustical claims, such as the area over which an ancient phalanx stretched, or noise from rattling hoplite armor, leading him to conclude that

> Under such circumstances it must have been impossible for a general, even if he had had the voice of a Stentor, to deliver a speech that could be heard by all the soldiers simultaneously [6].

Though propagation distance and background noise are certainly valid acoustical criteria to examine, the sentence above contains all the acoustical analysis the author thought necessary to include in his paradigm-shattering thesis. Of course, the problem with vague allusions to scientific truth to prove a point is that the other side can allude to science just as vaguely, which is indeed what happened in this case: a year later, the historian W. Kendrick Pritchett replied with a 100-page rejoinder defending the authenticity of the general's harangue, listing in detail different accounts of speeches to large crowds from ancient Greece and Rome, and also including more recent accounts, from Henry V at Agincourt to George Washington [7]. Pritchett mentioned these military examples as well as another from the preacher John Wesley, arguing that it is acoustically possible for a single speaker to reach such a large assembly.

This tendency to enlist science as an ally without careful numerical consideration often leads to an abuse or neglect of the specific issues in question. For example, Hansen's example of rattling hoplite armor, to which he returns several times in his essay, is based on a single reference to Alexander the Great in the midst of a great battle, rather than hoplites standing at attention *before* battle ([8], 4.13.37). In turn, Pritchett's reference to John Wesley uncritically accepted Wesley's estimate using an assumed crowd density of five people per square yard [9], while modern crowd estimation methods show that in large gatherings the highest densities achieved are less than half that value [10]. In such cases, quantitative references are somewhat superfluous; they are enlisted not to elucidate the past, but because the historians have already made up their minds and want to give their argument a veneer of scientific credibility (as some have asserted,"There are no statistics in ancient sources, just rhetorical flourishes made with numbers" [11]).

Humanities disciplines (including History, although it sometimes presents itself as Social Science instead) primarily *interpret* empirical facts which in themselves are not contested. The arguments employed in this process of interpretation are not (and in a sense cannot be) quantitative. Different historians come to different conclusions on the basis of the same collection of facts. The field which primarily seeks to collect empirical facts to be interpreted is not history but archaeology. In the past, archaeology focused on the excavation of physical artifacts, but recently has embraced computational methods to expand the range of acceptable empirical data [12]. This is a specific application of the movement generally known as the digital humanities, which seeks to use computational methods to shed light on uncontroversial but not immediately obvious facts, which may themselves be the basis for further humanistic interpretation and inquiry. The sub-field variously known as acoustic archaeology or archaeoacoustics uses computational acoustic simulation to uncover facts about the nature of sound in history, which is necessary to address the issue raised by Hansen and Pritchett's disagreement.

Both papers make reference to Julius Caesar's speeches to his army, both during the Gallic Wars and the Roman Civil War. The speeches during the Gallic Wars are smaller and less contested, but during the Civil War Caesar records giving two major speeches to large gatherings of soldiers, one following his army's defeat at the battle of Dyrrachium, and one immediately before the battle of Pharsalus, both in 48 BC [13]. This paper examines the historicity of these accounts through archaeoacoustic simulations of the speeches as Caesar describes them. Using this framework, it can be shown that, even without every historical detail preserved exactly, the plausibility or implausibility of certain historical speeches may be known with a high degree of certainty once we examine the acoustical evidence.

2. Background

After the controversy between Hansen and Pritchett, other historians writing on these speeches referenced the debate, but generally avoided making strong claims about whether these speeches actually occurred [14–16]. Again, this reticence may be a combination of not feeling comfortable with quantitative acoustics, as well as a general feeling that sound as a transient phenomenon is more or less lost to history once it is silenced. As it happens, there is a long tradition of working backwards through known physical laws to study sounds of the past, motivated by this same question: How many listeners can hear a single human voice?

2.1. Benjamin Franklin's Experiment

In 1739, the Methodist revivalist preacher (and friend of John Wesley) George Whitefield drew large outdoor crowds in London that were estimated as high as 80,000 people [17]. Across the Atlantic, Benjamin Franklin, who was at that time the publisher of The Pennsylvania Gazette in Philadelphia, had stopped printing the estimates of Whitefield's crowds because he thought they were exaggerated. Franklin described in his autobiography how he carried out an experiment to measure Whitefield's intelligible distance when the preacher came to Philadelphia. Using a semicircular approximation for the crowd shape, Franklin calculated that

> ··· [Whitefield] might well be heard by more than Thirty Thousand. This reconciled me to the newspaper accounts of his having preached to twenty five thousand people in the fields and to the ancient histories of generals haranguing whole armies of which I had sometimes doubted [18].

While Franklin used Whitefield's example as a conduit to explore the vocal ranges of ancient generals, the data from his experiment also provide information about Whitefield's vocal level, as the preacher was known for having one of the loudest voices of his generation [19]. Data from Franklin's experiment have been used to infer noise characteristics based on known noise sources and site geometry [20], study the maximum sound pressure [21] and directivity patterns for trained vocalists [22], and to simulate Whitefield's own maximum SPL and crowd size based on Franklin's data [23]. This work estimated that Whitefield's time-averaged on-axis L_{eq} could be as great as 90 dB_A, about 16 dB greater than the ANSI standard [24] for "loud speech" (about 74 dB when the ANSI spectrum is A-weighted [25]), and still significantly higher than that for "shouted speech" (about 82 dB_A). The corresponding ISO standard uses more discrete vocal levels, but similarly has its highest value of "Very Loud" speech as 78 dB after A-weighting is applied [25,26].

High vocal levels allow for animal communication at very long distances, producing an evolutionary sexual selection effect [27], which may vestigially influence human vocal capacity today. The computer simulations predicted that Whitefield could be heard intelligibly by a crowd of over 20,000 people without assuming overly optimistic acoustic conditions, although the reported crowd of 80,000 is acoustically implausible even under very favorable conditions [23].

This example shows that there exist regions of plausibility between being naively accepting or close-mindedly skeptical of all historical accounts of speeches to large crowds. Using what is already known about the human voice, sound propagation, and speech intelligibility, we can give a good approximation of how many people could hear a speaker on a specific occasion. While we cannot affix a precise crowd size to every historical account, we can shed a good deal of light on the historical account, which may inform the way we interpret the original text.

Even in situations where we do not have data as convenient as that which Franklin recorded, we can investigate a range of possible acoustic conditions (e.g., background noise). We can simulate the extreme "optimistic" and "pessimistic" ends of this range of conditions to better understand the historical situation. If, even under pessimistic conditions, the intelligibility (measured by the speech transmission index (STI)) is still acceptable throughout the crowd, we may consider the historical account plausible even if we cannot know the precise noise condition without the benefit of a time machine. Conversely, if under optimistic conditions the intelligibility is still too low throughout the simulated crowd, we may consider the account acoustically implausible. To apply this method to Caesar's speeches, we need to first consider his voice, the sites of his speeches, and the background noise present.

2.2. Caesar the Orator

Vocal training is an important factor to the maximum pressure achievable by a speaker [28], and Caesar received extensive training from childhood on. Though most famous for his achievements as a general and later as dictator, Julius Caesar was born in 100 BC into a high family in the Roman

Republic and was trained from childhood to perform the public ceremonies required of the family's inherited priesthood, and Caesar would later assume the title of *pontifex maximus*, the high priest of all Rome (the Latin title was later also given to the Bishop of Rome, i.e., the Pope) [29]. In 70 BC, Caesar traveled to the island of Rhodes to study oratory with the noted rhetoritician Apollonius Molon, who also trained Cicero in oratorical delivery [30]. In addition to his smaller speeches to his centurions or smaller military gatherings, Caesar also gave a noted speech in the Roman Senate advocating mercy for the Cataline conspirators in 63 BC [31].

In addition to the many examples of Caesar's experience with oratory, no less an orator than Cicero himself (who also wrote an entire text on the subject [32]) testified to the quality of Caesar's delivery. In a letter to Cornelius Nepos, Cicero wrote

> Do you know any man who can speak better than Caesar, even if he has concentrated on the art of oratory to the exclusion of all else? [33]

Despite his natural talent for oratory, Caesar chose to pursue the military instead of oratory as his chief vocation, and thus might not be expected to have perfected his delivery to the extent reported for speakers such as Demosthenes, who is reported to have undertaken specific exercises to perfect his elocution, lung control, and overall level [34]. In addition to vocal training, youth is also correlated with maximum vocal output, the level decreasing with increasing age [35]. In this regard, Caesar was no longer so young (52) at the time of the civil war in 48 BC, and thus simulations of his battle speeches cannot assume that he was near the highest vocal levels possible (90 dB$_A$), as in the case of Whitefield, who was only 24 at the time of Franklin's experiment. Because of this, a more moderate averaged SPL range of 74–80 dB$_A$ is assumed for Caesar in the computer models.

3. Simulation Method and Results

3.1. Dyrrachium

3.1.1. Environmental Factors

In 49 BC, fearing prosecution if he lost his military position, Caesar "crossed the Rubicon" into Italy without resigning his military command, initiating the Civil War between his followers, known as the Populares, and the Optimates, who followed Pompey the Great. Caesar followed Pompey's army into Greece, where the two battles both occurred. Pompey's forces managed to break through Caesar's lines at the city of Dyrrachium, leading to a rare decisive defeat for Caesar's army. After the battle he recounted

> Caesar was now forced to abandon his previous plans and believed that he needed to change his entire strategy for this war. Accordingly, he simultaneously withdrew his troops from all their fortified positions, thus ending the siege. He gathered his entire army in one place and addressed the troops in assembly, urging them not to be discouraged by what had happened: they should neither be frightened by their present experience nor consider this single setback–which, at any rate, was relatively minor–as equivalent to their many successful battles [13].

After the speech, Caesar reports that his army successfully withdrew from Pompey's larger force, setting up the later conflict at Pharsalus. The exact location of the speech is not recorded by Caesar or later historians, so it is unknown if there were any distinguishing surfaces or geometric properties which may have aided or impaired speech intelligibility. However, he had retreated from his former fortifications, so it is reasonable to assume it took place on the plain to the east, along the line of Caesar's retreat, a relatively flat area. According to Caesar's own account, he had about 15,000 troops at Dyrrachium, and his losses from the battle were about 1000, suggesting that his entire army at the time of his speech consisted of about 14,000 [13].

Hansen did not address Caesar's speech at Dyrrachium, as assemblies of armies not in formation do not meet his criteria for a "battlefield speech". He even noted that, since on such an occasion the troops could be drawn up closer together, such assemblies are at least plausible [6]. However, it is certainly a historical account of a single speaker addressing a crowd of 14,000, which is a difficult task for most untrained vocalists and thus merits further investigation.

Despite Hansen's repeated concerns about rattling armor, as mentioned before, the only text he cites refers to noise during battle, not before or after. In fact, Caesar's legions at Dyrrachium consisted of battle-hardened soldiers who were intensely loyal to him and adept at moving or being still in a coordinated fashion (in fact, this second quality would more or less win the war for Caesar at Pharsalus). Caesar mentioned that "much self-reproach came over the entire army because of the disaster" which suggests a sombre tone over the whole assembly as their general spoke to them [13]. Assuming that the soldiers could be still and attentive during such a speech, a moderate background noise level of 45 dB$_A$ seems reasonable based on the subjective descriptors of the occasion.

3.1.2. Simulation

The speech at Dyrrachium was modeled in CATT-Acoustic v9.1b [36,37]. Exact environmental data are not available, but the speech occurred in July and the weather was described as quite hot [13], so the model was assigned an air temperature of 27 °C, with 50% humidity, near the mid-summer averages for central Greece today, though to the author's knowledge no more exact temperature data are available for this period. Although geometric acoustic simulations (e.g., CATT) do not fully account for some wave effects [38,39], in this case, the model was needed only for propagation attenuation and STI calculation, so the geometrical model is sufficient for the level of historical precision that is known about the speech. CATT's diffraction prediction algorithm was enabled, allowing the simulation to account for the acoustic ground effect [40].

The model included a large map of a flat crowd, larger than that reported (200 m by 200 m), with assumed density of 2.7 people per m^2, corresponding to "strong" crowd conditions but not the extreme of "mosh-pit" conditions [10]. The modeled crowd's absorption coefficients are given in Table 1 using values measured in [41], and corresponding scattering coefficients were predicted based on an assumed average roughness depth of 0.1 m in CATT. Here and at Pharsalus, the measured Sabine absorption values were greater than 1 for the highest frequencies, but this is an artifact of the Sabine calculation. Since geometric acoustical simulations cannot accept values greater than 1, all such values were truncated to 0.99, as indicated in [37]. Above this, a source representing Julius Caesar was situated 1 m above the crowd, with listeners on all sides of him, assuming that his army would crowd around all sides to be able to hear him better.

Caesar's voice was simulated using a standard male vocal directivity pattern in CATT, since more specific information about Caesar's head shape or directivity pattern probably cannot be known. Caesar's voice was modeled with a steady-state L$_{eq}$ ranging from 74 dB$_A$ (standard "loud" speech) to 80 dB$_A$ (closer to "shouted" speech) at 1 m and a standard male vocal spectrum (which CATT calculates directly from the ANSI-1997 standard spectrum based on simulated level [37]). Background noise was modeled as 45 dB$_A$.

Table 1. Absorption coefficients by octave band center frequency (Hz) for standing soldiers in Dyrrachium model (2.7 persons per m^2, [41]).

Surface	125	250	500	1000	2000	4000
Audience area (dense)	0.24	0.47	0.94	0.99	0.99	0.99

After the simulations, CATT exported grid maps of the crowd, representing the Speech Transmission Index (STI) in each 2 m × 2 m square. From these, the minimum intelligible area (MIA), defined as the area over which STI $\geq$ 0.3 was calculated for each grid with a custom Matlab script. 0.3

was first used as a threshold of intelligibility in [23] because the phoneme-group score for various word groups drops precipitously below this value [42], leading STI measures below this to be characterized as "bad" intelligibility. Based on the source vocal level, the MIA is predicted to range from 2104 to 8096 m^2, as shown in Figure 1 (Caesar's position is shown by a yellow dot). Final predicted intelligible crowd sizes are 2.7 multiplied by the MIA, assuming a density of 2.7 persons per m^2, as shown in Table 2.

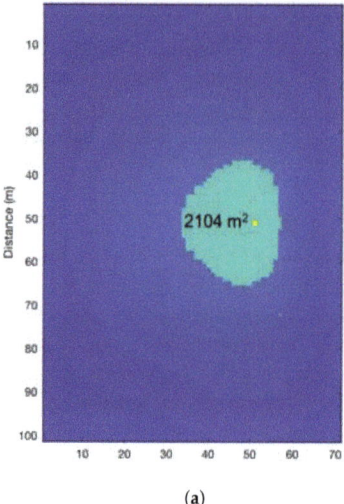
(a)

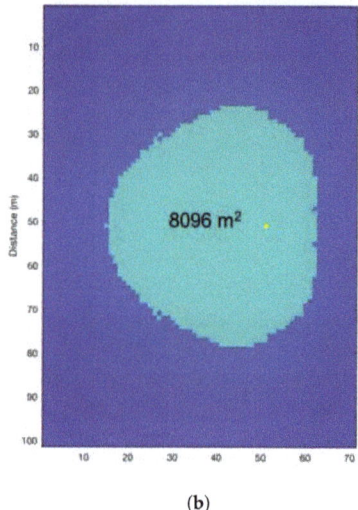
(b)

Figure 1. Minimally Intelligible Area (MIA) at Dyrrachium Based on Source Vocal Level: (**a**) 74 dB$_A$ source; and (**b**) 80 dB$_A$ source.

Table 2. Modeled crowd size, area, and density by vocal source level.

Vocal SPL (dB$_A$)	Background Noise Level (dB$_A$)	MIA (m^2)	Density (Persons/m^2)	Crowd Size
74	45	2104	2.7	5,681
80	45	8096	2.7	21,859

3.2. Pharsalus

3.2.1. Environmental Factors

After his speech at Dyrrachium, Caesar led his army across the Thessalian plain hoping to engage Pompey in the open field. Pompey, who had just won at Dyrrachium and had a much larger army (about 45,000 to Caesar's 22,000 at this point according to Caesar, who had received reinforcements from his commander Gnaeus Domitius Calvinus [13]), held higher ground and hoped for a defensive advantage. However, pressure from Pompey's associates to hurry up and finish the war led him to move out and engage Caesar's army. Caesar describes it thus:

> After· · · the signal for departure had already been given and the tents struck, it was noticed that a little earlier Pompey's battle line, exceeding what it had been used to doing each day, had moved forward farther from their rampart, so that it appeared possible to fight on ground that would not be unfavorable. Then Caesar, who was with his men as they were about to march out of the gate in formation, said to them, "We need to put off our march for now and turn our thoughts to battle, which is what we have been demanding all this time.

Let's take courage and be prepared to fight. It will not be easy to find another chance later." Then he immediately led out his troops, unencumbered and ready to fight [13].

Even Hansen the skeptic reckoned this speech may have actually happened, but argued that because it was so short it did not count as a true battlefield speech [6]. The exact site of the battle was unknown for nearly 2000 years, with many historians supposing that it happened south of the Epineus River, until F. L. Lucas suggested in 1921 that the battle be located on the north bank [43], which is now broadly accepted [44]. Because Pompey had come down out of the rougher hills to the west, it seems reasonable to model this as flat terrain as well.

However, unlike the controlled environment after Dyrrachium, the acoustic context of this speech is quite different. While in a gathering for the conveyance of information, we can assume Caesar's troops to be relatively quiet, in this instance, he specifically mentions that they were assembled and marching out the gate to challenge and exhaust Pompey. By Caesar's own account his troops were in battle formation and moving while he spoke to them, which would drastically increase the background noise. While we do not know the exact SPL this would produce, an optimistic figure is 55 dB$_A$, although the value could certainly be much greater.

3.2.2. Simulation

Caesar was simulated giving a speech to another 200 m × 200 m grid, with the same environmental factors as at Dyrrachium above. His vocal level was modeled as 80 dB (the quieter value of 74 dB being discarded for now to explore only the most optimistic scenario). The background noise was modeled as 55 dB$_A$ to account for the louder noise of an army on the march. In addition, since Caesar's army was in formation, the greatest density that can be assumed for them is lower than at Dyrrachium, about 1 person per m^2 [45]. Thus, Caesar's army was modeled using measured absorption values for a standing crowd with density of 1.17 persons per m^2, as measured in [46] and shown in Table 3.

Table 3. Absorption coefficients by octave band center frequency (Hz) for soldiers in Pharsalus model (1.17 persons per m^2, [46]).

Surface	125	250	500	1000	2000	4000
Audience area (dense)	0.20	0.35	0.70	0.99	0.99	0.99

Again, the STI map from CATT was exported and all intelligible grid squares' areas were summed in Matlab, with the difference this time that no space was allowed behind Caesar since he was presumably in front of his army when he addressed them (in general, cutting off the space behind a speaker, given the shape of the human voice's directivity pattern, will reduce the MIA for that scenario by about 20%). The simulated MIA under these conditions is shown in Figure 2.

It can be seen that the predicted MIA drops drastically under these conditions, to 596 m^2, as shown in Table 4. Thus, the greatest number of men that could have heard Caesar intelligibly at once is closer to 500 (the size of a single cohort) rather than his entire army of 22,000.

Table 4. Modeled crowd size, area, and density by vocal source level.

Vocal SPL (dB$_A$)	Background Noise Level (dB$_A$)	MIA (m^2)	Density (Persons/m^2)	Crowd Size
80	55	596	1.17	697

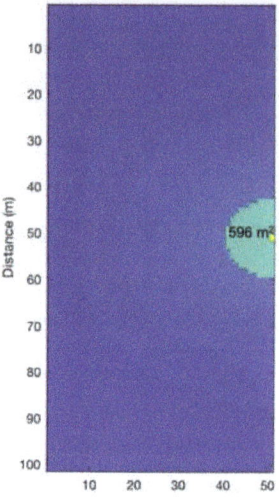

Figure 2. Minimally Intelligible Area (MIA) at Pharsalus.

4. Discussion

Based on these simulations, we may shed new light on the acoustical plausibility of these two speeches, which will affect our interpretation of Caesar's account. At Dyrrachium, we simulated rather pessimistic conditions—giving Caesar only a 6 dB boost above normal loud speech, while assuming background noise of 45 dB$_A$, over 10 dB greater than that measured at some ancient Greek amphitheaters [47]. It can be seen from the simulations that, even under these relatively pessimistic conditions, Caesar could have plausibly spoken to all his 14,000 soldiers at once (leaving room for more!), as long as he did not speak with a "normal" vocal level of 74 dB. Since the pessimistic case is plausible, we do not need to stretch the bounds of the simulation by probing how loud Caesar might have been, how quiet his men might have been, or assuming the existence of atypical environmental propagation, such as refraction patterns via a temperature inversion that would have carried his voice farther. The historical details must be sorted out by historians, but acoustically it must be said that this speech seems physically valid based on the descriptions we have of the event.

In contrast, before Pharsalus, even assuming the louder value for Caesar and a relatively quiet noise value for his army, his MIA value decreases by an order of magnitude. The maximum width of his intelligible area is less than 20 m, while the battlefront stretched up to 3 km once the armies were engaged [44]. Even assuming his army was somewhat more compact before he spoke to them, there does not appear to be any acoustically plausible scenario in which he could be heard intelligibly by his entire force if they were all in front of him, arranged less densely, and on the move. Since the optimistic case does not appear plausible, there is no need to simulate greater background noise values. In both cases, there is clearly some uncertainty about Caesar's exact STI value, vocal level, and background noise. However, by considering an implausible optimistic scenario or a plausible pessimistic scenario, it is possible to speak confidently about the general plausibility of an account even if we cannot know the precise intelligibility value at each point in the crowd.

Having made this acoustical point, however, it should not be inferred that Caesar's speech as written did not happen—it is quite possible that he did speak something similar to the soldiers and officers nearby when he realized Pompey's forces were extended far enough to engage them. The signal to engage could then have been conveyed by messengers, trumpet signal, or flag signals, all of which

Caesar used during the Civil War [13]. In fact, as Pompey's cavalry attempted to flank his right side, Caesar reacted as follows:

> However, once [Caesar] had a look at the enemy formation described above, he feared a flanking attack by the mass of enemy cavalry circling around his right wing; he therefore rapidly drew individual cohorts out of the third line of his formation, placed them as a fourth line to oppose Pompey's cavalry, and explained to them what he wanted them to do, making it clear that this day's victory would depend on the bravery of these cohorts. At the same time he commanded the third line not to move forward and engage with the enemy without explicit orders from himself: when he wanted this to happen, he would give the signal with a flag [13].

This action seems to be largely improvised, as Caesar recorded himself *explaining* and *commanding* his soldiers verbally of what they were about to do. Assuming that he could make himself heard by a single cohort of 500 men in formation, this seems reasonably in line with him riding behind his third line and speaking multiple times to multiple cohorts as he formed them into the fourth line which would counterattack Pompey's cavalry, resulting in Caesar's successful flank of Pompey, the route of Pompey's forces, and the end of the Civil War. The fact that Caesar ordered them all to coordinate this action by a flag signal again shows that he did not expect to be able to verbally communicate with his entire army, or even a portion of it, during the battle itself. However, his reported oratorical skill would have come in useful in this and many other noisy battles where Caesar's instincts and improvisational abilities led his forces to many such incredible victories and very few defeats.

Although this discussion has focused mainly on Caesar, clearly many such pitched battles between ancient armies occurred on fairly flat terrain without many significant reflecting surfaces, so these simulations may be seen as a proxy for any male speaker with the level and background noise conditions assumed here. This may be less useful for modern assemblies, which generally have electrical amplification, but it can be viewed as a lens through which to consider other ancient reported speeches before armies. However, for each situation, the orator's vocal ability, training, and age should be considered, along with the likely background noise of the crowd listening.

5. Conclusions

The simulations above show that Caesar's post-battle speech at Dyrrachium seems acoustically plausible, even without assuming optimistic conditions. In contrast, Caesar's speech before the battle at Pharsalus cannot plausibly have been heard intelligibly by most of his army in formation and on the march. This does not necessarily imply that we should adopt the extreme skepticism of Hansen, who doubted that commanders could even address 1000 soldiers at once in formation [6]. In other cases, when forces were drawn up to a static position before a battle, it may have been acoustically possible for their commanders to address all of them at once, depending on vocal level and background noise. Other cases such as accounts from Thucydides or Henry V's speech before Agincourt, seem to fall into this category, and require separate simulations to account for the particulars of each account. However, Caesar's speech at Pharsalus should be considered a local order to the troops immediately around him, rather than a grand address given to the entire army.

Funding: This research received no external funding.

Conflicts of Interest: The author declares no conflict of interest.

References

1. Thucydides. *History of the Peloponnesian War*; Bohn: London, UK, 1843.
2. Jacobs, H.A. To count a crowd. *Columbia Journal. Rev.* **1967**, *5*, 37–40.
3. Seidler, J.; Meyer, K.; Gillivray, L.M. Collecting Data on Crowds and Rallies: A New Method of Stationary Sampling. *Soc. Forces* **1976**, *55*, 507–519. [CrossRef]

4. Murphy, C.T. The Use of Speeches in Caesar's Gallic War. *Class. J.* **1949**, *45*, 120–127.
5. Miller, N.P. Dramatic Speech in the Roman Historians. *Greece Rome* **1975**, *22*, 45–57. [CrossRef]
6. Hansen, M.H. The battle exhortation in ancient historiography: Fact or fiction? *Historia* **1993**, *42*, 161–180.
7. Pritchett, W.K. The General's Exhorations in Greek Warfare. In *Essays in Greek History*; J. C. Gieben: Amsterdam, The Netherlands, 1994; pp. 27–109.
8. Rufus, Q.C. *History of Alexander*; Harvard University Press: Cambridge, MA, USA, 1946.
9. Wesley, J. *The Journal of John Wesley*; Moody Press: Chicago, IL, USA, 1951.
10. Watson, R.; Yip, P. How many were there when it mattered? *Significance* **2011**, *8*, 104–107. [CrossRef]
11. O'Neill, T. The Great Myths 5: The Destruction of The Great Library of Alexandria. 2017. Available online: https://historyforatheists.com/2017/07/the-destruction-of-the-great-library-of-alexandria/ (accessed on 8 February 2018).
12. Bordes, N.; Pailthorpe, B.; Hall, J.; Loy, T.; Williams, M.; Ulm, S.; Zhou, X.; Fletcher, R. Computational Archaeology. In Proceedings of the WACE 2004: Workshop on Advanced Collaborative Environments, Nice, France, 23 September 2004.
13. Raaflaub, K.A. (Ed.) *The Landmark Julius Caesar*; Pantheon Books: New York, NY, USA, 2017.
14. Ehrhardt, C.T.H.R. Speeches before Battle? *Hist. Z. Alte Gesch.* **1995**, *44*, 120–121.
15. Nordling, J.G. Caesar's Pre-Battle Speech at Pharsalus (B.C. 3.85.4): Ridiculum Acri Fortius ··· Secat Res. *Class. J.* **2005**, *101*, 183–189.
16. Zoido, J.C.I. The Battle Exhortation in Ancient Rhetoric. *Rhetorica* **2007**, *25*, 141–158. [CrossRef]
17. Dallimore, A. *George Whitefield: The Life and Times of the Great Evangelist of the Eighteenth–Century Revival*; Banner of Truth Trust: London, UK, 1970.
18. Franklin, B. *The Autobiography of Benjamin Franklin*, 2nd ed.; Yale University Press: New Haven, CT, USA; London, UK, 1964.
19. Stout, H. *The Divine Dramatist: George Whitefield and the Rise of Modern Evangelicalism*; William B. Eerdmans Publishing Company: Grand Rapids, MI, USA, 1991.
20. Boren, B.B.; Roginska, A. Analysis of noise sources in colonial Philadelphia. *Internoise* **2012**, *9*, 7543–7553.
21. Boren, B.B.; Roginska, A. Maximum Averaged and Peak Levels of Vocal Sound Pressure. In Proceedings of the 135th Audio Engineering Society Convention, New York, NY, USA, 17–20 October 2013.
22. Boren, B.B.; Roginska, A. Sound Radiation of Trained Vocalizers. In Proceedings of the Meetings on Acoustics: 21st International Congress on Acoustics, New York, NY, USA, 4 May 2013.
23. Boren, B.B. George Whitefield's Voice. In *George Whitefield: Life, Context and Legacy*; Jones, D.C., Hammond, G., Eds.; Oxford University Press: Oxford, UK, 2015.
24. American National Standards Institute. *American National Standard: Methods for Calculation of the Speech Intelligibility Index*; Acoustical Society of America: New York, NY, USA, 1997.
25. Rindel, J.H. *Odeon Application Note–Calculation of Speech Transmission Index in Rooms*; Odeon A/S: Lyngby, Denmark, 2014.
26. International Organization for Standardization. *ISO 9921: 2003 Ergonomics–Assesment of Speech Communication*; International Organization for Standardization: Geneva, Switzerland, 2003.
27. Forrest, T.G.; Raspet, R. Models of female choice in acoustic communication. *Behav. Ecol.* **1994**, *5*, 293–303. [CrossRef]
28. Mendes, A.P.; Rothman, H.B.; Sapienza, C.; Brown, W. Effects of vocal training on the acoustic parameters of the singing voice. *J. Voice* **2003**, *17*, 529–543. [CrossRef]
29. Kahn, A. *The Education of Julius Caesar*; Schocken Books: New York, NY, USA, 1986.
30. Plutarch. *The Parallel Lives*; Loeb Classical Library: Cambridge, MA, USA; London, UK, 1919.
31. Sallust. *Conspiracy of Catiline*; Harper and Brothers: New York, NY, USA; London, UK, 1899.
32. Cicero, M.T. *De Oratore*; Harvard University Press: Cambridge, MA, USA, 1967.
33. Suetonius, G. *The Twelve Caesars*; Penguin Books: London, UK, 1957.
34. Dobson, J.F. *The Greek Orators*; Methuen and Co.: London, UK, 1919.
35. Kent, R.; Kent, J.; Rosenbek, J. Maximum Performance Tests of Speech Production. *J. Speech Hear. Disord.* **1987**, *52*, 367–387. [CrossRef] [PubMed]
36. Dalenbäck, B.I.L. Room Acoustic Prediction Based on a Unified Treatment of Diffuse and Specular Reflection. *J. Acoust. Soc. Am.* **1996**, *100*, 899–909. [CrossRef]
37. Dalenback, B. *CATT-Acoustic v9*; CATT: Gothenburg, Sweden, 2011.

38. Rindel, J. The Use of Computer Modeling in Room Acoustics. *J. Vibroeng.* **2000**, *3*, 219–224.
39. Siltanen, S.; Lokki, T.; Savioja, L. Rays or Waves? Understanding the Strengths and Weaknesses of Computational Room Acoustics Modeling Techniques. In Proceedings of the International Symposium on Room Acoustics, Melbourne, Australia, 29–31 August 2010.
40. Forrest, T.G. From Sender to Receiver: Propagation and Environmental Effects on Acoustic Signals. *Am. Zool.* **1994**, *34*, 644–654. [CrossRef]
41. Adelman-Larsen, N.W.; Thompson, E.R.; Gade, A.C. Suitable reverberation times for halls for rock and pop music. *J. Acoust. Soc. Am.* **2010**, *127*, 247–255. [CrossRef] [PubMed]
42. Steeneken, H.J.M.; Houtgast, T. Basics of the STI measuring method. In *Past, Present and Future of the Speech Transmission Index*; van Wijngaarden, S.J., Ed.; TNO Human Factors: Soesterberg, The Netherlands, 2002; pp. 13–43.
43. Lucas, F.L. The Battlefield of Pharsalos. *Annu. Br. Sch. Athens Suppl.* **1921**, *24*, 34–53. [CrossRef]
44. Morgan, J.D. Palaepharsalus–The Battle and the Town. *Am. J. Archaeol.* **1983**, *87*, 23–54. [CrossRef]
45. James, S. 48 BC: The Battle of Pharsalus. In *Rome's Legions: Decoding Their Pythagorean Organization 753 BC to 410 AD*; Academia: Prague, Czechoslovakia, 2010.
46. Martellotta, F.; D'Alba, M.; Crociata, S.D. Laboratory measurement of sound absorption of occupied pews and standing audiences. *Appl. Acoust.* **2011**, *72*, 341–349. [CrossRef]
47. Bo, E.; Kostara-Konstantinou, E.; Lepore, F.; Shtrepi, L.; Puglisi, G.; Astolfi, A.; Barkas, N.; Mangano, B.; Mangano, F. Acoustic characterization of the ancient theatre of Tyndaris: Evaluation and proposals for its reuse. In Proceedings of the ICSV 2016–23rd International Congress on Sound and Vibration: From Ancient to Modern Acoustics, Atene, Greece, 10–14 July 2016.

© 2018 by the authors. Licensee MDPI, Basel, Switzerland. This article is an open access article distributed under the terms and conditions of the Creative Commons Attribution (CC BY) license (http://creativecommons.org/licenses/by/4.0/).

Article

Performance Space, Political Theater, and Audibility in Downtown Chaco

David E. Witt [1,*] and Kristy E. Primeau [2]

1 Department of Anthropology, State University of New York at Buffalo, Buffalo, NY 14261-0026, USA
2 Department of Anthropology, State University of New York at Albany, Albany, NY 12222, USA; kprimeau@albany.edu
* Correspondence: dwitt@buffalo.edu

Received: 6 November 2018; Accepted: 21 December 2018; Published: 27 December 2018

Abstract: Chaco Canyon, NM, USA, was the center of an Ancestral Puebloan polity from approximately 850–1140 CE, and home to a dozen palatial structures known as "great houses" and scores of ritual structures called "great kivas". It is hypothesized that the 2.5 km^2 centered on the largest great house, Pueblo Bonito (i.e., "Downtown Chaco"), served as an open-air performance space for both political theater and sacred ritual. The authors used soundshed modeling tools within the Archaeoacoustics Toolbox to illustrate the extent of this performance space and the interaudibility between various locations within Downtown Chaco. Architecture placed at liminal locations may have inscribed sound in the landscape, physically marking the boundary of the open-air performance space. Finally, the implications of considering sound within political theater will be discussed.

Keywords: archaeoacoustics; soundscapes; open-air performance space; political theater; Ancestral Puebloan; Chaco Canyon

1. Introduction

Chaco Canyon, San Juan County, NM, USA, was the center of an Ancestral Puebloan (also known as "Anasazi") polity from approximately 850–1140 CE. This location was home to a dozen palatial structures known as "great houses" and scores of ritual structures called "great kivas". Most of these were located in a 2.5 square kilometer region referred to as "Downtown Chaco", centered on the largest great house, Pueblo Bonito (Figure 1). We hypothesize that this downtown area served as an open-air performance space for both political theater and sacred ritual. Indeed, it is believed that ritual and politics were tightly bound together in Chaco, much like other early states, and similar to the nature of leadership within modern Pueblo communities.

Our purpose for this paper is to further explore the relationship between the built environment of Chaco and its soundscapes, a situation that we approached in various conference papers [1–3], and in a previous article, "Soundscapes in the past: Investigating sound at the landscape level" [4]. In our article, we reported that the physical relationship between modeled soundscapes and the locations of shrines throughout the wider landscape may be evidence of ritual performance space, where the shrines themselves marked the bounds of that space.

Our goals for this article are threefold. First, we briefly provide a literature review for those unfamiliar with the study of archaeoacoustics, particularly how it relates to the landscape scale. Secondly, we review the results of the initial study and then interpret those results with two linked bodies of anthropological theory: performance theory and political theater. Performance theory describes how activities gain their meaning in the context of group involvement [5], and political theater describes how elites utilize performances to present themselves as they want to be seen in order to legitimize their status [6]. These two interrelated theories can help researchers develop a stronger understanding of the nature of landscape experience at Chaco, illustrating how Chacoan elites may have guided the construction of specific landscape features to both anchor and bound socio-cultural performance space, and to serve as a stage for political theater for the legitimation of their roles. Finally, we show how both of these functions interacted simultaneously to construct and reify political power in the 10th and 11th century CE.

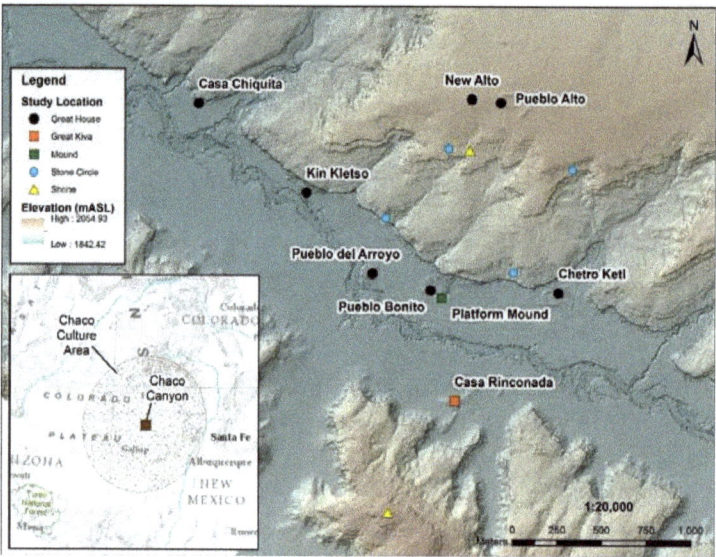

Figure 1. Map of "Downtown Chaco", Chaco Canyon, NM, USA.

Background

Archaeoacoustics is the study of the evidence of sound in the archaeological record. This can be achieved by studying the acoustical properties of artifacts, sites, or landscapes. An important method for understanding past people and cultures, archaeoacoustics provides an integral, albeit often ignored, component of the human experience. Most of the previous work on sound in the archaeological record has focused on the artifactual or site level [7–24], yet recent research has expanded to the landscape level [4,25–32].

Most recognized amongst landscape theories is phenomenology, an interpretive framework which explains that landscapes are places where memory, meaning, and identity interweave as integral parts of the lived experience [33–42]. Phenomenology, however, has been critiqued as methodologically weak, as it has traditionally relied upon qualitative, personal observations [34,35,43–49]. Our tools, described below, are being developed to answer this critique. Simultaneously, the tools also respond to claims that Geographic Information Systems (GIS) and other "abstracted experiences" are positivistic [39,50,51], by combining the strengths of both GIS and phenomenology to answer Tim Ingold's call that anthropologists adopt a greater awareness of the lived experience [52].

Our tools model the extent to which sounds, including those originating from the human voice and musical instruments, can be heard throughout the wider landscape. We hypothesize Chacoan elites to be practiced orators, able to speak for extended periods with a raised voice. For example, the historic Zuni (a Puebloan group) maintained a Priest of the Sun: "whose title, Pekwin, means, literally, Speaking Place . . . It is at the solstices that the sun is celebrated with great public ceremonies . . . In winter the public ceremonies are opened by the Pekwin's announcement made from the housetop at dawn. At this time he orders the people to make prayer sticks for their sun father and their moon mother" [53] (p. 512).

Within the American Southwest, musical instruments have been recovered from the pre-Hispanic period. These include bone flutes and whistles, wooden planks (i.e., "foot drums"), copper bells, and conch shell trumpets [54–58]. These instruments are linked to ritual and public performances, as illustrated by the ethnohistoric and archaeological records [57,59–62], as well as shown by use in modern contexts [63]. These performances take place within ritually charged locations such as enclosed kivas and open-air plazas. The recovery of conch shell trumpets and other instruments from similar ritual contexts in the archaeological record may illustrate similar use in the pre-Hispanic period [59,64].

The importance of sound in the past has recently been recognized by archaeologists working in the American Southwest, borrowing from researchers working in Europe and South America [15,30]. This recognition has primarily resulted from the work of Richard Loose, who has studied the acoustic properties of artifacts (such as conch shell trumpets), structures (such as the great kiva at the Aztec Ruins National Monument, San Juan County, NM, USA), and landscapes (such as Chaco Canyon and Casamero Pueblo, McKinley County, NM, USA) [65–69], but the topic is now receiving wider attention [4,25]. Within Chaco, the importance of sound is most clearly illustrated by Tse Biinaholtsa'a Yałti, Navajo for "Concavity in the Bedrock that Speaks". This feature is an alcove located on the north wall between Pueblo Bonito and Chetro Ketl, near the stone circle 29SJ1565. The alcove, which was modified by the removal of approximately 360 m^3 of bedrock, is associated with an altar and rock art panel, and is considered a portal to the dimension of Navajo deities [65,70]. The projected circle of the alcove's amphitheater, some 340 m in diameter, is "clearly a delineated space that is level, and conspicuously devoid of features and material culture" [70] (p. 206). Acoustical studies have been conducted by Richard Loose and colleagues, illustrating that this site creates a "virtual sound image" [68,69], resulting in a phenomenon that is described as filling the canyon floor with sound, and "a sensation of being 'bathed' in sound as standing waves of sound formed along the axis of the amphitheater" at certain frequencies [70] (p. 208). Other tests indicate the existence of other auditory phenomena, including echoes, reverberations, and the cancelling out of sound at various locations [65]. From this feature, it seems obvious that sound was intentionally manipulated as an aspect of landscape within Chaco Canyon and it likely played an important role in Chacoan rituals.

2. Modeling Methods

In 2016, we developed Soundshed Analysis Tool, beta version 0.9.2, part of an Archaeoacoustics Toolbox which models the spread of sound throughout a landscape [1–4]. Written in the Python programming language for ArcGIS 10.3, it is based upon SPreAD-GIS, a toolbox developed to model the propagation of engine noise within wildland settings [71,72]. The following year the acoustical modeling tool was updated [73,74], and the Archaeoacoustics Toolbox now includes preset versions of the Soundshed Analysis Tool which utilize elevation datasets with 1, 1.5, and 30 m resolutions. In addition, modeling at alternative resolutions is possible with minor adjustments to the Python script. For this analysis, our soundsheds feature a 1.5 m resolution based on LiDAR data.

Modeling the spread of sound in a GIS environment places an emphasis on the spatial location and extent of the soundshed, rather than a detailed acoustical reconstruction. This allows archaeologists to incorporate acoustics into their analyses of relationships between sites and features within the landscape, and study the cultural implications of those relationships. While noise analysis software can be cost-prohibitive or otherwise inaccessible to archaeologists, GIS is a prevalently used tool that most archaeologists can access and operate; we hope that, when complete, the Archaeoacoustics Toolbox will introduce many archaeologists to acoustical modeling as an open-source addition to readily available GIS software.

2.1. Model Inputs

Input variables for the model include environmental data and archaeologically derived cultural data, as illustrated in Table 1. Environmental inputs include an elevation dataset and information used to determine the physical characteristics of the spread of sound in air. These are typically gathered from the literature and include the percentage of relative humidity [75], the air temperature in degrees Fahrenheit [75], and the ambient sound pressure level (dB(A)) of the study location [76]. Cultural data describe the sound source. These consist of the location of the sound source, the output height (ft) of the person or instrument creating the sound (which can be derived from osteological data and/or the artifact assemblage), the sound pressure level of the source (dB(A)), the distance (ft) at which the sound pressure level of the source was measured, and the frequency (Hz) representing the fundamental tone or peak long-term average frequency at which the sound source was measured [66,77,78]. Specific modeling inputs used for this paper are provided in Table 2.

Table 1. Soundshed Analysis Tool v0.9.2 Input Variables.

Environmental Inputs	Cultural Inputs
Percentage of Relative Humidity	Location of Sound Source
Air Temperature (°F)	Height of Sound Source (ft)
Ambient Sound Pressure Level (dB(A))	Sound Pressure Level of Source (dB(A))
LiDAR-based DEM	Measurement Distance of Source (ft)
	Frequency of Source (Hz)

Table 2. Soundshed Analysis Tool v0.9.2 Modeling Inputs.

	Modeling Inputs	Elite Orator with a Raised Voice: Afternoon in June	Conch Shell Trumpet: Dawn in June
Environmental Inputs	Percentage of Relative Humidity	30%	30%
	Air Temperature	89.6 °F (32 °C)	55.4 °F (13 °C)
	Ambient Sound Pressure Level	20.7 dB(A)	20.7 dB(A)
Cultural Inputs	Height of Sound Source	5 ft (1.5 m)	6 ft (1.8 m)
	Sound Pressure Level of Source	84 dB(A)	96 dB(A)
	Measurement Distance of Source	3 ft (0.9 m)	4 ft (1.2 m)
	Frequency of Source	325 Hz	330 Hz

2.2. Modeling Steps

The Soundshed Analysis Tool is a geometric-type model which assumes sound is travelling through the air along straight-line paths. Currently, the model does not incorporate wave effects such as reverberation which require more processing power than a 32-bit GIS environment presently provides. The tool uses formulae of outdoor sound propagation, calculating free-field sound attenuation following ISO 9613-2 [79], atmospheric absorption loss following ANSI 1.26 [80], topographic loss following ISO 9613-2 [79], and barrier effects based on Maekawa's optical diffraction theory [81,82]. The results are output in soundshed rasters that indicate audibility over background noise levels, and provide a viewshed analysis for that site. Within this paper, "audibility" refers to the perception of sounds and does not necessarily implicate the intelligibility of speech. Rasters can be created for any frequency, however the examples presented herein match the fundamental tone or peak long-term average frequency of the sound source. Each study location is modeled independently, although a second tool being developed for inclusion in the Archaeoacoustics Toolbox can create cumulative soundsheds for sound sources propagating from multiple landscape locations simultaneously. Due to the environmental nature of the Chacoan landscape, vegetation attenuation and ground effects are not modeled in v0.9.2, however the script is currently under revision to include these calculations following ISO 9613-2 [79,83]. Absorption due to structural surfaces is also not modeled in the current tool.

3. Modeling Results

Using the above described tool and inputs, we modeled the spread of sound emanating from various sources at locations throughout Chaco Canyon [4]. Here, we continue the discussion by drawing attention to one specific location, the two platform mounds located immediately south of and in front of Pueblo Bonito, the largest great house in the Puebloan world, measuring approximately 90 by 150 m, or 1.2 ha in size (Figures 2 and 3). At this location, we modeled two scenarios: a practiced orator addressing a crowd, and an individual playing a conch shell trumpet.

Figure 2. Eastern platform mound at Pueblo Bonito (covered by vegetation). Man provided for scale.

Figure 3. Reconstruction of the Pueblo Bonito architecture. Reproduced with permission from Richard Friedman, in The Architecture of Chaco Canyon, New Mexico; published by University of Utah Press, 2007.

The platform mounds were important features of the landscape. The 3–4 m tall, rectangular mounds were constructed during the Classic Bonito Phase, from 1040–1100 CE, as indicated by ceramic dating [84,85]. The mounds were built up with adobe embankments, steps, and masonry retaining walls [84,86,87], as interpreted in Figure 3. The mounds contained artifacts that mostly reflect household refuse, with equivocal evidence for large scale feasting and specialized production [88,89]. However, the presence of relatively larger numbers of exotic goods such as turquoise, Narbona Pass chert, cacao residue, and macaw remains reinforce claims that Pueblo Bonito was a residence of Chacoan elites, and may indicate that ritual deposition occurred here [90].

Our scenarios were modeled at this location because the platform mounds were earthen architecture that were intentionally built. Researchers have hypothesized that the mounds are ritually charged due to their astronomical alignments, location, directionality, and ability to direct access to Pueblo Bonito [87,91–94]. While others have argued that the features, which were constructed of household refuse, resulted from the occupation of Pueblo Bonito, they are too large to be the result of only the relatively small population of the great house [95,96] and must have included imported materials. However, even if these mounds were merely domestic middens (or were meant to replicate domestic middens in some sense), they may still have been considered sacred places as middens were often the location of burials. Indeed, modern Puebloan people consider middens to be sacred for that reason [97,98]. Nevertheless, these mounds were important and may have served as performance stages, similar to Mesoamerican pyramids and Hohokam platforms [99]. Ruth Van Dyke states: "Standing atop them, with the great house and the north face of Chaco Canyon towering behind, ritual leaders would have been a very impressive sight. Ceremonies performed atop the mounds would have been highly visible to masses of people who, perhaps, did not have access into the great house itself" [97] (p. 130). For these reasons, we consider the mounds to be more than mere trash dumps (see also [100] for a discussion on the similar role of plazas in later Puebloan sites).

The modeling results are presented in Figures 4 and 5. The figures illustrate the amount by which the two sounds rise over ambient noise levels; hence, they also indicate a positive signal-to-noise ratio. While our study does not approach speech intelligibility as a specific topic of investigation, work by Alvarsson and his colleagues has shown that the signal-to-noise ratio is highly correlated to the Speech Intelligibility Index (SII) and may be used as a proxy of the SII outdoors [101].

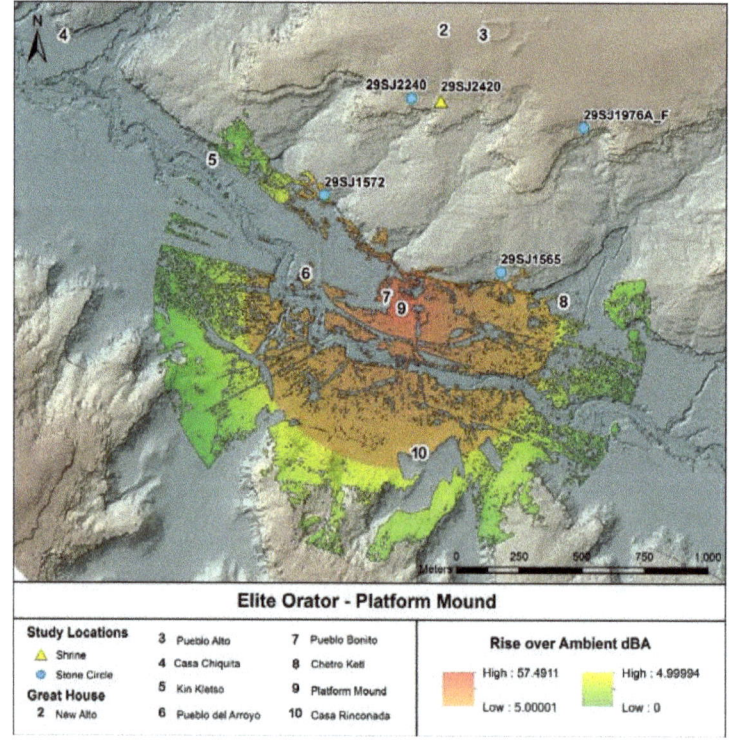

Figure 4. Soundshed of an elite orator speaking at 84 dB(A) with a peak long-term average frequency of 325 Hz.

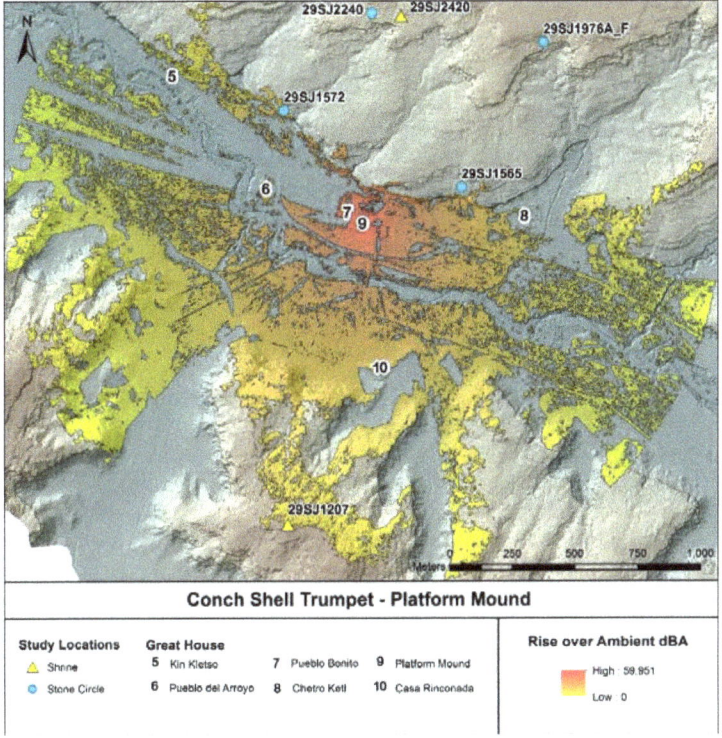

Figure 5. Soundshed of reconstructed conch shell trumpet at 96 dB and a fundamental tone of 330 Hz.

Figure 4 represents a person standing on the eastern mound (9), speaking loudly as if orating to a crowd. Studies have shown that male speakers accustomed to oration, such as actors or preachers, have reached a maximum vocal level of 90 dB(A) when addressing a crowd [102–104]. This individual, with a vocal level of 84 dB(A) and a peak long-term average frequency of 325 Hz, could be heard throughout Downtown Chaco. Using the signal-to-noise ratio as a proxy for SII, the 5 dB(A) contour, indicated by the abrupt shift between the orange and yellow shading, would equate to an approximate SII of 0.6 in a free field according to modeling conducted by Larm and Hongisto [105] (Figure 12 in the reference). Furthermore, Lazarus [106] (Figure 1 in the reference) and Jovičić [107] (Figure 5 in the reference) indicate that a signal-to-noise ratio of 5 dB(A) would equate to an approximate speech intelligibility (SI) score greater than 80 and 85 percent, respectively. Therefore, Figure 4 also indicates the extent to which an individual's speech may have been understood over environmental background noise given an absence of intervening noise sources, illustrating an approximate degree of speech intelligibility at the Pueblo Bonito (7), Pueblo del Arroyo (6), and Chetro Ketl (8) great houses, as well as the Casa Rinconada great kiva (10). Additionally, people at Kin Kletso (5) may have been able to hear the individual if they listened carefully for the sound of his voice, but likely would not have been able to understand what he was saying, as Kin Kletso is located beyond the contour indicating the SII of 0.6.

Figure 5 shows the spread of sound from a conch shell trumpet, an instrument that has been recovered in association with elite burials in Pueblo Bonito [59,66,108], and which was used in historic Puebloan rituals [59]. As illustrated by these figures, the sound of a reproduction trumpet, with an output of 96 dB and a fundamental tone of 330 Hz, spreads an additional 435 m beyond the output indicated by Figure 4, and individuals at the 29SJ1207 shrine overlooking the canyon from the south would have been able to hear the instrument. Additionally, individuals at the 29SJ1565 and 29SJ1572 stone circles on the north canyon rim may have heard both events. These shrines, in effect, inscribe and demarcate sound within the Chacoan landscape.

4. Discussion

We interpret these results through complementary aspects of performance theory and political theater. As mentioned previously, performance theory describes how activities gain their meaning in the context of group involvement [5]. Similarly, political theater describes how elites utilize performances to legitimize their status [6]. Thomas Luckman stated that "legitimation is making sense of power...to those who exercise power, to those who are subject to the exercise of power, or to both" [109] (p. 111). Inomata and Coben discuss the relationship between performance theory and political theater: "it is probably true that the development of large, centralized polities would have been impossible in any historical context without frequent public events, in which agents of political power presented themselves in front of a large number of spectators and the participants shared experiences through their bodily copresence" [5] (p. 11). Furthermore, "these events have profound implications for the understanding of any society, particularly in terms of the integration of communities and the establishment and maintenance of asymmetrical power relations, which are intricately intertwined with each other" [5] (p. 22). We believe that this is true for Chaco, and that the mounds in front of the largest great house served as a stage for these public events.

The involvement of the community (or, at least, various portions of the community) as audience within these events was a key requirement for the creation, reinforcement, and manipulation of power relations between elites and non-elites, as well as among different elites [110,111]. The mounds, as illustrated by our modeling, would have served as ideal locations for political theater. The audience would have included all within Downtown Chaco, not only the other leaders that occupied the various great houses in the vicinity, but also the commoners that lived within small sites throughout the area.

Yet Downtown Chaco was not merely the location of political theater. Whiteley stated that "ritual action, because of its intent to affect instrumentally the conditions of existence, is simultaneously political action" [112] (p. 68). We believe the converse may also be true at Chaco, especially considering how political power at the location has been so intimately tied up with ritual [113]. Therefore, the performance space of Downtown Chaco was also sacred space. Although the soundshed should be considered circumstantial evidence, the above illustrated locations of shrines and stone circles in relationship to the mounds' soundshed provide additional evidence for this claim.

However, the concepts of political theater and elite legitimation only begin to explore the meaning behind these platforms, and for this we return to performance theory. As mentioned earlier, these mounds contained more material than could be contributed by the occupants of Pueblo Bonito alone [96]. This material included trade items from throughout the Chacoan sphere, as well as much farther afield: Narbona pass chert, Chuska Gray Ware pottery, macaw remains, and cacao residue within rare ceramic vessel forms have been recovered from the mounds [88]. Yet as the papers within Crown's 2016 volume, "The Pueblo Bonito Mounds of Chaco Canyon", illustrate, the majority of the artifacts recovered from the mounds reflect normal, non-elite residential patterns. It is likely that debris from throughout the canyon's small house residences, as well as from visitors from throughout the American Southwest, was purposefully accumulated over the course of sixty years during occasional communal feasting and its accompanying performances of conspicuous consumption and ritual deposition.

As these platform mounds were constructed, the people of Chaco Canyon were in essence forming the stage for the political theater enacted by the elites of Pueblo Bonito. When seen in this light, instances of communal feasting take on a much more nuanced interpretation. Not only were these occasions organized by the elites (and thus served to formalize their role in comparison to others), but they physically shaped the political and sacred performance space utilized by those elites. As the mounds were created, not only did they become prominent features of the Chacoan landscape, but the extent of their soundscapes grew to encompass Downtown Chaco. If this soundscape is interpreted as an integral part of the legitimation of the Chacoan religious and political system, it was the performances of the Puebloan people themselves that reinforced the elite power structures.

5. Conclusions

As illustrated, specific features within a landscape can provide clues as to how people related to that place. While the platform mounds at Pueblo Bonito imply aspects of political theater and public performance, placing those features into a context of landscape archaeoacoustics highlights just how important a role they played within Chacoan society and culture. The mounds were constructed in an ideal location to serve as the stage for political theater that would have been observed by all within Downtown Chaco and perhaps farther afield. Their construction was an act of public performance, indicating that both elite and non-elite individuals participated in the creation of that stage and its resulting soundshed.

Sound is an integral part of the lived experience, and one that is becoming increasingly acknowledged by archaeologists. Chaco Canyon provides an example of just how important it may have been during the development of complex societies.

Author Contributions: Individual author contributions are as follows: conceptualization, D.E.W.; methodology, D.E.W., K.E.P.; software, K.E.P.; writing—original draft preparation, D.E.W., K.E.P.; writing—review and editing, D.E.W., K.E.P.; visualization, K.E.P.

Funding: This research received no external funding.

Acknowledgments: The authors would like to acknowledge Ruth Van Dyke and Kyle Bocinsky for providing GPS data, Rich Friedman for permission to use Figure 3, and Tommaso Mattioli for an invitation to present at a 2018 EAA session on Archaeoacoustics, which provided the impetus for this paper. David would like to thank his parents for a relatively undisturbed place in which he wrote the initial draft of this article during a vacation to visit family. Likewise, Kristy thanks Thomas Dyson for the same. David would also like to thank Ed Kandl for serving as a scale bar in Figure 2. Finally, the authors thank the three anonymous reviewers for their feedback, which has resulted in a stronger paper.

Conflicts of Interest: The authors declare no conflict of interest. As there is no outside funding, no additional person or organization had a role in the design of the study; in the collection, analyses, or interpretation of data; in the writing of the manuscript, or in the decision to publish the results.

References

1. Primeau, K.E.; Witt, D.E. Soundscapes in the Past: Towards a Phenomenology of Sound at the Landscape Level. Presented at the 81st Annual Meeting of the Society for American Archaeology, Orlando, FL, USA, 6–10 April 2016.
2. Primeau, K.E.; Witt, D.E. Soundscapes in the Past: A GIS Approach to Landscape Scale Archaeoacoustics. Presented at the Frontiers in Archaeological Sciences Symposium, Rutgers University, New Brunswick, NJ, USA, 23–25 October 2017.
3. Witt, D.E.; Primeau, K.E. Soundscapes in the Past: Interaudibility in the Chacoan Built Landscape. Presented at the 82nd Annual Meeting of the Society for American Archaeology, Vancouver, BC, Canada, 29 March–2 April 2017.
4. Primeau, K.E.; Witt, D.E. Soundscapes in the Past: Investigating Sound at the Landscape Level. *J. Archaeol. Sci. Rep.* **2018**, *19*, 875–885. [CrossRef]
5. Inomata, T.; Coben, L.S. Overture: An Invitation to the Archaeological Theater. In *Archaeology of Performance: Theaters of Power, Community, and Politics*; Inomata, T., Coben, L.S., Eds.; Altamira Press: Lanham, MD, USA, 2006; pp. 11–44.
6. Scott, J.C. *Domination and the Arts of Resistance: Hidden Transcripts*; Yale University Press: New Haven, CT, USA, 1990.
7. Cross, I.; Zubrow, E.B.W.; Cowan, F. Musical behaviours and the archaeological record: A preliminary study. In *Experimental Archaeology*; Mathieu, J., Ed.; British Archaeological Reports: Oxford, UK, 2002; pp. 25–34.
8. Cross, I.; Watson, A. Acoustics and the human experience of socially-organized sound. In *Archaeoacoustics*; Scarre, C., Lawson, G., Eds.; Oxbow Books: Oxford, UK, 2006; pp. 107–115.

9. D'Errico, F.; Lawson, G. The sound paradox: How to assess the acoustic significance of archaeological evidence? In *Archaeoacoustics*; Scarre, C., Lawson, G., Eds.; Oxbow Books: Oxford, UK, 2006; pp. 41–57.
10. Devereux, P. *Stone Age Soundtracks: The Acoustic Archaeology of Ancient Sites*; Vega: London, UK, 2002.
11. Eneix, L.C. (Ed.) *Archaeoacoustics: The Archaeology of Sound*; The OTS Foundation: Myakka City, FL, USA, 2014.
12. Jimenez, R.; Till, R.; Howell, M. (Eds.) *Music & Ritual: Bridging Material & Living Cultures*; Ekho Verlag: Berlin, Germany, 2013.
13. Scarre, C. Sound, place and space: Towards an archaeology of acoustics. In *Archaeoacoustics*; Scarre, C., Lawson, G., Eds.; Oxbow Books: Oxford, UK, 2006; pp. 1–10.
14. Watson, A.; Keating, D. Architecture and sound: An acoustic analysis of megalithic monuments in prehistoric Britain. *Antiquity* **1999**, *73*, 325–336. [CrossRef]
15. Kolar, M.A. Sensing sonically at Andean Formative Chavín de Huántar, Perú. *Time Mind* **2017**, *10*, 39–59. [CrossRef]
16. Azevedo, M.; Markham, B.; Wall, J.N. Acoustical archaeology—Recreating the soundscape of John Donne's 1622 gunpowder plot sermon at Paul's Cross. *Proc. Meet. Acoust.* **2013**, *19*, 015133. [CrossRef]
17. Markham, B.; Azevedo, M.; Wall, J.N. Recreating the soundscape of John Donne's 1622 gunpowder plot sermon at Paul's Cross. *J. Acoust. Soc. Am.* **2013**, *133*, 3581. [CrossRef]
18. Wall, J.N. Transforming the Object of our Study: The Early Modern Sermon and the Virtual Paul's Cross Project. Available online: http://journalofdigitalhumanities.org/3-1/transforming-the-object-of-our-study-by-john-n-wall/ (accessed on 24 December 2018).
19. Reznikoff, I. Sound resonance in prehistoric times: A study of Paleolithic painted caves and rocks. *J. Acoust. Soc. Am.* **2008**, *123*, 3603. [CrossRef]
20. Jahn, R.G.; Devereux, P.; Ibison, M. Acoustical resonances of assorted ancient structures. *J. Acoust. Soc. Am.* **1996**, *99*, 649–658. [CrossRef]
21. Iannace, G.; Trematerra, A.; Qandil, A. The Acoustics of the Catacombs. *Arch. Acoust.* **2014**, *39*, 583–590. [CrossRef]
22. Iannace, G.; Berardi, U. The Acoustic of Cumaean Sibyl. *Proc. Meet. Acoust.* **2017**, *30*, 015010. [CrossRef]
23. Iannace, G.; Marletta, L.; Sicurella, F.; Ianniello, E. Acoustic measurements in the Ear of Dionysius at Syracuse (Italy). In Proceedings of the 39th International Congress on Noise Control Engineering 2010, Lisbon, Portugal, 13–16 June 2010.
24. Kolar, M.A. Archaeological Psychoacoustics at Chavín de Huántar, Perú. Ph.D. Dissertation, Stanford University, Stanford, CA, USA, 2013.
25. Van Dyke, R.M.; de Smet, T. Chacoan Soundscapes. *Archaeol. Southwest Mag.* **2018**, *32*, 38–39.
26. Díaz-Andreu, M.; García Benito, C. Acoustics and Levantine rock art: Auditory perceptions in La Valltorta Gorge (Spain). *J. Archaeol. Sci.* **2012**, *39*, 3591–3599. [CrossRef]
27. Mattioli, T.; Farina, A.; Armelloni, E.; Hameau, P.; Díaz-Andreu, M. Echoing landscapes: Echolocation and the placement of rock art in the Central Mediterranean. *J. Archaeol. Sci.* **2017**, *83*, 12–25. [CrossRef]
28. Díaz-Andreu, M.; Atiénzar, G.G.; Benito, C.G.; Mattioli, T. Do You Hear What I See? Analyzing Visibility and Audibility in the Rock Art Landscape of the Alicante Mountains of Spain. *J. Anthropol. Res.* **2017**, *73*, 181–213. [CrossRef]
29. Mileson, S. Sound and Landscape. In *The Oxford Handbook of Later Medieval Archaeology in Britain*; Christopher, G., Alejandra, G., Eds.; Oxford University Press: Oxford, UK, 2018; pp. 713–727.
30. Mlekuz, D. Listening to landscapes: Modelling past soundscapes in GIS. *Internet Archaeol.* **2004**, *16*. [CrossRef]
31. Liwosz, C.R. Benchmarks: Ontological Considerations at Two Mojave Desert Petroglyph Labyrinths. Ph.D. Dissertation, University of California Santa Cruz, Santa Cruz, CA, USA, 2018.
32. Mattioli, T.; Díaz-Andreu, M. Hearing rock art landscapes: A survey of the acoustical perception in the Sierra de San Serván area in Extremadura (Spain). *Time Mind* **2017**, *10*, 81–96. [CrossRef]
33. Brück, J. Experiencing the past? The development of a phenomenological archaeology in British prehistory. *Archaeol. Dialogues* **2005**, *12*, 45–72. [CrossRef]
34. Cummings, V.; Whittle, A. *Places of Special Virtue: Megaliths in the Neolithic Landscape of Wales*; Oxbow Books: Oxford, UK, 2004.
35. Johnson, M.H. Phenomenological Approaches in Landscape Archaeology. *Annu. Rev. Anthropol.* **2012**, *41*, 269–284. [CrossRef]

36. Tilley, C. *A Phenomenology of Landscape*; Routledge: London, UK, 1994.
37. Tilley, C. *The Materiality of Stone: Explorations in Landscape Phenomenology*; Berg: Oxford, UK, 2004.
38. Tilley, C. *Body and Image: Explorations in Landscape Phenomenology 2*; Left Coast Press: Walnut Creek, CA, USA, 2008.
39. Tilley, C. *Interpreting Landscapes: Geologies, Topographies, Identities; Explorations in Landscape Phenomenology 3*; Left Coast Press: Walnut Creek, CA, USA, 2010.
40. Van Dyke, R.M. Phenomenology in archaeology. In *Encyclopedia of Global Archaeology*; Smith, C., Ed.; Springer: New York, NY, USA, 2014; pp. 5909–5917.
41. Hamilakis, Y. *Archaeology and the Senses: Human Experience, Memory, and Affect*; Cambridge University Press: New York, NY, USA, 2013.
42. Merleau-Ponty, M. *Phenomenology of Perception*; Routledge: London, UK, 1962.
43. Hamilton, S.; Whitehouse, R.; Brown, K.; Combes, P.; Herring, E.; Thomas, M.S. Phenomenology in practice: Towards a methodology for a 'subjective' approach. *Eur. J. Archaeol.* **2006**, *9*, 31–71. [CrossRef]
44. Eve, S. Augmenting Phenomenology: Using Augmented Reality to Aid Archaeological Phenomenology in the Landscape. *J. Archaeol. Method Theory* **2012**, *19*, 582–600. [CrossRef]
45. Gillings, M. Landscape Phenomenology, GIS and the Role of Affordance. *J. Archaeol. Method Theory* **2012**, *19*, 601–611. [CrossRef]
46. Llobera, M. Life on a Pixel: Challenges in the Development of Digital Methods Within an "Interpretive" Landscape Archaeology Framework. *J. Archaeol. Method Theory* **2012**, *19*, 495–509. [CrossRef]
47. Rennell, R. Experience and GIS: Exploring the Potential for Methodological Dialogue. *J. Archaeol. Method Theory* **2012**, *19*, 510–525. [CrossRef]
48. Trigg, D. Place and Non-place: A Phenomenological Perspective. In *Place, Space, and Hermeneutics*; Janz, B.B., Ed.; Springer: Cham, Switzerland, 2017; pp. 127–139.
49. Trigger, B.G. *A History of Archaeological Thought*, 2nd ed.; Cambridge University Press: Cambridge, UK, 2006.
50. Hacıgüzeller, P. GIS, critique, representation and beyond. *J. Soc. Archaeol.* **2012**, *12*, 245–263. [CrossRef]
51. Sui, D.Z. GIS and Urban Studies: Positivism, Post-Positivism, and Beyond. *Urban Geogr.* **1994**, *15*, 258–278. [CrossRef]
52. Ingold, T. *The Perception of the Environment*; Routledge: London, UK, 2000.
53. Bunzel, R.L. *Zuni Ceremonialism*; University of New Mexico Press: Albuquerque, NM, USA, 1992.
54. Brown, D.N. The Distribution of Sound Instruments in the Prehistoric Southwestern United States. *Ethnomusicology* **1967**, *11*, 71–90. [CrossRef]
55. Brown, E. Instruments of Power: Musical Performance in Rituals of the Ancestral Puebloans of the American Southwest. Ph.D. Dissertation, Columbia University, University Microfilms, Ann Arbor, MI, USA, 2005.
56. Brown, E. Musical instruments in the pre-hispanic southwest. *Park Sci.* **2009**, *26*, 46–49.
57. Brown, E. Music of the center place: The instruments of Chaco Canyon. In *Flower World: Music Archaeology of the Americas*; Stöckli, M., Howell, M., Eds.; Ekho Verlag: Berlin, Germany, 2014; Volume 3, pp. 45–66.
58. Brown, D.N. Ethnomusicology and the prehistoric southwest. *Ethnomusicology* **1971**, *15*, 363–378. [CrossRef]
59. Mills, B.J.; Ferguson, T.J. Animate objects: Shell trumpets and ritual networks in the greater southwest. *J. Archaeol. Method Theory* **2008**, *15*, 338–361. [CrossRef]
60. Taube, K. Gateways to Another World: The Symbolism of Supernatural Passageways in the Art and Ritual of Mesoamerican and the American Southwest. In *Painting the Cosmos: Metaphor and Worldview in Images from the Southwest Pueblos and Mexico*; Hays-Gilpin, K., Schaafsma, P., Eds.; Museum of Northern Arizona: Flagstaff, AZ, USA, 2010; pp. 73–120.
61. Hays-Gilpin, K.; Sekaquaptewa, E.; Newsome, E.A. Siitalpuva, "through the land brightened with flowers": Ecology and cosmology in mural and pottery painting, Hopi and beyond. In *Painting the Cosmos: Metaphor and Worldview in Images From the Southwest Pueblos and Mexico*; Hays-Gilpin, K., Schaafsma, P., Eds.; Museum of Northern Arizona: Flagstaff, AZ, USA, 2010; pp. 121–138.
62. Weiner, R.S. A Sensory Approach to Exotica, Ritual Practice, and Cosmology at Chaco Canyon. *Kiva* **2015**, *81*, 220–246. [CrossRef]
63. Van Dyke, R.M. The Chacoan past: Creative representations and sensory engagements. In *Subjects and Narratives in Archaeology*; Dyke, R.M.V., Bernbeck, R., Eds.; University Press of Colorado: Boulder, CO, USA, 2015; pp. 83–99.

64. Akins, N.J. The burials of Pueblo Bonito. In *Pueblo Bonito: Center of the Chacoan World*; Neitzel, J.E., Ed.; Smithsonian Books: Washington, DC, USA, 2003; pp. 94–106.
65. Loose, R.W. Tse'Biinaholts'a Yalti (Curved Rock That Speaks). *Time Mind* **2008**, *1*, 31–49. [CrossRef]
66. Loose, R.W. That old music: Reproduction of a shell trumpet from Pueblo Bonito. *Pap. Archaeol. Soc. N. M.* **2012**, *38*, 127–133.
67. Loose, R.W. Archaeoacoustics: Adding a Sound Track to Site Descriptions. *Pap. Archaeol. Soc. N. M.* **2010**, *36*, 127–136.
68. Loose, R.W. *A Report on Tse Biinaholtsa'a Yalti (Curved Rock that Speaks): An Open-Air Public Performance Theater at Chaco Canyon, New Mexico*; Manuscript on File; Chaco Culture National Historical Park: New Mexico, NM, USA, 2001.
69. Loose, R.W. Computer Analysis of Sound Recordings from Two Anasazi Sites in Northwestern New Mexico. *J. Acoust. Soc. Am.* **2002**, *112*, 2285. [CrossRef]
70. Stein, J.R.; Friedman, R.; Blackhorse, T.; Loose, R. Revisiting Downtown Chaco. In *The Architecture of Chaco Canyon, New Mexico*; Lekson, S.H., Ed.; The University of Utah Press: Salt Lake City, UT, USA, 2007; pp. 199–223.
71. Reed, S.E.; Mann, J.P.; Boggs, J.L. *SPreAD-GIS: An ArcGIS Toolbox for Modeling the Propagation of Engine Noise in a Wildland Setting*, version 1.2; The Wilderness Society: San Francisco, CA, USA, 2009.
72. Reed, S.E.; Boggs, J.L.; Mann, J.P. *SPreAD-GIS: An ArcGIS Toolbox for Modeling the Propagation of Engine Noise in a Wildland Setting*, version 2.0; The Wilderness Society: San Francisco, CA, USA, 2010.
73. Goodwin, G.; Richards-Rissetto, H.; Primeau, K.E.; Witt, D.E. Bringing Sound into the Picture: Experiencing Ancient Maya Landscapes with GIS and 3D Modeling. Presented at the Computer Applications and Quantitative Methods in Archaeology (CAA) International Conference, Tübingen, Germany, 19–23 March 2018.
74. Goodwin, G.; Richards-Rissetto, H.; Primeau, K.E.; Witt, D.E. Soundscapes and Visionscapes: Investigating Ancient Maya Cities with GIS and 3D Modeling. Presented at the 83rd Annual Meeting of the Society for American Archaeology, Washington, DC, USA, 11–15 April 2018.
75. Western Regional Climate Center. Chaco Canyon National Monument [sic], New Mexico, Monthly Climate Summary. Available online: http://www.wrcc.dri.edu/cgi-bin/cliMAIN.pl?nm1647 (accessed on 7 December 2018).
76. Ambrose, S. *Sound Levels in the Primary Vegetation Types in Grand Canyon National Park, July 2005*; NPS Report No. GRCA-05-02; Sandhill Company: Washington, DC, USA, 2006.
77. Hayne, M.J.; Rumble, R.H.; Mee, D.J. Prediction of crowd noise. In Proceedings of the Acoustics 2006, Christchurch, New Zealand, 20–22 November 2006; pp. 235–240.
78. Van Heusden, E.; Plomp, R.; Pols, L.C.W. Effect of ambient noise on the vocal output and the preferred listening level of conversational speech. *Appl. Acoust.* **1979**, *12*, 31–43. [CrossRef]
79. Organización Internacional de Normalización. *ISO 9613-2:1996, Acoustics: Attenuation of Sound during Propagation Outdoors. General Method of Calculation*; International Organization for Standardization: Geneva, Switzerland, 1996.
80. American National Standards Institute (ANSI). *ANSI S1.26-1995 Method for Calculation of the Absorption of Sound by the Atmosphere*; Acoustical Society of America: New York City, NY, USA, 1995.
81. Maekawa, Z. Noise reduction by screens. *Appl. Acoust.* **1968**, *1*, 157–173. [CrossRef]
82. Lamancusa, J.S. Outdoor sound propagation. In *Noise Control, ME 458 Engineering Noise Control*; Pennsylvania State University: University Park, PA, USA, 2009.
83. Primeau, K.E. Methodological Improvements in Landscape Archaeoacoustics: Exploring the Effects of Vegetation and Ground Cover. Presented at the 84th Annual Meeting of the Society for American Archaeology, Albuquerque, NM, USA, 10–14 April 2019.
84. Judd, N.M. *The Architecture of Pueblo Bonito*; Smithsonian Institution Press: Washington, DC, USA, 1964; Volume 147.
85. Windes, T.C. *Investigations at the Pueblo Alto Complex, Chaco Canyon, New Mexico, 1975–1979*; National Park Service: Sante Fe, NM, USA, 1987; Volume II, Pt. 2: Architecture and Stratigraphy.
86. Lekson, S.H. Great House Form. In *The Architecture of Chaco Canyon, New Mexico*; Lekson, S.H., Ed.; The University of Utah Press: Salt Lake City, UT, USA, 2007; pp. 7–44.

87. Stein, J.R.; Lekson, S.H. Anasazi Ritual Landscapes. In *Anasazi Regional Organization and the Chaco System*; Doyel, D.E., Ed.; University of New Mexico: Albuquerque, NM, USA, 1992; pp. 87–100.
88. Crown, P.L. (Ed.) *The Pueblo Bonito Mounds of Chaco Canyon*; University of New Mexico Press: Albuquerque, NM, USA, 2016.
89. Toll, H.W. Making and Breaking Pots in the Chaco World. *Am. Antiq.* **2001**, *66*, 56–78. [CrossRef]
90. Cameron, C.M. Pink Chert, Projectile Points, and the Chacoan Regional System. *Am. Antiq.* **2001**, *66*, 79–101. [CrossRef]
91. Ashmore, W. Building Social History at Pueblo Bonito: Footnotes to a Biography of Place. In *The Architecture of Chaco Canyon, New Mexico*; Lekson, S.H., Ed.; The University of Utah Press: Salt Lake City, UT, USA, 2007; pp. 179–198.
92. Marshall, M.P. The Chacoan Roads: A Cosmological Interpretation. In *Anasazi Architecture and American Design*; Morrow, B.H., Price, V.B., Eds.; University of New Mexico: Albuquerque, NM, USA, 1997; pp. 62–74.
93. Sofaer, A. The Primary Architecture of the Chacoan Culture: A Cosmological Expression. In *Anasazi Architecture and American Design*; Morrow, B.H., Price, V.B., Eds.; University of New Mexico Press: Albuquerque, NM, USA, 1997; pp. 88–132.
94. Van Dyke, R.M. Sacred Landscapes: The Chaco-Totah Connection. In *Chaco's Northern Prodigies: Salmon, Aztec, and the Ascendancy of the Middle San Juan Region after AD 1100*; Reed, P.F., Ed.; The University of Utah Press: Salt Lake City, UT, USA, 2008; pp. 334–348.
95. Wills, W.H. Ritual and Mound Formation during the Bonito Phase in Chaco Canyon. *Am. Antiq.* **2001**, *66*, 433–451. [CrossRef]
96. Windes, T.C. Gearing Up and Piling On: Early Great Houses in the Interior San Juan Basin. In *The Architecture of Chaco Canyon, New Mexico*; Lekson, S.H., Ed.; The University of Utah Press: Salt Lake City, UT, USA, 2007; pp. 45–92.
97. Van Dyke, R.M. *The Chaco Experience: Landscape and Ideology at the Center Place*; School for Advanced Research Press: Santa Fe, NM, USA, 2008.
98. Cameron, C.M. Sacred Earthen Architecture in the Northern Southwest: The Bluff Great House Berm. *Am. Antiq.* **2002**, *67*, 677–695. [CrossRef]
99. Lekson, S.H. *A History of the Ancient Southwest*; School for Advanced Research Press: Santa Fe, NM, USA, 2008.
100. Chamberlin, M.A. Plazas, Performance, and Symbolic Power in Ancestral Pueblo Religion. In *Religious Transformation in the Late Pre-Hispanic Pueblo World*; Glowacki, D.M., Keuren, S.V., Eds.; University of Arizona Press: Tucson, ZA, USA, 2012; pp. 130–152.
101. Alvarsson, J.J.; Nordström, H.; Lundén, P.; Nilsson, M.E. Aircraft noise and speech intelligibility in an outdoor living space. *J. Acoust. Soc. Am.* **2014**, *135*, 3455–3462. [CrossRef]
102. Boren, B. Whitefield's Voice. In *George Whitefield*; Oxford University Press: Oxford, UK, 2016; pp. 167–189.
103. Boren, B. The Maximum Intelligible Range of the Human Voice. Ph.D. Dissertation, New York University, New York, NY, USA, 2014.
104. Boren, B.; Roginska, A.; Gill, B. Maximum Averaged and Peak Levels of Vocal Sound Pressure. In Proceedings of the 135th Audio Engineering Society Convention, New York, NY, USA, 17–20 October 2013.
105. Larm, P.; Hongisto, V. Experimental comparison between speech transmission index, rapid speech transmission index, and speech intelligibility index. *J. Acoust. Soc. Am.* **2006**, *119*, 1106–1117. [CrossRef] [PubMed]
106. Lazarus, H. Prediction of Verbal Communication is Noise—A review: Part 1. *Appl. Acoust.* **1986**, *19*, 439–464. [CrossRef]
107. Jovičić, S.T. A relation between speech intelligibility and distribution of speech pressure about the head. *Appl. Acoust.* **1991**, *34*, 51–59. [CrossRef]
108. Judd, N.M. *The Material Culture of Pueblo Bonito*; Smithsonian Institution Press: Washington, DC, USA, 1954; Volume 124.
109. Luckman, T. Comments on Legitimation. *Curr. Sociol.* **1987**, *35*, 109–117. [CrossRef]
110. Barker, R. *Legitimating Identities: The Self-Presentations of Rulers and Subjects*; Cambridge University Press: Cambridge, UK, 2004.
111. Brown, J.; Elliot, J.H. *A Palace for a King: The Buen Retiro and the Court of Philip IV*; Yale University Press: New Haven, CT, USA, 1980.

112. Whiteley, P.M. *Deliberate Acts: Changing Hopi Culture through the Oraibi Split*; University of Arizona Press: Tucson, AZ, USA, 1988.
113. Yoffee, N. The Chaco "Rituality" Revisited. In *Chaco Society and Polity: Papers from the 1999 Conference*; Cordell, L.S., Judge, W.J., Piper, J.-E., Eds.; New Mexico Archaeological Council: Albuquerque, NM, USA, 2001; pp. 63–78.

© 2018 by the authors. Licensee MDPI, Basel, Switzerland. This article is an open access article distributed under the terms and conditions of the Creative Commons Attribution (CC BY) license (http://creativecommons.org/licenses/by/4.0/).

Article

Towards Italian Opera Houses: A Review of Acoustic Design in Pre-Sabine Scholars

Dario D'Orazio [1,*,†] **and Sofia Nannini** [2]

1. Department of Industrial Engineering, DIN University of Bologna, 40126 Bologna, Italy
2. Department of Architecture and Design, DAD Politecnico di Torino, 10125 Torino, Italy; sofia.nannini@polito.it
* Correspondence: dario.dorazio@unibo.it; Tel.: +39-051-2090549
† Current address: Viale Risorgimento 2, 40126 Bologna, Italy.

Received: 31 December 2018; Accepted: 22 February 2019; Published: 1 March 2019

Abstract: The foundation of architectural acoustics as an independent science is generally referred to Sabine's early studies and their application. Nevertheless, since the 16th Century, a great number of authors wrote essays and treatises on the design of acoustic spaces, with a growing attention to the newborn typology of the Opera house, whose evolution is strongly connected to the cultural background of the Italian peninsula. With roots in the Renaissance rediscovery of Vitruvius's treatise and his acoustic theory, 16th- to 19th-Century Italian authors tackled several issues concerning the construction of theatres—among them, architectural and structural features, the choice of the materials, the social meanings of performances. Thanks to this literature, the consolidation of this body of knowledge led to a standardisation of the forms of the Italian Opera house throughout the 19th Century. Therefore, the scope of this review paper is to focus on the treatises, essays and publications regarding theatre design, written by pre-Sabinian Italian scholars. The analysis of such literature aims at highlighting the consistencies in some 19th-Century minor Italian Opera houses, in order to understand to what extent this scientific and experimental background was part of the building tradition during the golden age of the Italian Opera.

Keywords: opera house; cultural heritage; shape optimisation; room acoustics

1. Introduction

Almost all Italian Opera houses with outstanding room acoustical reputations have been built before what is considered the beginning of architectural acoustics as an independent science, which dates back to 1898 [1,2]. The acoustics of historic Opera houses can be considered the result of a Darwinian type evolution, with architects designing halls inspired by the success of former "good" halls, frequently destroyed by fires. According to previous works [3–5], it is possible to identify three main room acoustic design approaches which were the starting point of this evolutionary process: the theory of "circulation of sound", the proto-geometrical acoustics and the "echo theory".

The theory of "circulation of sound" was based on the Renaissance rediscovery of Vitruvius's writings [6]. The Latin author argued that the voices of actors should be unobstructed in order to create favorable room-acoustics [5]. Such method thus influenced the building of auditoria with round shapes, rounded proscenium arches, without obstructions which "slowed the circulation" or "broke the voice" [3–5].

The proto-geometrical acoustics—from the 17th Century—was based on the assumption that the trajectory of sound was analogous to the sound rays reflected from a surface. Therefore, sound could be conveyed by modifying the plan shape. The similarity between light and sound was first assumed by Giuseppe Biancani (1566–1624), who defined the so-called "echometria" [7]. Assuming that rays spread evenly over a plan after being reflected from parabolic or elliptically shaped walls [5], scholars discussed the best plan for the cavea: elliptical, bell- and horseshoe-shaped—the latter becoming the standard shape for 19th-Century theatres. Moreover the raising of proscenium arches, directly bound both to scenography and acoustics, increased the early reflections—or, better, the first reflection; lastly, the shape of the ceilings was never designed as concave in order not to concentrate reflections.

The "echo-theory" is a recent guideline based on geometrical acoustics which dates back to the end of the 18th Century [5,8]. This theory focused on a "quantification of the perception threshold between direct and reflected sounds: when a first order reflection exceeded this threshold, an echo would be perceived, which was considered detrimental for the acoustics of the hall" [5]. Instead of uniform radiation pattern, a directional one was assumed and the "echo-theory" guideline was grounded on simple measurements, which employed a speaking person and a human observer judging audibility or intelligibility as a function of distance and direction. Because of such voice directivity and propagation guideline, the size of audience areas in Opera houses was limited.

Analyses through virtual reconstructions [9–13] provided further information on the acoustic quality of historical theatres, also taking into account the occupied conditions. The results generally highlighted situations that are more reverberating than the current standards. If we compare these conditions to the limited available literature on the current state of such theatres, following the ordinary measuring procedures, it is possible to define local building characteristics that are deeply connected to the politic fragmentation of Italy [14–20]. A collation of such data was suggested by Prodi et al. [20]: this work sought the correlation between acoustic and geometric descriptors of the environment (volume, seats, volume/audience ratios, etc.). Moreover, some authors [21,22] suggested the adjustment of some descriptors used in international theatres to the small-medium typology of Italian Opera houses. Lastly, Garai et al. [23] examined the coupling effects between fly tower and cavea.

Thanks to the reading of several original texts, spanning between the 16th and the 19th Century, this paper aims at adding another contribution to the scientific discussion on the pre-Sabine theories. Indeed, these texts show a particular attention on materials and other features that can be interpreted by means of architectural acoustics.

2. The Late-Renaissance Theatre

Until the end of the 17th-Century, the acoustic spaces were treated using the Vitruvian categories, taken from Vitruvius's well known treatise De Architectura (On Architecture). The treatise was written around the 1st Century B.C. and rediscovered in the 15th Century, becoming one of the bases of Renaissance humanist culture. Vitruvius's theory on theatre design was based on the propagation of spherical sound waves—the so called "circulation" [4,5]:

- Dis-sonantes, when the circulation is dissipated by destructive interference provided by the architectural elements (hard and sharp corners, etc . . .);
- Con-sonantes, when the environment increases the circulation;
- Circum-sonantes, when curved surfaces create a reverberation effect;
- Re-sonantes, when echo phenomena occur.

One of the first music rooms of the modern age was the Odeo Cornaro in Padua, designed by Giovanni Maria Falconetto (1468–1534) around 1524–1530 for the nobleman and patron Alvise Cornaro (1484–1566). This project was included by the architect and theorist Sebastiano Serlio (1475–1554) in his Sette libri dell'architettura (Seven Books of Architecture) [24]. Serlio's drawing of the Odeo Cornaro shows a central symmetric plan with four niches on the sides and it is covered by a dome (see Figure 1). This small room fully agreed with the features of Vitruvian circum-sonantes spaces, as the round niches increased the diffusivity (As Serlio writes, "Et li quattro niccij per la sua ritonditá concava ricevono le voci, & le ritengono" (And the four niches, due to their round concavity, receive the voices and retain them) [24].). Indeed, this space was oriented to music listening rather than mere speech. Serlio's great interest in theatre design can be particularly understood by reading the last pages of his second book on perspective, where a specific paragraph is dedicated to theatres and scenographies. By blending his readings of Vitruvius's acoustic theory and the laws of geometrical perspective, Serlio included a section and a plan of what might have been the model of his timber theatre designed for the courtyard of the palazzo Porto in Vicenza in 1539. On one hand, the first terraces of the cavea are semicircular and then acquire an almost elliptical shape [25]; on the other hand, the stage shows a sloped surface for a better perspective view from the audience (see Figure 2).

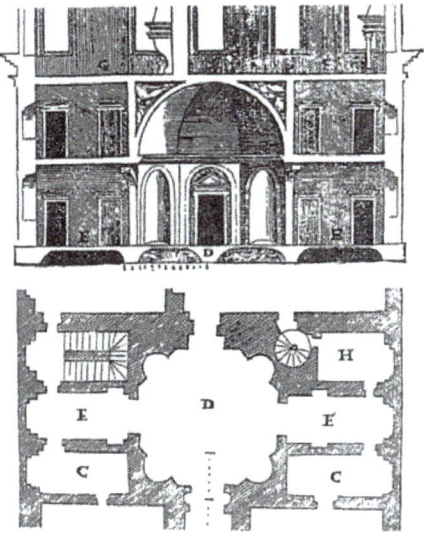

Figure 1. Plan of Odeo Cornaro in Padua (Hall D in the figure) (after [4]).

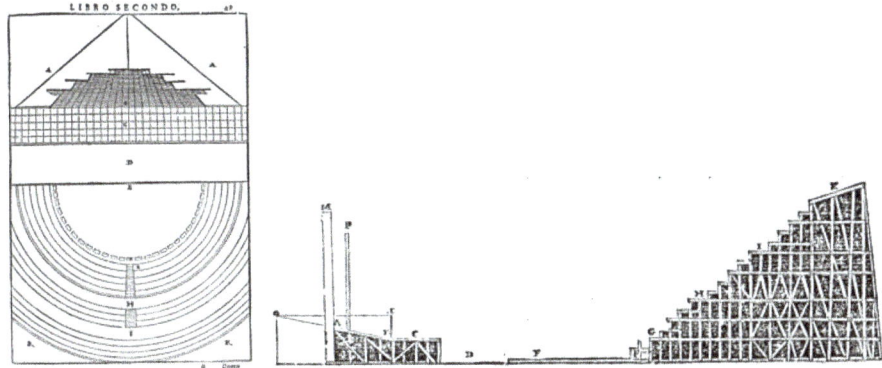

Figure 2. Plan and section of Sebastiano Serlio's timber theatre, in the book dedicated to Perspective (after [24]).

However, it is important to keep in mind that Vitruvius's acoustic and musical theory was based on considerations that mainly regarded the propagation of the human voice, as this was the key element of ancient Graeco-Roman theatre. Therefore, the acoustics of Greek and Roman theatres was generally revolving around the speech intelligibility of the actors [26–31]. Clearly, the evidence of these theatres' exceptional acoustics is due to their amphitheatrical shapes, and thus to the knowledge of the propagation of circular sound waves. However, it is not possible to affirm that the theory of the Latin author—despite his extensively quoting of Greek sources such as Aristotle's pupil Aristoxenus—was also applied to former Greek and Hellenistic theatres. Nevertheless, Vitruvius's rediscovered writings became the core of the Renaissance architectural theory, thus necessarily influencing also its acoustic outcomes [32,33].

A few decades after, with the aim of increasing the intelligibility of the speech and still following the models of Graeco-Roman theatres, late–Renaissance architects started experimenting with con-sonantes spaces. Such is the case of the Teatro Olimpico in Vicenza, by Andrea Palladio (1508–1580), inaugurated in 1585 with Sophocle's Oedipux Rex, with choruses by Andrea Gabrieli (1533–1585) [4,13]. The theatres of this period were characterised by steps and showed a reverberation higher to the present ones. In fact, the Odeo Cornaro shows a measured reverberation time at mid frequencies of 2.4 s in an unoccupied condition and a simulated reverberation time in an occupied condition of 1.3 s [34]; the Teatro Olimpico shows a measured reverberation time at mid frequencies of 3.3 s in unoccupied conditions and a simulated reverberation time of 2.2 s in occupied conditions [13]. Such acoustic conditions allowed a good intelligibility of the early writing styles of melodrama, as the Gabrieli's choruses in the Teatro Olimpico, as confirmed by historical sources [34,35] (see Figure 3).

At the same time, the development of acoustics was strongly connected to political and religious issues prompted by the event of the Protestant Reformation and the Council of Trent (1545–1563), which opened to the Counter-Reformation. On one hand, the Lutheran culture was oriented to the act of reading, due to the availability of printed Bibles thanks to Gutenberg's innovation. On the other hand, the reformed Catholic culture gave more importance to the act of listening, due to the changes in ritual offices. For example, coffered wooden ceilings in churches were preferred by Jesuits due to the decrease in reverberation time and increase in speech intelligibility [4]. An increasing interest in the propagation of sound was collected by the Jesuit Giuseppe Biancani (1566–1624), who published in 1620 the essay Sphaera Mundi [7], whose third paragraph is titled Echometria, idest Geometrica tractatio de Echo (Echometry, i.e., the Geometric Treatise of Echo). Biancani's Echometria opened a new branch of scientific knowledge blending sound and optics, which will be further called Acustica (Acoustics).

In this cultural environment, the building of the theatres corresponded to the birth of a new form of representation: the so-called Melodrama, which integrated various types of performance: mélos meaning singing, drama meaning acting. Indeed, the word Opera in Latin is the plural of opus, which means "act, performance". In the last decades of the 16th Century, the performance acquired a semi-public dimension—open to the Ruler and the court—and its own independent architecture. In the spaces designed for melodrama, one of the most significant aspects is the structural and typological background, deriving from the form of the Roman Basilica [36]. In fact, they were composed of a rectangular body ending with an apse, becoming, respectively, the audience and the stage. The theatre plan was typically divided into areas with different functions: the cavea for the audience, the stage and the rooms behind it reserved for the staff and for the artists. This is the case of Teatro all'antica (Ancient-Style Theatre) in Sabbioneta (1588–1590), by Vincenzo Scamozzi (1548–1616), which became the first building fully designed to be a free-standing theatre (see Figure 4).

Figure 3. Section and view of Andrea Palladio's Teatro Olimpico in Vicenza (1585–1590) [37].

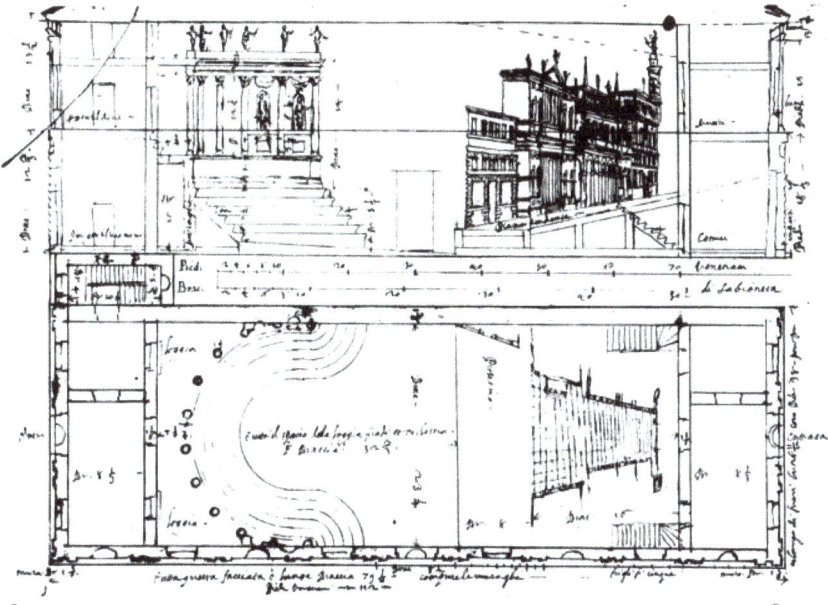

Figure 4. Vincenzo Scamozzi's draft of Teatro all'antica in Sabbioneta (1588–1590) [38].

3. The Baroque Theatre

At the beginning of the 17th Century, a new element entered the theatre design, which would influence both scenography and acoustics: the proscenium arch. This element, placed between the auditorium and the stage, was a direct consequence of the late-Renaissance studies on perspective—especially Serlio's aforementioned paragraph on perspective and scenography [39]. When it comes to its architectural use, the first and most famous architect constantly experimenting with proscenium arches was Inigo Jones (1573–1652). Inspired by his Italian architectural visits, he started introducing proscenium arches in his scenographies already in the first decade of the 17th Century, thus bringing the Italian perspective studies to Great Britain [40,41].

At the same time, the Italian architect Giovan Battista Aleotti (1546–1636) introduced such new spatial disposition in two of his works: the Teatro degli Intrepidi in Ferrara (1605) [9], now demolished, and Farnese Theatre in Parma (1610) [11] (see Figure 5). In both structures, he built a proscenium arch which materialised the difference between what is real and what is a theatrical fiction. The dimensions of the scenic arch increased with the progressive standardisation of its form, together with the increasing complexity of Baroque scenery as in the works of Giacomo Torelli (1604–1678) [42], Nicola Sabbattini (1574–1654) [43] and Joseph Furttenbach (1591–1667) [44]. The role of this element was both visual and acoustical: on one hand, it provided a separation between the space of the representation and the space of the audience; on the other hand, it enhanced the first reflections from the stage and the orchestra, thus increasing the intelligibility in the stalls.

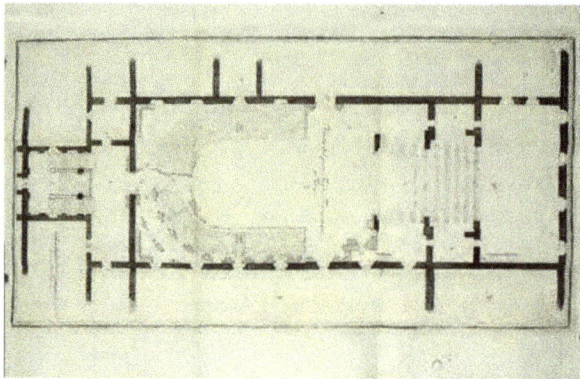

Figure 5. Plan of the Teatro Farnese in Parma (1610) (after [45]).

During the 17th Century, the most significant evolution of the theatre shape took place. It was due to the transformation of the acting performances, the appearance of the masques, and the renovation of the subjects. An important date, which represented a turning point in the theatre history, was 6th March 1637, when all social classes attended the inauguration of the theatre in campo San Cassian in Venice, by paying a ticket [46]. The paying audience and the new theatrical genre, the melodrama, consequently led to a redefinition of the theatre shape. In fact, the payment of a ticket for the performance made it possible to plan the theatrical seasons and the related investments, thus allowing for building permanent structures and to buy theatrical machineries [47]. The space for the audience became a field of architectural experimentations, with the main purpose of incrementing the seating capacity and improving the visual conditions of the attending public. At the same time, the upper classes began to claim for independent and private spaces, leading to the building of wooden partitions on the different tiers, creating the so-called "boxes". This spatial division reflected the social subdivision into classes. Moreover, such spatial configuration introduced a new kind of performance, where the main actors were the spectators themselves. Not only were the boxes rich and sophisticated showcases for the upper class, but they were also a window from which the richest could observe the common people sitting on the uncomfortable benches of the audience [46].

The first essay dealing with the so-called "Italian theatre" was written by the architect Fabrizio Carini Motta (16??–1699) in 1676 [48]. The author summarised all the architectural characteristics of the theatre and it recognised its different typologies: the 16th-Century typology with steps, with boxes or galleries, with boxes joined or not joined with the proscenium. He identified four main parts called "theatre square", "scene square", "boxes area" and "stage area". Concerning the so-called horseshoe typology, Carini Motta proposed two different models changing the ratio between the width and the length of the scene square, as shown in the author's drawing of Figure 6. The architect also dealt with the design of the partition between boxes: if the partitions were built perpendicular to the audience perimeter, the people sitting behind were not able to see the stage. Considering the possibility of choosing a focal point in the middle of the proscenium arch and building diagonal partitions by pointing to that focus, he noticed that there was unused space due to the presence of acute angles following Vitruvius's recommendation, allowing the public sitting behind to enjoy a better view of the stage area. Influenced by Carini Motta's observations, the Swedish architect Nicodemus Tessin The Younger (1654–1728) introduced some of these elements—rounded off back walls and a royal box—in the design for a theatre at the court of Denmark in the last years of the 17th-Century [5,49]. Furthermore, Carini Motta theorised a new kind of partition, shown in the last drawing of Figure 6. This was a mix between the aforementioned solutions, but it was unfortunately very difficult to realise.

If a longer 'scene square' was needed, it was possible to set the wanted length with a line parallel to EF and to draw the biggest circle tangential to that line (see the last draw in Figure 6). According to Carini Motta's theory, all theatre dimensions had to be multiples or fractions of the theatre width. For instance, the height of the audience ceiling should have been higher than 2/3 of the theatre width but smaller than the theatre width itself. On the other hand, he did not give specific indications about the height of the ceiling above the stage area: he only stated that it should have been higher than the audience vault.

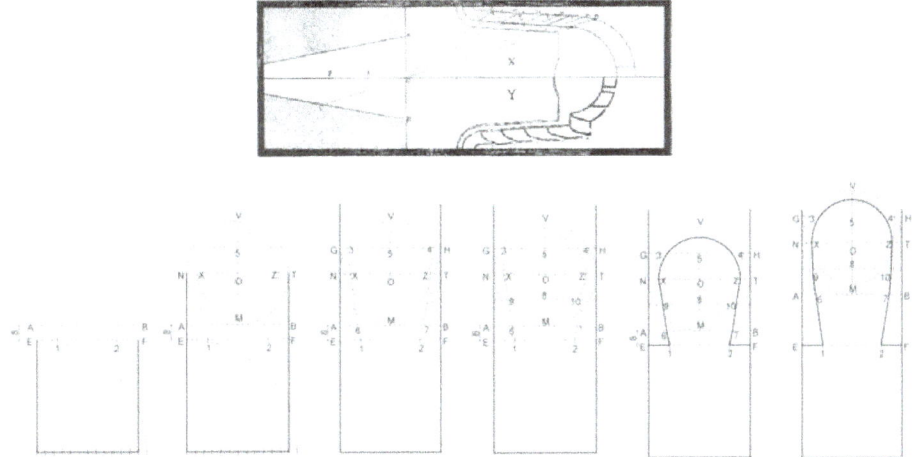

Figure 6. Fabrizio Carini Motta's original drawing (1676) [48] and redrawings [50]. The last figure refers to the possibility of obtaining a longer "scene square" if needed—there are more details in the text.

4. The Evolution of Theatrical Architecture during the 18th Century

During the 18th Century, the form and the functions of the theatre were optimised to host the melodrama: the stage was extended towards the audience—through the proscenium—in order to enhance the intelligibility of the soloists, the orchestra was placed in front of the stage; the boxes' arrangement was adapted for reasons of visibility. The mixtilinear shape of the Italian theatre was developed by many architects, who experimented with new solutions in order to improve the acoustical conditions and the possibility to have exact visuals (see Figure 7).

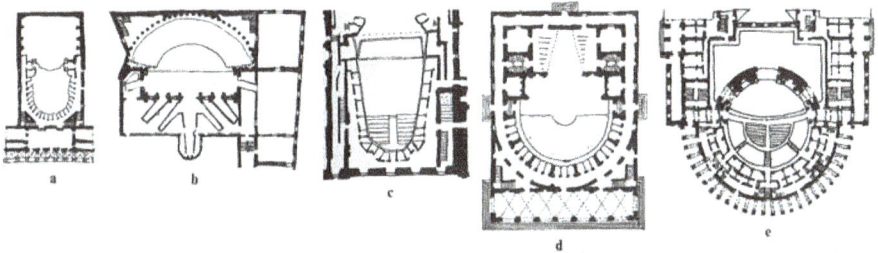

Figure 7. Theatre plan comparison: (**a**) bell shape: Teatro Comunale of Bologna (1763); (**b**) semielliptical shape: Teatro Olimpico of Vicenza (1585); (**c**) open U-shape: Molière hall in the Palais Royal of Paris (1640); (**d**) semi-circular shape: a design by Enea Arnaldi (1762); (**e**) circular shape: project by Vincenzo Ferrarese (1771) (after [36]).

Despite being extremely active in theatre and scenography design [51], the members of the famous Galli Bibbiena family did not write acoustic treatises with which we could directly retrace their thoughts. Nevertheless, some authors referred in their writings to the works by the Galli Bibbiena architects. For example, Francesco Algarotti (1712–1764) ground his study Saggio sopra l'opera in musica (Essay on the Opera, 1764) [52] on the works of Ferdinando Galli Bibbiena (1657–1743) and his sons Giuseppe (1696–1757) and Antonio (1697–1774). The former designed the Margravial Opera House in Bayreuth (1750) and several theatres in Europe; the latter designed the Teatro Comunale in Bologna (1763), the Teatro Scientifico in Mantua (now Teatro Bibiena,1769) and theTeatro dei Quattro Nobili Cavalieri in Pavia (now Teatro Fraschini, 1773) (see the plans in Figure 8).

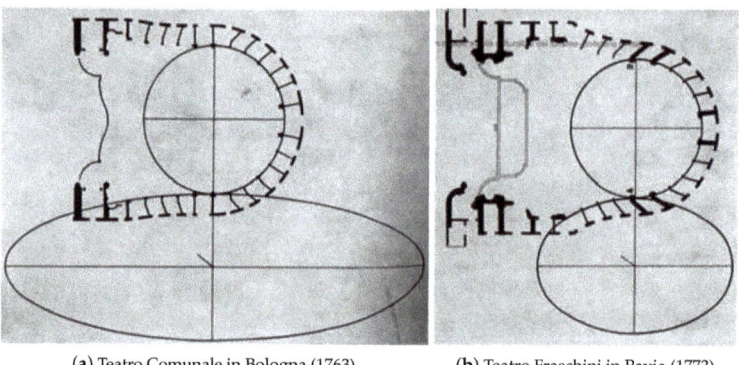

(a) Teatro Comunale in Bologna (1763)   (b) Teatro Fraschini in Pavia (1773)

Figure 8. Plans of two bell-shaped theatres by Antonio Galli Bibbiena [53].

Algarotti agreed with some building techniques used by Galli Bibbiena, e.g., the mixed use of masonry and wood in the theatre: masonry for the bearing structure in order to prevent fire and timber for indoor cladding in order to "equalise" the sound. He proposed the use of timber as cladding, analysed the case of a room—considered as a finite volume—with different boundary conditions (walls, tapestries, timber). Moreover, he linked their effects both on the vocal timbre and on the sound level spatial decay. He also noted that a room with flat walls made the voice too "sharp", while a room covered by tapestry made the voice too "silent". Instead of this, a room cladded by timber made the voice loud and pleasant. Therefore, he stated that the all the wooden parts inside the theatre had to be aged. Building solutions of the Teatro Comunale in Bologna, built by Antonio Galli Bibiena in the same years of Algarotti's essay, reflected this approach. Moreover, Algarotti discussed about the maximum dimensions of a theatre, which should be limited taking into account the sound strength of the voice, quoting Vitruvius's fifth book [6]. Concerning the shape of the audience, Algarotti denied the bell-shape for sake of visibility, although it is ideal for the strength of the voice. He proposed the use of the elliptical shape, being the semicircular shape an ideal solution but unachievable due to the too large width of proscenium. Furthermore, he avoided the decorations with sharp elements, referring again to Vitruvius's dissonantes category. In order to optimise the hall volume, he also proposed that all the tiers of boxes had the same height: the architectural praxis was to increase the width of boxes in the higher tiers in order to emphasise the cavea perspective. Algarotti's essay was extensively quoted in the work by Antonio Planelli (1737–1803), titled Dell'opera in musica. Trattato del cavalier Antonio Planelli dell'ordine gerosolimitano (On the Musical Opera. Treaty by Knight Antonio Planelli of the Jerosolimitan Order, 1772), in the section related to the construction of theatres [54].

Another late 18th-Century author, who analysed the effects of boxes in Italian theatres, was Francesco Riccati (1718–1790) [55–58]. He compared the Italian theatre with the one developed in the European courts, characterised by open galleries, but he also pointed out the benefits, in terms of social connections, that the boxes ensured to their owners. In particular, he did not agree with the presence of not proportioned openings in theatre cavea, which was in contrast with the architectural rules and with an idea of harmony and elegance of the structure. According to Riccati, a theatre should satisfy at least two main requirements: it should ensure the possibility to see and to hear properly. In his book Della costruzione de' teatri secondo il costume d'Italia vale a dire divisi in piccole logge (About the Construction of Theatres According to the Italian Style, i.e., with Small Boxes, 1790) [55], he dealt with both such themes (The essay opens with the following words: "Fra tutte le produzioni dell'Architettura Civile la più disastrosa, la più difficile, la più contumace ad assoggettarsi alle teoriche della scienza, ed ai precetti dell'arte, ella, non v'ha dubbio, è la costruzione de' Teatri giusta il costume d'Italia" (Among all the outcomes of Civil Architecture, the most disastrous, the most difficult, the most insubordinate towards the theories of science, and the norms of art, it is, without doubt, the Italian construction of theatres).). First of all, he identified the curved shape as the best solution for the cavea because it allows for aligning a high number of boxes along the cavea itself, ensuring a sufficient visibility to all of them. Secondly, among the different kinds of curved shapes, he suggested the use of the "divergent" shapes. In fact, the adoption of a concave curve ensured the visibility of the stage to the major part of the public, especially if the part of the curve overlooking the proscenium was made as large as possible. According to the author, the horseshoe shape originated from the degeneration of this attempt. In particular, he disagreed with this particular shape as it enlarged the part of the curve facing the stage way too much, leading to the necessity of narrowing the part of the curve closer to the stage itself. In this way, the boxes located at the sides of the cavea were necessarily disadvantaged. An appropriate shape for the audience was, in Riccati's opinion, the bell shape, which was often adopted by the architects Galli Bibbiena in their theatres. Indeed, this shape could ensure a good visibility to the whole public. Riccati strongly criticised the use of the elliptical shape, which was instead proposed by Luigi Rizzetti (17??–18??), in a piece titled Memoria intorno alla più perfetta costruzione di un Teatro (Essay on the Most Perfect Construction of a Theatre) [59], quoted by Riccati in his treatise and later on published as the small pamphlet Risposta del sig. Conte Luigi Rizzetti alle accuse date al Teatro da lui proposto (Response to the Allegations Made to the Theatre He Proposed, 1792) [60]. Rizzetti's choice of this shape was probably due to a property of the ellipse: the rays coming from a focus reflect one another. For this reason, Rizzetti placed the stage border on a focus of the ellipse, thinking that it would support the sound diffusion. Riccati observed that the public sitting in the boxes at the stage sides would not be able to have comfortable visuals to it because it would directly face the other side of the cavea. Probably influenced by the political and social issues that revolved around the concept of an "ideal" theatre (see next paragraph), Riccati claimed that the best shape was the circular one, although it was possible to adopt it only in small theatres, otherwise the stage—which would have been placed on the diameter of the circle—would become too wide.

Riccati thus proposed an elliptical shape obtained by different circle sectors. These sectors were drawn fixing the circle centre on the diameter of the wanted ellipse. In Figure 9, an ellipse is shown built up with three circular sectors (CF, FG, GD). The sectors CF and GD are drawn centering, respectively, the compasses on the points N and O, located on the shorter diameter out of the ellipse perimeter. The sector FG is drawn centering the compasses on the longer diameter, on the point E, located on the cross point of the segment NF and OG. The whole construction is ruled by the dimension of the segment BE. Increasing this dimension, it is possible to obtain an ellipse with a larger apex, which can contain a consistent amount of boxes facing the stage.

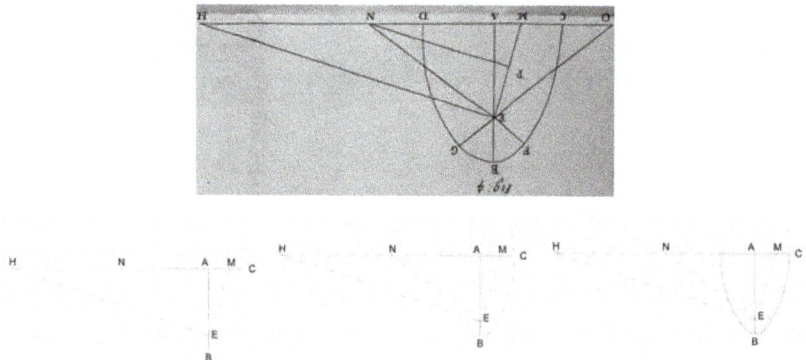

Figure 9. Francesco Riccati's guidelines to determine the elliptical shape of a theatre with circular sectors. Riccati's original drawing (1790) and re-drawings—more details are in the text.

The elliptic shape was used also by Cosimo Morelli (1732–1812) [61]; at the time, one of the most influential architects of the Papal States [62]. Due to visibility properties of elliptical shape, Morelli designed a three-part stage for the theatre of Imola (1780)—see Figure 10. He compared the width of the proscenium of an elliptic-shaped plan (segment GH in Figure 10) with that of the proscenium in the bell-shaped plan (segment EF) and the horse-shoe shaped one (segment DD). Indeed, the proscenium would be wider using the elliptical-shaped design. According to the architect, this configuration allowed also a larger stage for the scenes, a better visibility from the boxes, and a higher occupancy in the stalls.

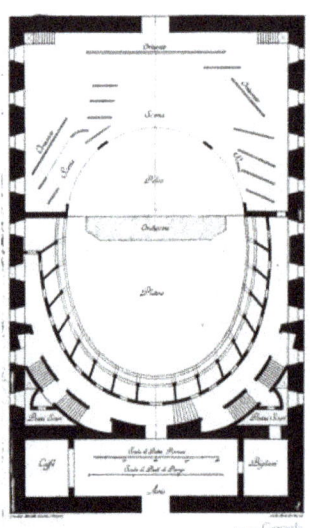 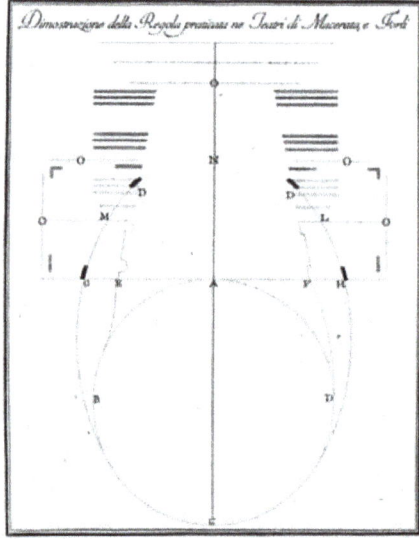

Figure 10. Plan of the Theatre in Imola designed by Cosimo Morelli [61].

The essay by the Venetian author and nobleman Andrea Memmo (1729–1793), titled "Semplici lumi tendenti a render cauti i soli interessati nel teatro da erigersi nella parocchia di S. Fantino in Venezia" (Simple Hints for Those Who Are Interested in the Theatre to Build in the Parish of S. Fantino in Venice, 1790) is an interesting account of the scientific knowledge on the theatrical construction in those years [63]. Far from the political and social ideologies that had been surrounding the idea of theatre in the second half of the 18th Century, Memmo focused on the building characteristics of a space for music and performances. He first pointed out the directivity of the sound sources. In his opinion, the Vitriuvian design of circular-shaped theatres was not optimised to the voice directivity, which decreases with the increase of the angle from the principal direction. Furthermore, the timbre of the voice depends both on the direct sound and the reverberated field. The author thus noted that reflected acoustic rays should converge to a central line and not to a point. In this case, a phenomenon of resonance (circumsonare) can be avoided. He focused on the necessity of balancing sound reflexion and diffusion, with the proper control of concave and convex shapes. At the same time, the direct and the reverberated fields had to be balanced: in particular, the latter could be equalised with a proper choice of the materials. In fact, Memmo was aware of the different acoustic absorption of porous and rigid materials, and he also referred to the problem of the different kinds of wood and their ageing. Furthermore, he stressed the importance of vaults made of light materials, in order to have low frequency absorption due to membrane-resonance effects. Quoting the essay of the Neapolitan engineer Vincenzo Lamberti (17??–18??), who dedicated his work to Memmo himself [64], Memmo referred to the bell shape—named pelcinona by Lamberti—as the optimal shape for a theatre. It is also interesting to notice that the author made no specific distinction between the acoustic and the lighting technique, which on the contrary were usually divided between the concerns of the physicists and those of the scenographers. By reading this essay, it is possible to claim that the technical and scientific knowledge around the building of theatres was extremely high in late 18th-Century publications, despite the lack of written works by the architects who were active in the field. This confirms the refined results by contemporary theatres, such as Teatro alla Scala in Milan (1776–1778) by Giuseppe Piermarini (1734–1808) and the already mentioned theatres by the architects Galli Bibbiena. A similar knowledge was established also by George Saunders (1762–1839), who wrote A Treatise on Theatres (1790) [65]. In this book, Saunders designed a theatre whose dimensions were based on audible threshold of a human voice and prescribed the use of wood in order to absorb the right amount of sound. Saunder's design of theatre was semicircular and showed some similarities with Enea Arnaldi's treatise (see the next section).

These last two sections have dealt with still existing Italian Opera houses, but it is important to keep in mind that several others were demolished or burned down. These events contributed, as stated in the Introduction, to a "Darwinian" evolutionary process of the optimisation of the shape and of the technical solutions of the theatres that were rebuilt after their demolition. For a complete chronology of the Opera houses of this period, see [66].

5. The Ideal Theatre

Throughout the 18th-Century, a widespread theoretical debate exploded on the social role of the theatre. Several French authors and theorists, such as Voltaire (1694–1778) and Diderot (1713–1784) [67] among the others [68–80], wrote on the educational importance of theatre performances. At the same time, the model of the classical, circular theatre was taken at the ultimate formal example [81,82], thus mirroring the necessity to erase the strict social division that characterised the partition between the audience and the boxes. Indeed, equality in listening and seeing had to match social and political equality, as a forerunner of the ideologies that led to the French Revolution. For a precise analysis of the connections between Italian theatres and French theoretical texts, see [83,84]. If in France the only architectural outcome of these theories was the theatre of Claude-Nicolas Ledoux (1736–1806) in Besançon (1775–1784) [81], these experimentations had no physical results in the Italian context. Nevertheless, architectural theorists like Enea Arnaldi (1716–1794) and Francesco Milizia (1725–1798) published on the topic of the "ideal theatre" and, most probably, they directly influenced some of the French designs [83,84].

In 1762, the architect Enea Arnaldi published in Vicenza the essay titled L'idea di un teatro nelle principali sue parti simile a' teatri antichi all'uso moderno accomodato (The Idea of a Theatre Being

Similar in its Parts to the Ancient Theatres and Adjusted to the Modern Use) [85]. His ideal theatre merged the characteristics of the ancient models—of Vitruvian origins—together with the Italian typology. In fact, he added a series of semi-circular steps to the audience and gave a cylindrical shape to the orchestra pit (see Figure 11).

Figure 11. Section of the theatre proposed by Enea Arnaldi [85].

Shortly after, in 1773, the theorist Francesco Milizia wrote an essay about the Italian theatre [86]. He compared the so-called modern theatre with the ancient one, thus complaining about its degraded conditions. The author observed that modern theaters were built with flammable materials and they were affected by humidity problems, whereas the main features of such constructions were supposed to be robustness, comfort and convenience. Among all European theatres, he only mentioned as "solid" those of Turin, Naples, Bologna and Berlin. Examining the shape of the auditorium, Milizia also highlighted the inappropriate form adopted for the Italian theatre. He compared the semi-circular ancient theatre, where all the spectators were able to see and hear properly, with the shape of the modern theatre and its boxes, that did not allow satisfying visuals to the whole public. He wrote: "the geometry itself demonstrates that, in a circle, all the angles of the circumference, which have as their basis the same diameter, are equal" [86]. For this reason, the author recommended assuming a circular shape and to place the stage on the main diameter, as the only way to guarantee the same perspective on the performance to each person in the audience. In addition, Milizia criticised the presence of boxes, which, in his opinion, divided the volume of the theatre in too many small spaces. On this, he wrote that "the boxes cut and reverberate the resonant air", thus confusing the perceived sound. In order to strengthen the criticism of the modern theatre shape, Milizia necessarily quoted Vitruvius. In fact, the Latin author affirmed that the inclination of the ancient theatre was chosen following musical proportion and it permitted a visual contact between the listeners and the actors. On this issue, Milizia published the project of an ideal theatre, called Teatro Ideale (1773). Taking example from Arnaldi's theory, he designed a more radical project, introducing French revolutionary themes into theatre design. In the Teatro Ideale, the architect replaced the separated boxes, reminders of the class divisions, with open galleries. The ideal theatre was characterised by a semi-circular shape, semi-circular ceiling, cavities and resonators, and galleries. On one hand, Milizia's assumptions were not acoustically appropriate and this was shown by the numerical simulation of the Teatro Ideale, realised from the original drawings [12]. In fact, considering the overall values of the acoustic descriptors the Teatro Ideale appears as a reverberant theatre—$T_{30} = 2.6\,\text{s}$ at 1000 Hz in unoccupied conditions—more than the existing Italian theatres of that period [15,19,20]. On the other hand, Milizia's theoretical contribution was extremely important when it comes to the social role of the theatre, which was seen as a public space. Indeed, his struggle for a more democratic spatial configuration of the audience could be seen as a forerunner of Richard Wagner (1813–1883) and his "democratisation of acoustics" [87].

While the 18th-Century theatre was still a place for aristocracy, Milizia saw in theatre performances as a means to convey the social image of a more equal and educated society, able to mirror the idea of a nation—the Italian one—still scattered into different states (see Figure 12).

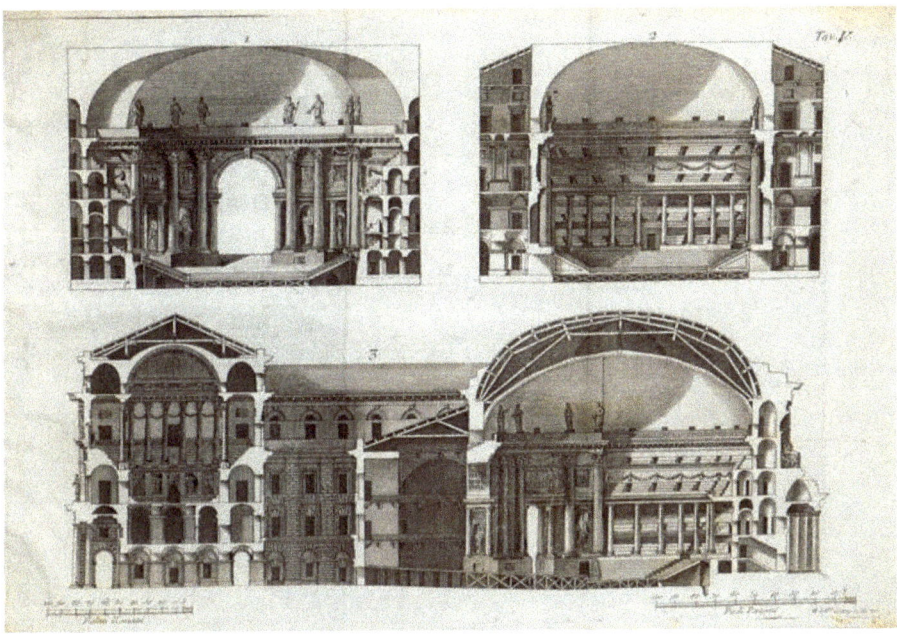

Figure 12. Design of the "Ideal Theatre" by Francesco Milizia and Vincenzo Ferrarese [86].

Milizia's writings strongly influenced the work of the architect Giovanni Antonio Antolini (1753–1841) who designed the grand "Foro Bonaparte" (1801) in Milan, in the area surrounding the Sforza Castle. Within the unbuilt architectural complex of the "Foro", Antolini introduced Milizia-Ferrarese circular-shaped theatre—that he named the "school for morals" [88]—equally divided between the stage and the semi-circular steps for the audience (see Figure 13).

Another follower of Milizia's theories was Tommaso Carlo Beccega (17??–18??), who published in 1817 an essay titled Sull'architettura greco-romana applicata alla costruzione del teatro moderno italiano e sulle macchine teatrali (On the Application of the Graeco-Roman Architecture to the Construction of the Modern Italian Theatre and to the Theatrical Machines) [89]. The basis of this essay was Vitruvius's theory and, at the same time, he tried to develop a theatre standard, whose shape was a semi-circle with elongated extremities (see Figure 14). Recalling the Vitruvian principle of comfort, he deducted the stage inclination, the orchestra pit position—lower than audience in the stalls—the proscenium arch shape and proportion, the construction of the dividers between boxes. He also claimed that the theatre height should have been equal to the cavea width, in order to avoid excessively deep halls. He fixed the maximal length of the audience, which should not exceed 45 m, the recognized audibility threshold.

Figure 13. Giovanni Antonio Antolini's design of a theatre in the Foro Bonaparte in Milan (after [83]).

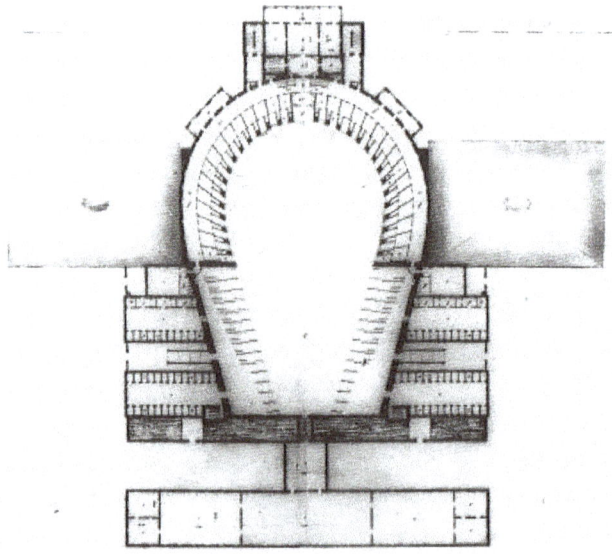

Figure 14. Ideal theatre shape according to Tommaso Carlo Beccega (1817) (after [83]).

6. The 19th-Century Theatre

The evolution of the 18th-Century into the 19th-Century theatre has not been diffusely discussed in architectural essays. However, it is clear that the theaters built during the 19th Century presented similar characteristics to the buildings belonging to the preceding century. By the mid-19th Century, the form of theatre was standardised and the vast majority of Italian opera houses was built after this period, in order to satisfy the request of each small town to have its theatre and the request of bigger cities to build increasingly larger ones.

At the beginning of the 19th Century, Antonio Niccolini (1772–1850) published a pamphlet entitled Alcune idee sulla risonanza del teatro (Some Ideas on Theatre Resonance) [90], in which he attributed the theatre deafness to its proportions and, exactly as Milizia did, to the use of the boxes [83]. Differently from Milizia, Niccolini understood that the modern theatre characteristics derived from social and economical needs; therefore, he elaborated solutions in order to improve the acoustic of these theatres, taking into account the circumstantial necessities. Niccolini's research followed a completely different path than those of his predecessors: he investigated the stage dimensions—and compared to the dimensions of the audience—and the vault continuity as deafness causes. In his opinion, the shape of the theatre and the used materials were irrelevant. The first element significantly contributing to the theatre resonance was a wide stage opening, which conveyed an air flow from the stage area to the audience (see the Niccolini's plan of Teatro San Carlo in Figure 15)—where the air was more rarefied because of the presence of public—thus reinforcing the sound propagation. The second element was the vault which in his opinion should ensure the uniformity of the sound path.

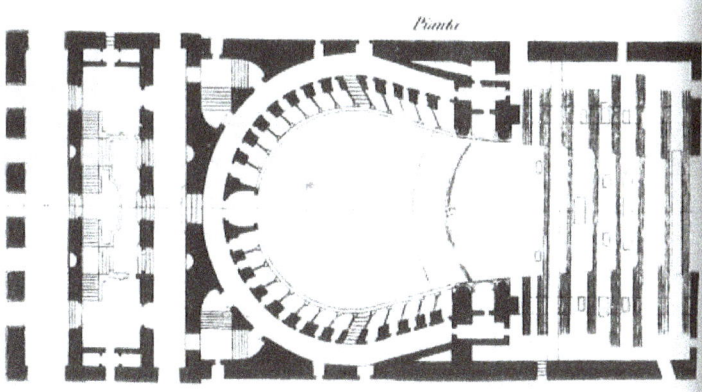

Figure 15. Antonio Niccolini's design of San Carlo theatre in Naples (1816) [90].

The architect and scenic painter Paolo Landriani (1757–1839), in his Osservazioni sui difetti prodotti nei teatri dalla cattiva costruzione del palco scenico e su alcune inavvertenze nel dipingere le decorazioni (Observations on the Flaws Deriving by the Poor Construction of the Stage and on Some Oversights in Decoration Paintings) [91], provided a description of a common method used to determine the theatre shape in that period. The usual shape adopted for theatres was a curve composed by a semi-circle and two lines, with a focal point at a distance of two diameters of the aforementioned semi-circle. The whole length of the theatre was usually similar to the length of the circle diameter; the proscenium width was determined by the intersection of the aforementioned lines with a line tangent to the circle. It could also happen that the length of the audience was longer than the circle diameter: therefore, the proscenium width could be determined as explained in Figure 16. It was the author's firm belief that the acoustic quality of a theatre depended on: the horseshoe shape commonly adopted; the lack of interruption or projections in the cavea boundary; the wooden parapets of the boxes; the elliptical vault above the audience, which should be as flat as possible; the proscenium opening, which should be wider on the audience side; a slightly inclination of the proscenium arch ceiling towards the stage, so that the actors' voices could spread easily into the auditorium as a ray.

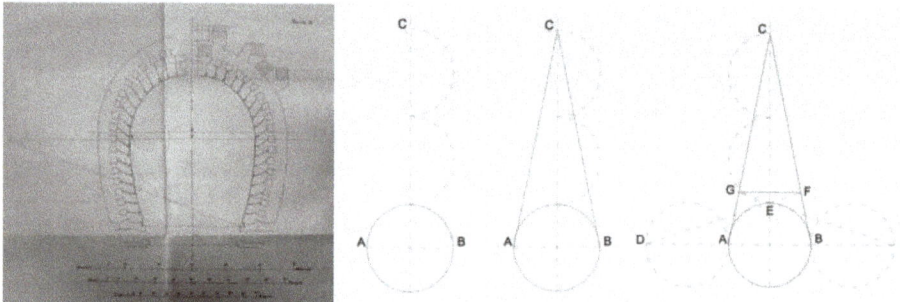

Figure 16. Paolo Landriani's guidelines to determine the proscenium width. Landriani's original drawing (1828) and redrawings [50]. See details in the text.

As stated before, theatres were prone to fires and therefore they were often rebuilt. Just like Niccolini, also the Venetian architect Giovanni Battista Meduna (1800–1886) and his brother Tommaso (1798–1880) had to deal with the reconstruction of a theatre. In their case, they rebuilt the theatre La Fenice in Venice, already debated by Memmo in his essay: the theatre was destroyed by fire in 1836 and it was reopened in 1837 [92]. Meduna's method is still "classic", regarding both the shape of the hall and the use of the materials. In particular, the Meduna brothers used the properties of timber in order to create resonant volumes under the stalls [93].

Already in the first decades of the 19th Century, the shape and the building techniques for theatre construction were almost consolidated: therefore, contemporary scholars focused on manuals or historical overviews [94–97], which progressively accounted for a theatre form with standard shapes and building typologies. It is not a coincidence if, in the same years, some Italian composers—such as Gaetano Donizetti (1791–1848), Vincenzo Bellini (1801–1835) and Giuseppe Verdi (1813–1901)—wrote some of the most iconic Italian operas [98]. The popular fame of these operas is strongly connected to the need of building theatres in every town, thus "scaling" the geometries on reduced volumes and occupancy [16–18].

From the second half of the 19th Century onwards, there was a progressive change in the building typologies of theatres. For example, the use of steel (such as in the Politeama theatre in Palermo, designed and built between 1865 and 1891) [99] and the use of cast-iron in the bearing structures, thus replacing timber or masonry, had resulting effects on the acoustic of such spaces [100,101].

The last moment in which Italian Opera houses were extensively built can be located in time just before the arrival of cinema. In those years, some great Sicilian theatres were built: like the Bellini theatre in Catania (inaugurated in 1890) and the Massimo theatre in Palermo (inaugurated in 1897), which is still today the biggest theatre of the Italian peninsula [102,103]. These last theatres were deeply influenced by geometries and spatial schemes that could not be described as merely Italian anymore, but "European", as they were affected by French (Opéra Garnier, 1875) and Austro-Hungarian (Wiener Staatsoper, 1869) designs. In fact, they do not only show boxes, but also galleries; they have large foyers that could host the new bourgeoisie, together with the high classes.

A reverse exemple—that is, an Italian architect working in an European architectural background, as the architects Galli Bibbiena had done in the former Century—was the architect Antonio Canoppi (1769–1832). In 1830, he published his Opinion d'A.C. sur l'architecture en général et en spécialité sur la construction des théâtres modernes (The Opinion of A.C. on Architecture in General and in Particular on the Construction of Modern Theatres) [104]. He was an Italian architect who had been working in Russia since 1805 as theatre director for the Russian Empire. He designed two theatres in St. Petersburg (see Figure 17), whose occupancy were, respectively, of 2200 and 1000 people. Thanks to studies on optics, perspective and acoustics, he fixed the length of the cavea—in the larger theatre—at

25.8 m and the proscenium width at 12.9 m. The dimensions of the stage were very large, in order to host monumental scenographies.

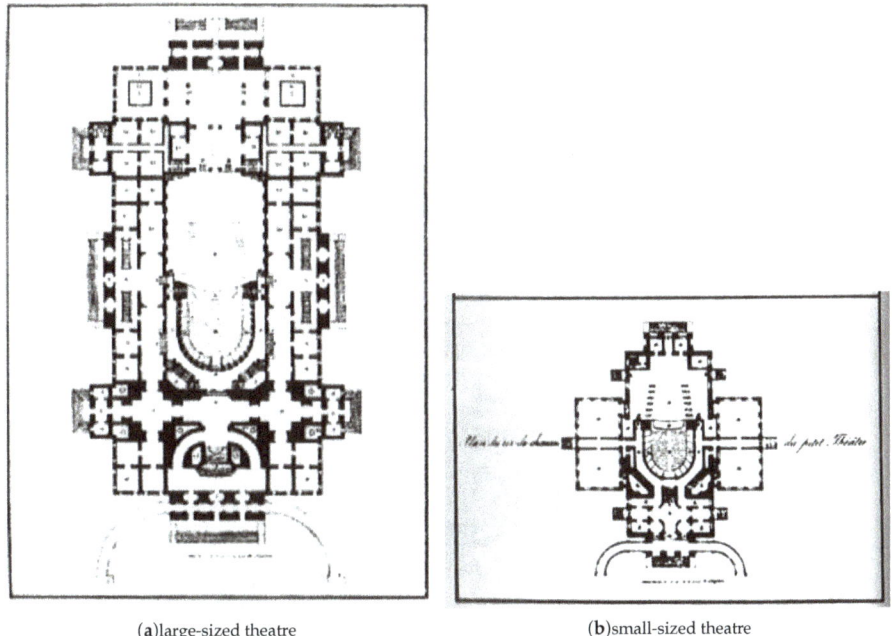

(a) large-sized theatre   (b) small-sized theatre

Figure 17. Antonio Canoppi's drawings of two theatres in St. Petersburg (after [83]).

In order to better understand the chronological evolution of each hall shape, together with their analysis in literature, Figure 18 represents a timeline overlapping the construction year and shape of each mentioned theatre, together with the life spans of the debated authors.

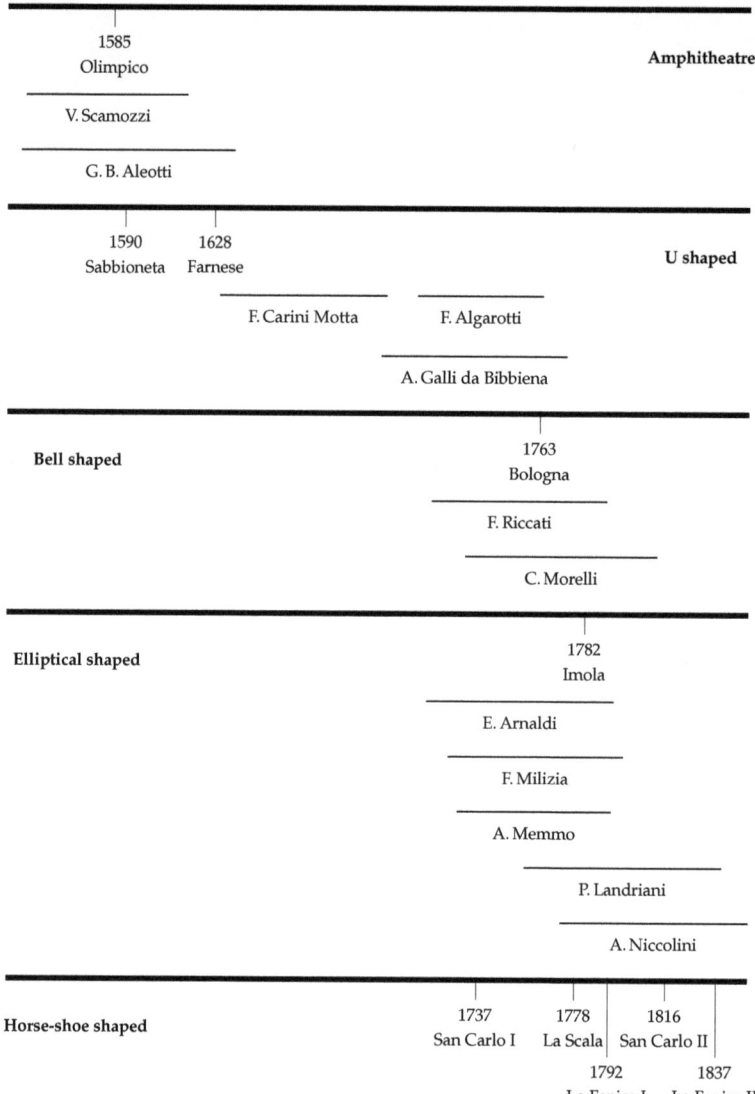

Figure 18. Evolution during time of the different shapes for Opera houses. In particular, some of theatres mentioned in the text have been marked on each timeline, together with the life span of mentioned authors.

7. Evidences in Small–Sized Theatres of the 19th Century

Five small-sized Italian Opera houses have been selected as case studies because of their similar dimensions and social function. All of these theatres were designed and built in the 19th Century: Municipal Theatre of Cervia (CER), Municipal Theatre of Cesenatico (CES), Municipal theatre of Russi (RUS), Dragoni theatre of Meldola (DRA), Petrella Theatre of Longiano (PET); see further details in Table 1. Such theatres were chosen due to the availability of the original designs in the city archives (see Figure 19) and a complete documentation about the restoration works of the last several decades [36,105–112].

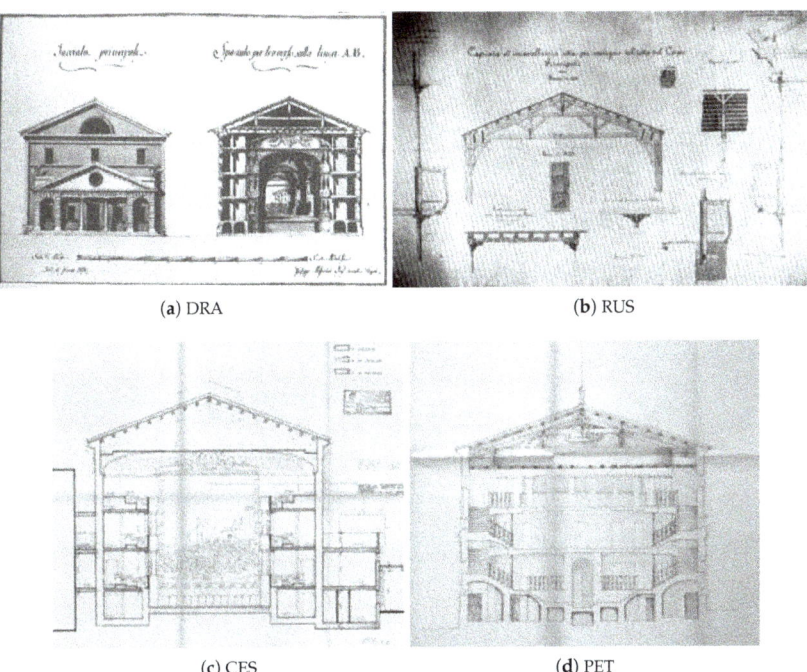

(a) DRA (b) RUS

(c) CES (d) PET

Figure 19. Some original drawings of the small–sized opera houses under study. Such drawings are available at the local city archives and local publications [36,105–112] and represent one of the key sources for the historical research on such buildings, collected in [50].

Table 1. Architectural features of the small Italian historical Opera houses under study. $T_{30,M}$ is the measured value of Reverberation time at mid-frequencies averaged over the whole audience. Other measured room criteria are fully available on [19].

	Theater, City	Seats	V_{hall}	V_{stage}	Scenic Arch	$T_{30,M}$ (s)	year
DRA	Dragoni, Meldola	318	1140	1080	6.8 × 7.3	0.83	1838
RUS	Comunale, Russi	305	900	1370	8 × 7.5	0.97	1887
CES	Comunale, Cesenatico	271	870	1320	7.4 × 8.6	0.90	1865
CER	Comunale, Cervia	224	730	1140	8.1 × 7.4	0.84	1862
PET	Petrella, Longiano	241	630	1390	7.5 × 9.1	1.07	1870

A subset of these theatres can be made considering their shape: for example, CES and MEL show an elongated shape. Nevertheless, the two theatres show a significant difference, as CES has two boxes included in the proscenium arch (one on each side). Its shape is thus characterised by a linear part linked with the horseshoe shape. Among the others, CER and RUS show exactly the same bending. PET, instead, is very similar in its bottom part, even if it is smaller than the other two, but with a linear evolution closer to the stage. Considering the stage area, all theatres show more or less a rectangular shape, with the same irregularity due to the presence of service rooms. CER is an exception: because of urban necessities, the stage area is characterised by a strong asymmetry and the rear wall of the stage is not parallel to the proscenium arch.

The aforementioned guidelines by Carini Motta, for the design of horse-shoe theatres, may be applied to these case studies. Following the author's instructions, an elongated shape with a linear part close to the proscenium arch can be obtained: therefore, this method has been applied to the theatres of CES and DRA, which present similar features Figure 20. The shape obtained following Carini Motta's method, with the dimensions of CES, is an offset of the real curve defining the audience boundary of the theatre. Considering the dimensions of DRA, the shape obtained is only a little too rounded on the sides of the audience, but, in general, it corresponds quite well to the real theatre shape. It also tried to apply Carini Motta's method to a different geometry, such as CER, yet with unsatisfactory results.

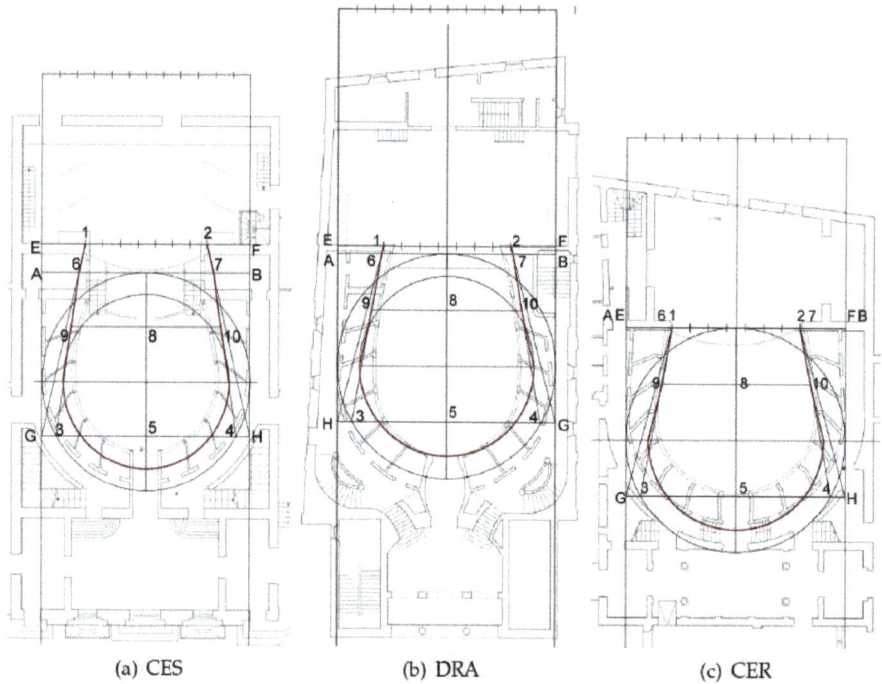

(a) CES   (b) DRA   (c) CER

Figure 20. Attempts to apply Carini Motta's guidelines for drawing horseshoe-shaped theatres.

In Figure 21, Landriani's method to design the proscenium arch was applied to the small-sized theatres under study. This method has been first applied to the theatres with a more rounded shape, such as CER and RUS. For both theatres, the shape obtained is quite close-fitting to the real one. However, Landriani's method to define the proscenium arch width did not give satisfying results. The figures of CER and RUS show that the shape obtained following Landriani's guidelines is bigger if compared to the boundary of CER audience, while it is a bit smaller in the part close to the stage in the case of RUS. This means that RUS is defined by a more rounded shape. The same method was also applied to the theatres with a more elongated shape (that is, DRA, CES, PET). In DRA, the shape obtained fits quite well with the rear part of the audience, but it becomes too narrow close to the stage. The same observation also counts for PET; instead, Landriani's method applied to CES produced a tight-fitting shape. As for the theatres of CER and RUS, Landriani's instructions to define the proscenium arch width did not give satisying results.

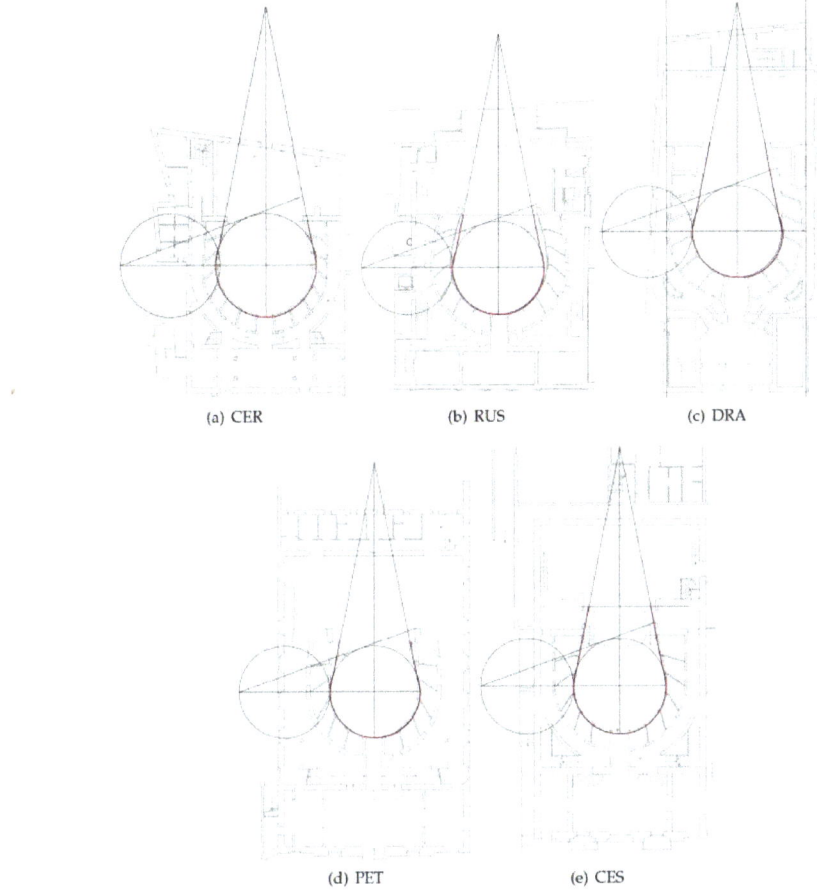

Figure 21. Application of Paolo Landriani's guidelines for dimensioning the proscenium width.

Such design features may influence the acoustic criterion of the hall. Despite of refurbishment differences in each theatres—changes in velvet, chairs and other elements and materials—the acoustic behaviour at the present day may still be dependent from design. The theatres under study were measured in a previous campaign of theatres in Romagna [19]. The positions of the omnidirectional sound source [113] on the stage were chosen similarly in all theatres: the first position on the longitudinal axis of the stage at 1 m from the edge (fore-stage); the second one at the centre of the area closed by the curtains (centre-stage). ISO 3382 [114] criteria were extracted from the measured impulse responses [115].

The values of reverberation time are quite low, if compared with other Italian Opera houses of bigger dimensions—swinging between 1.2 s (La Scala Theatre in Milan), those classified as "type B" by Prodi et al. [20], and 1.7 s (Teatro Comunale in Bologna), those classified as "type A". As the reverberation time appears to be quite constant on all the audience, due to the complex connection between geometry and positions, it may be more useful to analyse a value that depends on the positions of the source and the receiver, such as the sound strength [114].

Table 2 summarises the measured values of G_M, using the two source positions described above (one in the proscenium and one at the centre of the stage), and also the averaged values on all the receivers of the same categories (all positions in the stalls, boxes and galleries). In general, results show how the position of the soloist on the stage influences the sound strength in the audience. G_M values in the fore-stage are higher that the ones in the centre-stage. This may be due to the higher acoustic coupling when the soloist is placed in the fore-stage. This situation was well known by Opera authors, who often stressed in their librettos the right place of each aria, in order to balance the sound level of each soloist, the choir and the orchestra [87].

Table 2. Measured values of Sound Strength at mid-frequencies G_M in the theatres under study (dB).

Sources:	Proscenium			Center-Stage		
Receivers:	Stalls	Boxes	Gallery	Stalls	Boxes	Gallery
DRA	10.2	7.3	5.5	8.1	6.2	4.5
RUS	11.0	8.6	7.3	8.7	7.4	6.8
CES	11.4	9.7	9.4	9.3	8.4	8.4
CER	16.0	12	12.8	13.0	12.0	11.9
PET	15.7	10.4	11.5	11.8	9.0	10.3

The use of a velvet band within the cavea of DRA—introduced by the renovation works of 1970s [36,105,108]—is for sure the cause of a low G value in the stalls. However, in this theatre, the shape factor of the hall is different from the others, as DRA is much higher than what the cavea surface would allow. This influences a low G value in the gallery.

CER and CES, as stated above, are very similar. The choice of drawing the stage further back, in CES, in order to insert a small orchestra pit, has a negative influence on the sound strength (about 3 dB are lost, for each couple source-receiver). In CES, the presence of the original design for the proscenium shows a sound strength homogeneity in the audience. In fact, apart from the values in the stalls for a source on the proscenium, that are necessarily influenced by its proximity, G values are more or less constant in all audience, gallery included. This behaviour mirrors the scope of a good design of a theatre built for a community, which is a city theatre.

If RUS is too wide if compared to Landriani's method, PET fits perfectly into these guidelines. The latter was restored by preserving the original materials, as the former was renovated into a small contemporary prose theatre. PET also shows a great result of acoustic coupling between the fly tower and the cavea. G values are constant, with the exception of those in the stalls, due to the proximity effect.

To conclude, the measures of sound strength and its spatial distribution show that the design of a theatre hall is mainly based on the shape of the theatre itself, more than on the materials used (this is valid for small-sized Opera houses with a horseshoe shape, such as the aforementioned case studies). Restoration works may influence negatively on the acoustic behaviour of the halls, yet the influence of the original design still plays an important role.

8. Conclusions

This study had the aim of summarising the copious literature on theatre design related to pre-Sabinian scholars, from the 15th- and 16th-Century rediscovery of Vitruvius's treatise to 19th-Century standardised applications of guidelines and instructions by several Italian authors. The aim was to draw evidence of such literature in the design of Italian Opera houses, with a particular attention to small-sized Opera houses to be found in minor Italian towns.

In a contemporary acoustic point of view, the pre-Sabinian Italian scholars pointed out all the critical aspects of theatre design: Aleotti the relevance of the scenic arch [9], Carini-Motta [48] the separation of the acoustic volumes of the stage and the audience, Milizia [86] the coupling of the boxes, Riccati [55] the homogeneity of direct sound to the listeners, Rizzetti [60] the diffusion of the sound, Niccolini [90] the energetic transmission between stage and cavea and the relationship between reverberation and acoustic mean free path, and Beccega [89] the geometric ratios of the theatre and the echo-flutter phenomena. The dualism optics-acoustics might have often been useful to these authors, the knowledge in optics at the times being more advanced than that in acoustics.

Some of the topics that emerged from this study are at the core of the acoustic properties of Italian Opera houses. The majority of existent and active Italian theatres was bult between 1750 and 1880 and is characterised, for example, by a vast use of timber, whose effects have only partially been studied so far. Recent renovation works and the adaptation to fire-escape regulations have reduced the occupancy of such theatres. Moreover, their acoustic quality does not mirror the acoustic conditions at the time of their construction and full functionality in the 19th Century. In the authors' opinion, this may not necessarily be seen as a problem as the theatre has always been a flexible building, adaptable to the changes of the audience and to the functional necessities of the performances. The knowledge of such literature could instead be of help in the case of refurbishment of a historical theatre or of a new building with new materials, which still aims at preserving the acoustic characteristics of Italian Opera houses.

The birth of architectural acoustics as an independent science dates back to Wallace Clement Sabine's work. Interestingly, some Vitruvian categories were still used by Sabine himself [1], regarding speech intelligibility. Nevertheless, some of the main issues of architectural acoustics had already been consciously tackled by pre-Sabinian Italian scholars in their treatises. Although the concept of reverberation was not explicitly known, this literature shows a particular attention to the effects of materials, related to both intelligibility and abatement. In fact, timber was used not only for its ability to radiate sound energy (the analogy with the musical instruments is often recalled), but also for its capacity of absorbing part of the sound energy and thus equalising the sound. Moreover, issues related to the echo-flutter were tackled, both focusing on the elliptical shape and on the maximum dimensions of the hall. Furthermore, the Italian Opera house is a system of acoustic coupling of volumes. The fly tower was born for scenographic reasons—it is missing in the first performance spaces, but it soon became a critical element for the correct acoustic coupling of sound sources on the stage. On one hand, the fly tower cannot be too big for scenographic issues; on the other hand, some geometries like the bell and the ellipse enlarge the dimension of the scenic arch in order to host a sufficient number of people in the audience. This need is perhaps the reason why the horseshoe-shape eventually took hold.

To conclude, by quoting one of the authors analysed in the paper, this sentence by Antonio Niccolini—designer of the new hall of the Teatro San Carlo in Naples (1817)—may be still valid today: "L'indole della propagazione del suono è tuttora nel nostro teatro reputata da molti come un mistero ignoto, e da molti altri erroneamente definita" (Until the present day, the nature of sound propagation is considered by many as an unknown mystery, and it is erroneously defined by many others).

Author Contributions: Conceptualization and methodology, D.D.; Investigation, D.D. and S.N.; Writing—Original Draft, D.D. and S.N.

Funding: This research received no external funding.

Acknowledgments: The authors acknowledge Elena Bo for her contribution about ancient theatre and the work of Vitruvius, Lorenzo Fecchio for his suggestions on the works by Sebastiano Serlio and Andrea Palladio, and Valentina Sartini for bibliographic research and for the use of Figures 20 and 21.

Conflicts of Interest: The authors declare no conflict of interest.

References and Notes

1. Sabine, W. *Collected Papers on Acoustics*; Harvard University Press: Cambridge, MA, USA, 1922.
2. Kwon, Y.; Siebein, G.W. Chronological analysis of architectural and acoustical indices in music performance halls. *J. Acoust. Soc. Am.* **2006**, *128*, 654–663.
3. Barbieri, P. The acoustics of Italian opera houses and auditoriums. *Recercare* **1998**, *X*, 263–328.
4. Barbieri, P. The state of architectural acoustics in the late renaissance. In *Architettura e Musica Nella Venezia del Rinascimento*; Howard, D., Moretti, L., Eds.; B. Mondadori: Milan, Italy, 2006.

5. Postma, B.N.J.; Jouan, S.; Katz, B.F.G. Pre-Sabine room acoustic design guidelines based on human voice directivity. *J. Acoust. Soc. Am.* **2018**, *143*, 2428. [CrossRef] [PubMed]
6. Vitruvius. *I dieci libri dell'architettura [...] tradotti e commentati da Mons. Daniel Barbaro*; Ten Books of Architecture); de' Franceschi: Venezia, Italy, 1567. (In Italian)
7. Biancani, G. *Sphaera Mundi seu Cosmographia, ac Facili Methodo Tradita*; Sebastiano Bonomi for Geronimo Tamburini: Bologna, Italy, 1620. (In Latin)
8. Postma, B. A history of the use of time intervals after the direct sound in concert hall design before the reverberation formula of Sabine became generally accepted. *Build. Acoust.* **2013**, *20*, 157–176. [CrossRef]
9. Fabbri, P.; Farina, A.; Fausti, P.; Pompoli, R. The second life of the Teatro degli Intrepidi by Giovan Battista Aleotti through the new techniques of virtual acoustics. In Proceedings of the 2nd International Conference on Acoustic and Musical Research, Ferrara, Italy, 19–21 May 1995. (In Italian)
10. Cammarata, G.; Fichera, A.; Pagano, A.; Rizzo, G. Acoustical prediction in some Italian theatres. *Acoust. Res. Lett. Online* **2001**, *2*, 61. [CrossRef]
11. Prodi, N.; Pompoli, R. The acoustics of three Italian historical theatres: The Early days of modern performance spaces. In Proceedings of the Tecniacustica 2000, Madrid, Spain, 16–20 October 2000.
12. Tronchin, L. Francesco Milizia (1725–1798) and the Acoustics of his Teatro Ideale (1773). *Acta Acust. United Acust.* **2013**, *99*, 91–97. [CrossRef]
13. Weinzierl, S.; Sanvito, P.; Schultz, F.; Büttner, C. The acoustics of renaissance theatres in Italy. *Acta Acust.* **2015**, *101*, 632–641. [CrossRef]
14. Iannace, G.; Ianniello, C.; Maffei, L.; Romano, R. Objective measurement of the listening condition in the old Italian opera house "Teatro di San Carlo". *J. Sound Vib.* **2000**, *232*, 239–249. [CrossRef]
15. Farina, A. Acoustic quality of theatres: Correlations between experimental measures and subjective evaluations. *Appl. Acoust.* **2001**, *62*, 899–916. [CrossRef]
16. Pompoli, R. (Ed.) *Proc. of Teatri d'opera dell'Unità d'Italia (Proc. of Opera houses of the Italian Unification)*; Associazione Italiana di Acustica: Venezia, Italy, 2011.
17. Cirillo, E.; d'Alba, M.; Martellotta, F. Acoustic characterisation of Apulian historical theatres. In Proceedings of the 38 Congress of Italian Acoustic Association, Rimini, Italy, 8–10 June 2011. (In Italian)
18. Ceniccola, G. Architetture in scena. Teatri storici in Campania tra XVIII e XX secolo: Conoscenza e nodi critici nel progetto di conservazione (Performance Architecture. Historical Theatres in Campania between 18th and 20th Century: Knowledge and Criticalities in the Conservation Project). Ph.D. Thesis, University of Naples "Federico II", Naples, Italy, 2011.
19. Garai, M.; Morandi, F.; D'Orazio, D.; De Cesaris, S.; Loreti, L. Acoustic measurements in eleven Italian opera houses: Correlations between room criteria and considerations on the local evolution of a typology. *Build. Environ.* **2015**, *94*, 900–912. [CrossRef]
20. Prodi, L.; Pompoli, R.; Martellotta, F.; Sato, S. Acoustics of Italian Historical Opera Houses. *J. Acoust. Soc. Am.* **2015**, *138*, 769–781. [CrossRef] [PubMed]
21. De Cesaris, S.; Morandi, F.; Loreti, L.; D'Orazio, D.; Garai, M. Notes about the early to late transition in Italian theatres. In Proceedings of the 22nd International Congress on Sound and Vibration (ICSV22), Florence, Italy, 12–16 July 2015.
22. Morandi, F.; De Cesaris, S.; D'Orazio, D.; Garai, M. Energy criteria in Italian historical opera houses: A survey over 11 theatres. In Proceedings of the 9th International Conference on Auditorium Acoustics, Paris, France, 29–31 October 2015; Volume 37.
23. Garai, M.; De Cesaris, S.; Morandi, F.; D'Orazio, D. Sound energy distribution in Italian opera houses. *Proc. Mtgs. Acoust.* **2016**, *28*, 015019.
24. Serlio, S. *Sette libri dell'architettura di Sebastiano Serlio Bolognese (Seven Books of Architecture)*; Francesco de'Franceschi senese: Paris, France, 1545. (In Italian)
25. Mazzucato, T. Idea del espacio escénico y lugares para la representación teatral entre los siglos XV y XVI. Modelos de teatro a la manera de Italia (Idea of the Scenic Space and Places for Theatre Performance between the 15th and the 16th Century. Models for the Italian Theatre). *Studia Aurea* **2009**, *3*, 139–172. (In Spanish)
26. Canac, F. *L'acoustique des Theatres Antiques: Ses Insegnements (The Acoustics of Ancient Theatres: Its Lessons)*; Edition du Centre National de la Recherche Scientifique: Paris, France, 1967. (In French)
27. Maconie, R. Musical Acoustics in the Age of Vitruvius. *Music. Times* **2005**, *146*, 75–82. [CrossRef]

28. Farnetani, A.; Prodi, N.; Pompoli, R. On the acoustic of ancient Greek and Roman Theatres. *J. Acoust. Soc. Am.* **2008**, *124*, 157–167. [CrossRef] [PubMed]
29. Chourmouziadou, K.; Kang, J. Acoustic evolution of ancient Greek and Roman theatres. *Appl. Acoust.* **2008**, *69*, 514–529. [CrossRef]
30. Mo, F.; Wang, J. The Conventional RT is Not Applicable for Testing the Acoustical Quality of Unroofed Theatres. *Build. Acoust.* **2013**, *20*, 81–86. [CrossRef]
31. Amadei, D. L'ordine e la geometria nel teatro antico. Diffusione e fortuna del De Architectura di Vitruvio. Caso studio: Il teatro romano di Fanum Fortunae (Order and Geometry in the Ancient Theatre. Diffusion and Fame of Vitruvius's De Architectura. ase Study: The Roman Theatre of Fanum Fortunae). Ph.D. Thesis, Polytechnic University of Marche, Ancona, Italy, 2015. (In Italian)
32. Mullin, D.C. The Influence of Vitruvius on Theatre Architecture. *Educ. Theatre J.* **1966**, *18*, 27–33. [CrossRef]
33. Torello–Hill, G. The exegesis of Vitruvius and the creation of theatrical spaces in Renaissance Ferrara. *Renaiss. Stud.* **2014**, *29*. [CrossRef]
34. Sanvito, P.; Weinzierl, S. L'acustica del Teatro Olimpico di Vicenza (The Acoustics of the Teatro Olimpico in VIcenza). *Odeo Olimpico* **2013**, *23*, 463–492. (In Italian)
35. Pigafetta, F. *Due Lettere descrittive l'una dell'ingresso a Vicenza della Imperatrice Maria d'Austria dell'anno MDLXXXI l'altra della Recita nel Teatro Olimpico dell'Edippo di Sofocle nel MDLXXXV (Two Letters on the Arrival of Maria of Austria in 1581 and on the Performance at the Teatro Olimpico of Sofocles's Oedipus in 1585)*; Crescini, V., Ed.; Raccolta di Lettere Sulla Pittura, Scultura ed Architettura: Padova, Italy, 1830; pp. 25–31. (In Italian)
36. Farneti, F.; Van Riel, S. *L'architettura teatrale in Romagna 1757–1857 (Theatrical Architecture in Romagna)*; Uniedit: Firenze, Italy, 1975. (In Italian)
37. Scamozzi, O.B. *Le fabbriche e i disegni di Andrea Palladio (Andrea Palladio's Works)*; Giovanni Rossi: Vicenza, Italy, 1796. (In Italian)
38. Scamozzi, V. *L'idea della architettura universale (About of Universal Architecture)*; Expensis Auctoris: Venezia, Italy, 1615. (In Italian)
39. Stone Peters, J. *Theatre of the Book 1480–1880. Print, Text, and Performance in Europe*; Oxford University Press: Oxford, UK, 2000.
40. Crabtree, S.; Beudert, P. *Scenic Art for the Theatre*; Focal Press: London, UK, 1998.
41. Peacock, J. *The Stage Designs of Inigo Jones: The European Context*; Cambridge University Press: Cambridge, UK, 1995.
42. Bjurstrom, P. *Giacomo Torelli and Baroque Stage Design*; Almqvist & Wiksell: Stockholm, Sweden, 1961.
43. Sabbatini, M. *Pratica di Fabricar Scene e Machine Ne' teatri (On Making Scenes and Machines for Theatres)*; CNR IRCrES: Ravenna, Italy, 1638. (In Italian)
44. Lazardzig, J.; Rössler, H. (Eds.) *Technologies of Theatre: Joseph Furttenbach and the Transfer of Mechanical Knowledge in Early Modern Theatre Cultures*; Verlag Vittorio Klostermann: Frankfurt, Germany, 2016.
45. Donati, P. *Descrizione del Gran Teatro Farnesiano di Parma e notizie storiche sul medesimo di Paolo Donati parmigiano architetto teatrale e accademico di Bologna e professore della Reale Accademia di Firenze (Description of the Farnese Theatre in Parma and Historical Information . . .)*; Blanchon: Parma, Italy, 1817.
46. Mariano, F. *Il Teatro Nelle Marche: Architettura, Scenografia e Spettacolo (The Theatre in the Marche: Architecture, Scenography and Show)*; Banca delle Marche: Ancona, Italy, 1997. (In Italian)
47. Quagliarini, E. *Costruzioni in Legno nei Teatri All'Italiana del '700 e '800: Il Patrimonio Nascosto Dell'architettura Teatrale Marchigiana (Wooden buildings in Italian Historical Opera Houses: The Hidden Heritage of Theatrical Architecture)*; Allinea: Firenze, Italy, 2008. (In Italian)
48. Carini Motta, F. *Treaty about the Structure of the Theatres and of the Scenes*; Guastalla, 1676; Republished; Il Polifilo: Milano, Italy, 1972. (In Italian)
49. Donnelly, M.C. Theaters in the Courts of Denmark and Sweden from Frederik II to Gustav III. *J. Soc. Arch. Hist.* **1984**, *43*, 328–340. [CrossRef]
50. Sartini, V. Comparative Analysis of Five Small Sized Italian Theatres of the 19th Century. Master's Thesis, University of Bologna, Bologna, Italy, 2012.
51. Galli Bibiena, F. *L'architettura Civile Preparata su la Geometria e Ridotta Alle Prospettive (Civil Architecture: Geometry and Perspective)*; Paolo Monti: Parma, Italy, 1711. (In Italian)

52. Algarotti, F. *Saggio Sopra L'opera in Musica (Essay on Opera)*, 2nd ed.; Library of Congress: Washington, DC, USA, 1764. (In Italian)
53. Giordano, L. Il Teatro dei Quattro Cavalieri e la presenza di Antonio Galli Bibiena a Pavia (The "Teatro dei Quattro Cavalieri" and the work of A. Galli Bibiena in Pavia). *Bollettino D'arte* **1975**, *60*, 88–102. (In Italian)
54. Planelli, A. *Dell'opera in Musica. Trattato del Cavaliere Antonio Planelli Dell'ordine Gerosolimitano (About the Opera)*; D. Campo: Napoli, Italy, 1772. (In Italian); re-published and edited by F. Degrada; Discanto edizioni: Fiesole, Italy, 1981. (In Italian)
55. Riccati, G. *Della Costruzione De' Teatri Secondo il Costume d'Italia: Vale a Dire Divisi in Piccole Logge (About the Construction of Theatres According to the Italian Style, I.e., With Small Boxes)*; Remondini from Venice: Bassano, Italy, 1790. (In Italian)
56. Barbieri, P. *Giordano Riccati Fisico Acustico e Teorico Musicale (Giordano Riccati Acoustician and Musical Theorist)*; Leo S. Olschki: Florence, Italy, 1990; pp. 279–304. (In Italian)
57. Bagni, G.T. *Vincenzo, Giordano, Francesco Riccati e la matematica del settecento (V., G.; F. Riccati and Mathematics in the 18th Century)*; Teorema: Treviso, Italy, 1993. (In Italian)
58. Bortolozzo, R. *L'universo ben Temperato Dei Riccati: Cosmologia e Musica in Una Famiglia di Illuministi Trevigiani (The Well Tempered Universe of the Riccati Family ...)*; Il Cardo: Venezia, Italy, 1995. (In Italian)
59. Rizzetti, F. Memoria intorno alla più perfetta costruzione di un Teatro. as quoted in Riccati, 1790.
60. Rizzetti, L. Risposta del sig. Conte Luigi Rizzetti alle accuse date al Teatro da lui proposto (Response [...] to the allegations made to the theatre he proposed). In *Collection of Scientific and Literary Brochures*; Stamperia Coleti: Ferrara, Italy, 1792. (In Italian)
61. Morelli, C. *Pianta e Spaccato Del Nuovo Teatro d'Imola (Plan and Section of the New Theatre of Imola)*; Casaletti: Roma, Italy, 1780. (In Italian)
62. Matteucci, A.M.; Lenzi, D. *Cosimo Morelli e L'architettura Delle Legazioni Pontificie (Cosimo Morelli and the Architecture of the Papal States)*; University Press Bologna: Imola, Italy, 1977. (In Italian)
63. Memmo, A. *Semplici lumi tendenti a render cauti i soli interessati nel teatro da erigersi nella parocchia di S Fantino in Venezia [...] (Simple Hints for Those Who Are Interested in the Theatre to Build in the Parish of S. Fantino in Venice)*. 1790. (In Italian)
64. Lamberti, V. *La regolata costruzion de' teatri secondo il costume d'Italia, vale a dire divisi in picciole logge ... (Italian Construction of Theatres, i.e., Divided in Boxes ...)*; V. Orsini: Napoli, Italy, 1782. (In Italian)
65. Saunders, G. *Treatise on Theaters*; I. and J. Taylor: London, UK, 1790.
66. Available online: https://www.theatre-architecture.eu (accessed on 8 February 2019).
67. D'Alembert, J.; Diderot, D. "Theatre" in *Encyclopédye ou Dictionaire raisonnè des Sciences, des Arts et del Métiers*; Briasson: Paris, France, 1751–1772. (In French)
68. Bondin, N. *Teatro Antico. Ragionamento Sopra la Forma e la Struttura (Ancient Theatre. Reasoning on Form and Structure)*; s.e.: Venezia, Italy, 1746. (In Italian)
69. Cochin, C.-N. *Lettre sur les salles de spectacle, Mercure de France (Letter on Musical Halls)*. 1760. (In French)
70. Dumont, G.M. *Suite de projects détaillés de salles de spectacle, avec del principes de construction, tant pour la mécanique del theatres que pour das décoration ... (Set of Detailed Projects of Performance Halls, with Construction Principles, both for Theatre Mechanics and Decoration)*. Paris, France, 1773. (In French)
71. De La Dixmerie, N.B. *Lettres sur l'etat present de nos spectacles, avec del vues nouvelles sur chacun d'eux, particulierement sur la Comedie Francaise et l'Opera (Letters on the Present Situation of Our Performances, with New Opinions, Especially on the Comedie Française and l'Opéra)*; Duchesne: Amsterdam, The Netherlands; Paris, France, 1765. (In French)
72. Chaumont, D.J. (*Veritable construction exterieure d'un theatre d'Opera, a l'usage de France, relative a celle donnee l'annee dernere, pour la Construction interieure (Exterior Construction of a French Style Opera House, Related to the One Published Last Year, Related to the Interior)*; De Lormel: Paris, France, 1766. (In French)
73. Monginot, G.M. *Exposition des principes qu'on doit suivre dans l'ordonnance des theatre modernes*; C.A. Jombert: Paris, France, 1769. (In French)
74. Damun, J. *Prospectus du noveau theatre tracé sur les principes des Grecs et des Romains (Project of a New Theatre Based on Greek and Roman Principles)*. Paris, France, 1772. (In French)
75. Roubo le Fils, A.J. *Traité de la construction des theatres et des machines theatrales (Treatise on the Construction of Theatres and Theatre Machines)*; Cellot et Jombert: Paris, France, 1777. (In French)

76. Patte, P. *Description du theatre de la ville de Vicence en Italie, chef-d'oeuvre d'Andrea Palladio (Description of the Theatre of Vicenza, in Italy, Work by Andrea Palladio)*. Paris, France, 1780. (In French)
77. Noverre, J.G. *Observations sur la construction d'une salle d'Opera (Observation on the Construction of an Opera House)*; Cellot: Paris, France, 1781. (In French)
78. Patte, P. *Essai sur l'architecture théatrale, ou, De l'ordonnance la plus avantageuse à une salle de spectacles, relativement aux principes de l'optique et de l'acoustique: Avec un examen des principaux téatres de l'Europe, et une analyse des écrits les plus importans sur cette matiere (Essay on Theatre Architecture, or on the Most Advantageous Design of a Performance Hall, and an Analysis of the Most Important Writings on This Topic)*; Moutard: Paris, France, 1782. (In French)
79. Bouellet. *Essai sur l'art de construire les theatres, leurs machines et leur mouvements (Essay on the Art of Building Theatres, their Machines and their Movements)*; Ballard: Paris, France, 1801. (In French)
80. Grobert, J.-F.-L. *Exécution dramatique en rapport avec le materiel de la Salle et de la Scene (Dramatic Execution rEgarding the Materials of the Hall and the Scene)*; Schoell: Paris, France, 1809. (In French)
81. Ledoux, C.N. *L'architecture considerée sour le rapport de l'Art, des Moeurs et de la Legislation . . . (The Architecture Considered through the Connections with Art, Traditions and Law)*. Paris, France, 1804. (In French)
82. Boullée, E.L. *Architecture, Essai sur l'art (Architecture, Essay on Art)*. London, UK, 1953. (In French)
83. Tamburini, E. *Il luogo teatrale nella trattatistica italiana dell'800 (The Theatrical Space in Italian 19th-Century Treatises)*; Bulzoni Editore: Roma, Italy, 1984. (In Italian)
84. Barbieri, P.; Tronchin, L. L'impostazione acustica dei teatri nei progetti del primo neoclassicismo Italiano (1762–1772), (The Acoustical Structure of Theatres in the First Italian Neoclassical Projects). In *Francesco Milizia e il teatro del suo tempo Architettura, Musica, Scena, Acustica*; Russo, M., Ed.; Collana Studi e Ricerche n. 2: Trento, Italy, 2011; pp. 137–161, ISBN 978-88-8443-396-1. (In Italian)
85. Arnaldi, E. *Idea di un teatro nelle principali sue parti simile a' teatri antichi all'uso moderno accomodato*; Veronese: Vicenza, Italy, 1762. (In Italian)
86. Milizia, F. *Trattato completo, formale e materiale del teatro (About the Theatre)*; Pasquali: Venezia, Italy, 1773–1794. (In Italian)
87. D'Orazio, D.; De Cesaris, S.; Morandi, F.; Garai, M. The aesthetics of the Bayreuth Festspielhaus explained by means of acoustic measurements and simulations. *J. Cult. Herit.* **2018**, *34*, 151–158. [CrossRef]
88. Westfall, C.W. Antolini's Foro Bonaparte in Milan. *J. Warbg. Courtauld Inst.* **1969**, *32*, 366–385. [CrossRef]
89. Beccega, T. *Sull'architettura greco-romana applicata alla costruzione del teatro moderno italiano e sulle macchine teatrali (On the Application of the Graeco-Roman Architecture to the Construction of the Modern Italian Theatre and to the Theatrical Machines)*. Alvisopoli, Italy, 1817. (In Italian)
90. Niccolini, A. *Alcune idee sulla risonanza del teatro (Some Ideas on the Resonance of the Theatre)*; Masi: Napoli, Italy, 1816. (In Italian)
91. Landriani, P. *Osservazioni sui difetti prodotti nei teatri dalla cattiva costruzione del palco scenico e su alcune inavvertenze nel dipingere le decorazioni (Observations on the Flaws Deriving by the Poor Construction of the Stage and on Some Oversights in Decoration Paintings)*; Regia tipografia di Milano: Milan, Italy, 1828. (In Italian)
92. Meduna, T.; Meduna, G. *Il Teatro La Fenice in Venezia . . . (On the La Fenice Theatre . . .)*; Antonelli: Venezia, Italy, 1849. (In Italian)
93. Cocchi, A.; Garai, M.; Tronchin, L. Influenza di cavità risonanti poste sotto la fossa orchestrale: Il caso del teatro Alighieri di Ravenna (The Influence of Resonating Cavities under the Orchestra Pit: The Case of the Alighieri Theatre in Ravenna). In *Teatri storici. Dal restauro allo spettacolo*; Nardini Editore: Firenze, Italy, 1997; pp. 135–153.
94. D'Apuzzo, N. *Cenno intorno ai teatri moderni e sopra gli archi di trionfo degli antichi (Notes on Modern Theatres and Ancient Triumphal Arches)*. 1817. (In Italian)
95. De Grazia, V. *Discorso sull'architettura del teatro moderno (On the Architecture of the Modern Theatre)*. 1825. (In Italian)
96. Ferrario, G. *Storia e descrizione de' principali teatri antichi e moderni (History and Description of the Main Ancient and Modern Theatres)*. Forni, Italy, 1830. (In Italian)
97. de Cesare, F. *La scienza dell'architettura applicata alla costruzione, alla distribuzione, alla decorazione degli edifici civili (The Science of Architecture on the Construction, Distribution and Decoration of Civil Buildings)*. 1885. (In Italian)

98. D'Orazio, D.; De Cesaris, S.; Garai, M. Recordings of Italian opera orchestra and soloists in a silent room. *Proc. Meet. Acoust.* **2016**, *28*, 015014. [CrossRef]
99. Barbera, P. *Giuseppe Damiani Almeyda artista architetto ingegnere (Giuseppe Damiani Almeyda Artist, Architect, Engineer)*; Pielle Edizioni: Palermo, Italy, 2008.
100. Boniotto, E.; Bovo, M.E.; Di Bella, A.; Frinzi, G.; Granzotto, N.; Rinaldi, C.; Zecchin, R. L'acustica nel restauro dei teatri storici: Il caso del Teatro Civico Di Schio (The Acoustics in the Restoration of Historical Theatres. The Case of the City Theatre in Schio). In Proceedings of the Congress of Italian Association of Acoustics, Ischia, Italy, 10–12 May 2006.
101. Silingardi, V.; Rinaldi, C.; Granzotto, N.; Barbaresi, L.; di Bella, A. A restoration based on the result of a public debate: The case of Civic Theatre of Schio. *Rivista Italiana di Acustica* **2017**, *41*, 1–14.
102. Basile, G.B.F. *Sulla costruzione del Teatro Massimo Vittorio Emanuele (On the Construction of Teatro Massimo Vittorio Emanuele)*; Tip. Dello Statuto: Palermo, Italy, 1883. (In Italian)
103. Fundaró, A.M. *Il Concorso per il Teatro Massimo di Palermo—Storia e progettazione (Competition for the Teatro Massimo in Palermo: History and Design)*; STASS: Palermo, Italy, 1974
104. Canoppi, A. *Opinion . . . sul l'architecture en général et en spécialité sul la construction des Théatres modernes (On Modern Theatres and Ancient Triumphal Arches)*. 1830. (In French)
105. Bortolotti, L.; Masetti, L. *Teatri storici. Dal restauro allo spettacolo (Historical Theatres. From Renovation to Performance)*; Nardini: Fiesole, Italy, 1977. (In Italian)
106. Savini, F. *Cenni storici e vicende del Teatro Comunale di Cesenatico (Historical Information on the City Theatre of Cesenatic)*. Cesenatico, Italy, 1979.
107. Fabbri, P. *Teatri di Russi, dal vecchio al nuovo Comunale (The City Theatre in Russi, from the Old to the New)*; Longo Editore: Ravenna, Italy, 1979.
108. Van Riel, S. (Ed.) *Il Teatro di Meldola: Storia e Restauro (The Theatre in Meldola: History and Renovation)*; Alinea: Firenze, Italy, 1982.
109. Castagnoli, S. Il teatro "E. Petrella" di Longiano dalla fabbrica al restauro (The Theatre Petrella in Longiano, from Design to Renovation). In *Studi Romagnoli*; XXXVI; La Fotocromo Emiliana: Bologna, Italy, 1985. (In Italian)
110. Bortolotti, L. *(a cura di), Le stagioni del teatro. Le sedi storiche dello spettacolo in Emilia Romagna*; Grafis Industrie Grafiche: Bologna, Italy, 1995.
111. Ceredi, C.; Piraccini, O. (Eds.) *Teatro Comunale di Cesenatico (City Theatre in Cesenatico)*. Comune di Cesenatico, Italy, 2004.
112. Vasumi Roveri, E. *I teatri di Romagna. Un sistema complesso (The Theatres of Romagna: A Complex System)*; Compositori Editore: Bologna, Italy, 2005. (In Italian)
113. D'Orazio, D.; De Cesaris, S.; Guidorzi, P.; Barbaresi, L.; Garai, M.; Magalotti, R. Room acoustic measurements using a high SPL. In Proceedings of the 140th AES Convention, Paris, France, 4–7 June 2016.
114. ISO 3382-1. *Acoustics—Measurement of Room Acoustic Parameters. Part 1: Performance Spaces*; ISO: Geneva, Switzerland, 2009.
115. Guidorzi, P.; Barbaresi, L.; D'Orazio, D.; Garai, M. Impulse responses measured with MLS or Swept–Sine signals applied to architectural acoustics: An in-depth analysis of the two methods and some case studies of measurements inside theaters. *Energy Procedia* **2015**, *78*, 1611–1616. [CrossRef]

© 2019 by the authors. Licensee MDPI, Basel, Switzerland. This article is an open access article distributed under the terms and conditions of the Creative Commons Attribution (CC BY) license (http://creativecommons.org/licenses/by/4.0/).

Article

The Proscenium of Opera Houses as a Disappeared Intangible Heritage: A Virtual Reconstruction of the 1840s Original Design of the Alighieri Theatre in Ravenna

Dario D'Orazio [1,*,†], Anna Rovigatti [2] [1]

1   Department of Industrial Engineering, University of Bologna, 40136 Bologna, Italy
2   Acoustics Air and Emissions, Atkins, London SW1E 5BY, UK
*   Correspondence: dario.dorazio@unibo.it; Tel.: +39-051-209-0549
†   Current address: Viale Risorgimento 2, 40126 Bologna, Italy.

Received: 1 July 2019; Accepted: 23 August 2019; Published: 1 September 2019

Abstract: In a Historical Opera House (HOH), the proscenium is the foreground part of the stage. Until the end of the 19th Century, it was extended through the cavea, being the orchestra placed at the same level of the stalls, without an orchestra pit. Soloists often moved in the proscenium when they sung, in order to increase the strength of the voice and the intelligibility of the text. The Alighieri theatre in Ravenna, designed by the Meduna brothers, the former designers of Venice's "La Fenice" theater, is chosen as a case study. During a refurbishment in 1928, the proscenium of the stage was removed in order to open the orchestra pit, which was not considered in the original design. The original design and the present one are compared by using numerical simulations. Acoustic measurements of the opera house and vibro-acoustic measurements on a wooden stage help to reach a proper calibration of both models. Results are discussed by means of ISO 3382 criteria: the proscenium increases the sound strength of the soloists but reduces the intelligibility of the text.

Keywords: opera house; cultural heritage; room acoustics; shape optimisation; archeoacoustics

PACS: 43.55.Gx; 43.55.Ka

1. Introduction

During last three decades, the interest of scholars was focused on the acknowledgment of acoustics of HOH as *intangible* cultural heritage, using this term with the meaning accepted by the academic community [1,2]. The so-called "Charter of Ferrara"—a document delivered in 1999—stated: "Preserving the acoustical heritage of historical opera houses means first of all being fully aware of it, identifying its presence and getting to know it. Then, it implies making an inventory of it and, finally, introducing legal protection measures to avoid its spoilage" [3]. Basing on the Charter of Ferrara—but not only—in the last 20 years, a huge set of acoustical data of Historical Opera Houses (HOHs) have been accumulated by scholars, but only few studies are available on international references [4–12].

Acoustic impulse responses are measured by setting sound sources and receivers in some relevant positions, respectively in the orchestra pit and on the stage for the source and in the stalls and in the boxes for the receiver [13] and extracting the room criteria defined by ISO 3382-1 [14]. HOHs need periodic refurbishments, which imply effects from an acoustic point of view [15,16]. The refurbishment of an HOH is a debated topic: on one hand, it would return the hall at its originally working condition; on the other hand, the safety and performance requirements vary in time. The acoustics of an HOH needs to change during all its life and this variability must be taken into account when we consider HOH acoustics an intangible cultural heritage [17]. Nevertheless, during the 19th Century, a dramatic change occurred in many opera houses, which influenced more than others the acoustics: the introduction of the orchestra pit by removing the proscenium of the stage.

The proscenium was introduced in the 17th-Century theatre in order to improve the singer intelligibility during the *recitativo* and to bring the singer close to cembalist—or other keyboardist—who played the *continuo* (see Table 1 for the meaning of Italian musical terms). Moreover, the shapes of the proscenium were optimised by pre-Sabine designers in order to make the voice of soloist louder during the *aria* [18].

The orchestra pit was introduced by Richard Wagner (1813–1883) in the Bayreuther Festspielhaus (1876)—the wooden theatre designed by architect Gottfried Semper (1803–1879) to host Wagner's *Ring* [19]. They proposed a *mystic gulf*, covered by a double proscenium arch, in order to divide the scene from the audience. In the mystic gulf, the orchestra was placed under the level of the audience, in a pit covered by a shell. A second shell was added in a second time, in order to decrease further the sound strength of the orchestra. The plan for a covered pit was not a Wagner original idea: the concealment of the orchestra was proposed in essays in 1775 and 1817 [18], and achieved in the Theatre of Besançon, designed by Claude-Nicholas Ledoux (1736–1806) [20], now demolished.

After three centuries from the birth of the Italian opera house—which dated back to Palladium's Teatro Olimpico (1580)—several reasons contributed to creating another volume, coupled to the main hall, for the orchestra:

- When the orchestra plays in the pit, it is also possible to dim the lights in the cavea, increasing the focusing of audience on the scenes. During the 19th Century, before the Festspielhaus opening—some show companies, such as the *Meiningen Company*, dimmed the auditorium during the performance [21] . Wagner's main goal in Festspielhaus was to prevent the audience reading the libretto during the drama, enhancing the visual and the listening experience of the audience: for this reason, the sunken pit was intended to cover the lights used by musicians [22].
- In the 19th Century, orchestras were changing, by adding more brass and, consequently, more strings and woodwinds were needed to balance the tone [23]. Consequently, the sound level of the orchestra was louder than the soloist and the pit was needed also to decrease the orchestral sound strength [24,25].
- *Recitativo* gradually disappeared during the 19th-Century opera, so soloist and conductor did have more need for proximity, which allows— from the point of view of the soloist— to follow the *continuo* accompaniment, often played by the conductor himself.

Wagner's proposals for Bayreuther Festspielhaus were followed by almost all the composers, and not only by composers: some of today's practice—e.g., to clap hands at the end of each act only—were introduced during the early representations of the *Ring*. After the Wagnerian revolution in conceiving, playing and attending an opera, almost all the HOHs needed to be updated to the new wave. In the early 20th Century, the pre-existent HOHs removed the proscenium in order to open the orchestra pit. In the "La Scala" theatre in Milan, the pit was opened in 1902, two years after the death of Giuseppe Verdi (1813–1900): it means that most of the operas were written taking into account the proscenium as "natural gain" of singers' voices. It can be confirmed by original *libretto*, which included the design of the wings and the displacement of the soloist [26].

Due to these reasons, the aim of the present work is to compare the acoustics for an HOH with and without the proscenium, by using the virtual simulation of the discontinued configuration. It could be useful to better understand how the audience attended the opera in the 19th Century. Furthermore, the present works may help to understand the range value of some room criteria during the golden age of the Italian opera.

Some data of the present work were provided by the MSc theses of Laura Reggiani [27] and Marco Rinaldi [28].

Table 1. List of the musical terms in Italian used in this paper.

aria	solo vocal piece with instrumental accompaniment
libretto	text written for and set to music in an opera
recitativo	style of delivery in which a singer is allowed to adopt the rhythms of ordinary speech
continuo	(or bass continuo) a part for keyboard consisting of a succession of bass notes with figures that indicate the required chords

2. Methodology

As mentioned in the abstract, during a refurbishment in 1928, the proscenium of the stage was removed in order to open the orchestra pit, which was not considered in the original 1840 design. Acoustic differences between the original design of Alighieri theatre and the present one were studied by using numerical simulations. Two acoustic simulation models were done: the original design and the present one.

The present model was calibrated with measurements done in September 2014. The calibration process compares the measured and the simulated room criteria, considering values in each octave band, averaged over three regions of receivers (stalls, boxes and gallery). The analysis of this process takes into account both the spatial variation of the sound field and the uncertainty due to the simulation chain [29].

The proscenium, as mentioned in the abstract, was the foreground part of the stage, and it was made of wood. The wooden stage is fixed above the air cavity: it works as an acoustic absorber at low frequencies and, at the same time, it shows some re-radiation properties above its coincidence frequency. In order to study the acoustic behaviour of the historic wooden stage, further vibro-acoustic measurements were done. Laboratory tests on specimens helped to evaluate the influence of the aging on fir wood. Then, in situ measurements on the stage of the Alighieri Theatre confirmed the results of specimens, helping the vibro-acoustic behaviour of the historical stage.

After the calibration procedure, two models of the theatre were built: the original and the present one. A virtual sound source was simulated in each model, which was placed in the position used by the soloist during arias, respectively, in the proscenium and on the fore-stage. Room criteria were simulated over the same receivers used in the calibration procedure. Finally, results were compared.

The reliability of the method used in the present study may be evaluated through the point of view of the so-called archeo-acoustics. A taxonomy of the procedures used in the previous studies [30–40] is shown in Table 2.

Table 2. Summary of the calibration method for several previous studies on virtual reconstruction for historic buildings that are no longer existing.

Room under Study	Calibration Reference	Calibration Approach
Public/Ritual Spaces [30]	Architectural drawings	Collected historical records.
Fogg Art Museum [31]	impulse response measurement (1973) and reverberation time data (1912)	Calibrated model using reverberation time data and photo, reconstruction former room states by modifying model and final comparison of original state to Sabine's data.
Concert Hall, Gewandhaus [32]	Historical reverberation time data (1933)	Material properties based on detailed construction documentation, seating material determined from photo inspection, adjusted to reverberation time data.

Table 2. *Cont.*

Room under Study	Calibration Reference	Calibration Approach
Acheron Necromancy, Olympia echo hall, Temple of Zeus [30]	Archival plans	Material properties based on common materials of the time.
Finnish concert halls [33]	Reviews of acoustical qualities (2014)	Reproduce acoustical characteristics by adjusting material and scattering properties.
Súleymaniye Mosque [34]	impulse response measurement	Calibrated model based on the field test results.
Santa Maria de la Murta [35]	Archival plans and images	Material properties determined from historical images and similar studies.
Mosque, Cordoba [36]	Previous studies and archival materials	Calibration based on the current model results.
Festspielhaus, Bayreuth [37]	Archival plans and sections	Material properties determined from historical images and similar studies.
Greek Theatre of Syracusae [38]	Impulse response measurement and previous studies	Material properties based on detailed construction documentation and photo inspection.
Palais du Trocadero [39]	Historical documentations and historical acoustic data (1911, 1906, 1907)	Historic photos of the space, documents concerning the construction and design and acoustic data.
Lazarica Church [40]	Impulse response measurement	Calibrated model using on-site measurement results.
Present work	Original plans and sections	Material properties based on acoustic and vibro-acoustic measurements of the current status of the theatre.

2.1. The Alighieri Theatre in Ravenna

In 1838, Tomaso Meduna (1798–1880) was asked to design the new opera house in Ravenna. Tomaso Meduna, formerly the architect of "La Fenice" Theater in Venice, accepted the job along with his brother Giovanni Battista (1800–1886). The works began in 1840 and the theater was inaugurated in 1853 [41]. In the 1840 design—in the following cited as the "original" one—the stage was protruded in the cavea (the so-called *proscenium*), to a corresponding surface until the line of the first boxes (see Figure 1). Furthermore, the orchestra was placed at the same level of the listeners in the stalls, corresponding to the second and the third row of chairs. The floor of the stalls, which was entirely made of pine wood, was supported with pine wooden truss holding up larch architraves and brick pillars. A river cobblestone was placed under the floor [42]. In 1928, the orchestra pit was built, cutting the proscenium. From 1959 to 1967, the theatre remained closed, in order to replace the wooden structures affected by termites: the original wooden structures were replaced by newer ones made by concrete and steel. The theatre reopened in 1967. The wooden stage was replaced a second time in 1970 [43]. In the summer of 2015, the wooden stage was renewed for a third time. The theatre is now used for opera and symphonic music by the resident "Luigi Cherubini" Young Orchestra, found and conducted by Riccardo Muti.

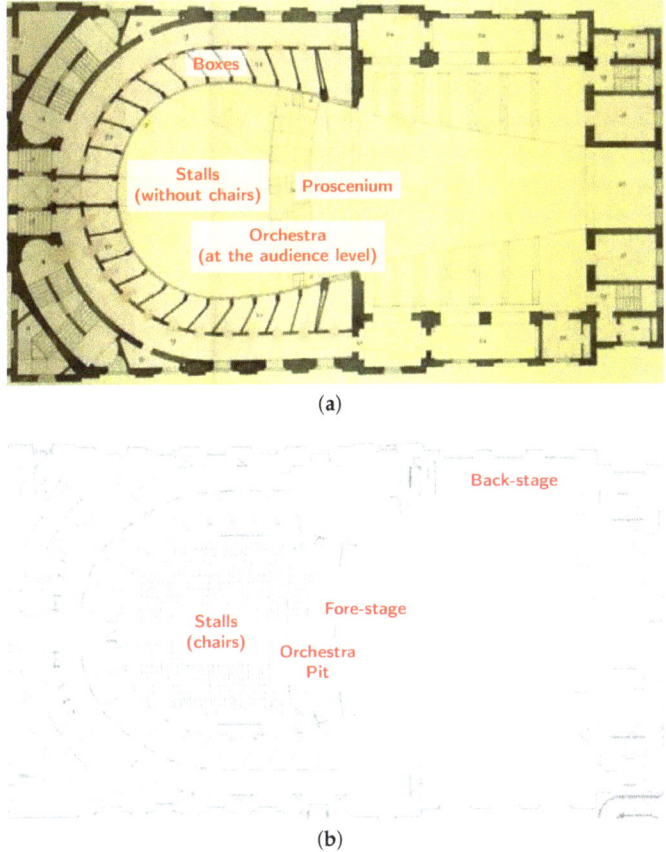

Figure 1. Plans of Alighieri Theatre. Comparison between the original Meduna's design (**a**) and the actual configuration (**b**). Focusing on the fore-stage, the proscenium of Meduna's plan was replaced by an orchestra pit; on the back-stage, some fitting rooms were dismissed in order to enlarge the stage.

2.2. Measurements and Simulation Set-Up

On the basis of available plans and sections of the present state of the Alighieri Theatre, a geometrical acoustic model was set up for numerical simulation purposes. This model was calibrated according to the acoustic parameters extracted from the Room Impulse Responses (RIRs), which have been measured during the acoustic measurements in September 2014 and vibro-acoustic measurements on the stage in November 2014.

The workflow of acoustic measurements followed the ISO 3382-1 [14] recommendations. Three positions of sound sources were used in the stage (two) and inside the orchestra pit (one), following the recommendation of the Ferrara charter [13]. RIRs were measured for each sound source in each position of stalls, boxes and gallery, for a total amount of about 2500 measurements [10]. Exponential Sine Sweeps (ESS) were used as excitation signals through a custom high-SPL sound source [44]. The excitation signal was 128 K length with a sample rate of 48 kHz, allowing for measuring IRs that have enough signal-to-noise ratio in each octave band. A proprietary software was used in order to optimise the time-windowing of the ESS and, consequently, the frequency response of the measured impulse responses [45].

2.3. Assumptions and Numerical Calibration

In order to understand the acoustic improvements generated by the last renovations and to investigate the acoustic behavior of the original design—the one with the *proscenium*—a geometrical model of it was created for numerical simulation aims. In Table 3, geometrical data of both acoustic models are reported.

The acoustic simulations were provided by hybrid ray tracing software Odeon v.12 (Lyngby, Denmark). The software uses two different calculation methods for early and late reflections and it switches from one to the other according to the transition order (TO) that the user indicates. The early reflections calculation is carried out with the deterministic image source method, while the late reflections calculation is driven by the statistical ray tracing method. A transition order equal to 2 was used. The number of late rays used in the calculations was enough to allow an adequate accuracy, according to the investigation of the reflection density parameter that should be higher than at least 25 reflections/ms for reliable results [46]. A length of the simulated impulse response of 2500 ms was selected to be greater than the estimated reverberation time. The virtual sound source was placed on the fore-stage, at 1 m from the edge; in the original design model, the sound source was placed on the proscenium, at 1 m from the edge. Twelve virtual receivers were placed in the stalls, eight on two column boxes—two in each tier—and two in the gallery, as suggested by a Ferrara-charter [13].

Table 3. Geometrical data of the acoustic simulation models corresponding to the original and present configurations.

Configuration	Volume (m^3)	Number of Surfaces	Number of Late Rays
Original	10,455	7680	10,584
Present	10,455	7676	10,564

The model of the theatre (see Figure 2) was calibrated according to the acoustic measurements and the reliability of the calibration process was evaluated following simulation procedures based on the previous studies [29,47]. The calibration workflow mainly consisted of an iterative process which involved the acoustic material proprieties, as absorption and scattering coefficients, and thus the selection of these values is responsible for the reliability of the simulations. A data collection for credible values of absorption coefficients was carried out based on the scientific literature on this matter, according to both the material typologies and historical context [48,49]. Some values were slightly adapted according to the specific case and conditions, e.g., elements which were simplified or even not modeled as the technical equipments in the fly tower. In fact, some values of absorption coefficients have been slightly modified in order to adapt them to the specific case of the theatre, while remaining within reasonable ranges considering comparable materials in the literature. Values of plaster inside the fly tower and hall results higher than the referenced ones due to the presence of furniture—as scenography furniture for the stage and general furniture for the hall—during the measurement session. The most significant adaption of values of absorption coefficients was done for the case of the boxes where the geometrical details as seats and objects inside were not modeled in order to avoid the increase of number of surfaces, which are not necessary for the simulation process [46]. In particular, the absorption and scattering coefficients assigned to the boxes are significantly high due to the lack of interior modeled surfaces. Considering the lack of complex small details in comparison to the main volumes, particular scattering proprieties were assigned to fix it. The scattering and absorption coefficients of all the materials involved in the geometric acoustic (GA) simulations are reported in Table 4.

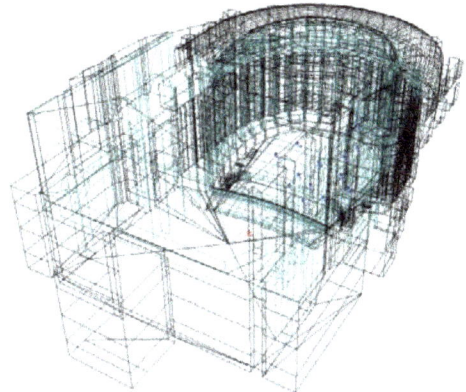

Figure 2. Wireframe of the theatre model. More details are provided in Table 3.

Table 4. Absorption and scattering (s) coefficients for all the materials involved in the simulation. The scattering coefficient is referred to 707 Hz.

Materials	Surface %	Absorption						s	Ref.
		125 Hz	250 Hz	500 Hz	1000 Hz	2000 Hz	4000 Hz		
Marble	1	0.02	0.02	0.03	0.05	0.05	0.05	0.05	[48]
Plaster (Fly Tower)	23	0.12	0.14	0.15	0.15	0.16	0.16	0.05	adapted [49]
Plaster (Hall)	16	0.08	0.10	0.10	0.10	0.10	0.07	0.05	adapted [49]
Boxes	33	0.20	0.22	0.30	0.33	0.33	0.33	0.50	adapted [49]
Wood (Stage)	10	0.18	0.12	0.10	0.09	0.08	0.07	0.05	adapted [48]
Wood (Hall)	3	0.04	0.04	0.07	0.06	0.06	0.07	0.05	[48]
Stage grid	4	0.15	0.15	0.25	0.40	0.50	0.50	0.20	adapted [10]
Wings	6	0.15	0.20	0.20	0.40	0.40	0.42	0.05	adapted [48]
Velvet	1	0.15	0.35	0.55	0.77	0.70	0.60	0.05	adapted [49]
Wooden chairs	1	0.12	0.11	0.10	0.09	0.08	0.07	0.70	[48]
Seats	2	0.40	0.55	0.70	0.80	0.80	0.70	0.70	adapted [49]

The following acoustic parameters were considered in the calibration process: Early Decay Time EDT, Reverberation Time T_{30}, Center Time t_s and Clarity C_{80} [50]. Giving the couples of sound sources/receivers used in the calibration process, the work was completed once the difference between the measured values and the simulated one was less than twice Just Noticeable Differences (JNDs). For sake of brevity, the calibration results are reported in Figure 3, where EDT and C_{80} values were plotted—for each octave band—showing for sound source the averaged values over three groups of receivers—respectively, stalls, boxes, and gallery.

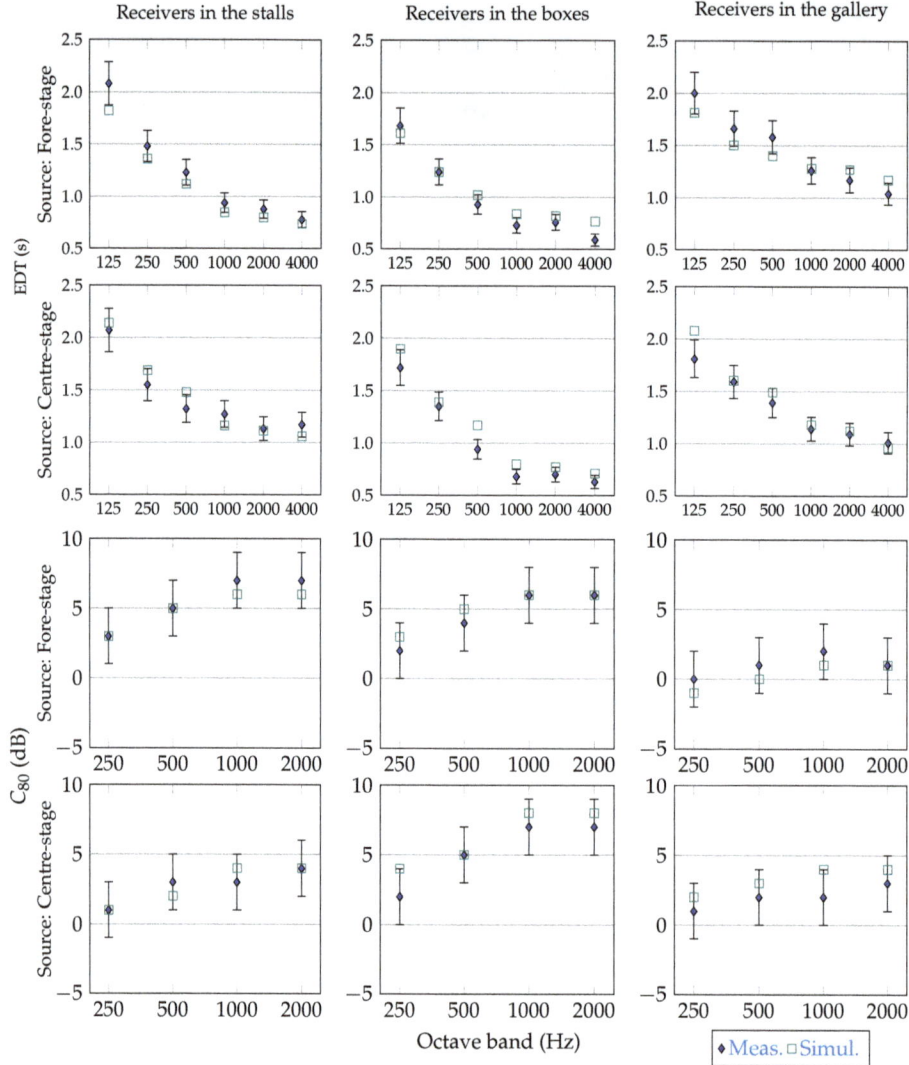

Figure 3. Calibration of Alighieri Theatre: comparison between the simulated values (Simul.) and the measured values (Meas.) of EDT and C_{80} for two significant sound sources. The values are subdivided into the three categories (stalls, boxes and gallery) and averaged over the corresponding receivers. The error bars correspond to twice the Just Noticeable Difference (JND), which was assumed as 1 dB for C_{80} and 5% of the measure value for EDT.

2.4. Vibroacoustic Measurements of the Historical Wooden Stage

A crucial aspect of the work concerns the stage. In an HOH, the stage was typically made of light wooden fir, in order to fix the scenes during operas. Any acoustic reason was taken into account by pre-Sabinian scholars regarding the stage design [18]. Previous works related to wooden stage-air cavity are purely numerical [51,52], or related to the fork-effect of cellos and double-bass only [53] or the early work of Beranek, which showed an increment of radiation in the 125 Hz octave band [54]. Instead, the present work needed vibro-acustic measurements for archeo-acoustical reasons. The question was: "Are the historical stage properties similar to the present ones?".

In 2014, the direction of the Alighieri Theatre stated the renovation of the wood of the stage and the authors were asked to study the effects of wood aging on stage acoustics. The original stage was made of fir boards, arranged in square blocks, measuring 17.30 m by 17.30 m and sloping "nel rapporto di vent'uno di base per uno di altezza" (the ratio is 21/1 in height, corresponding to a slope of 2.6 degrees). Samples of not-aged and two-year aged firs were analyzed in a laboratory and compared with a historical sample of the stage.

A set of measurements was carried out in the laboratory to determine several relevant parameters related to the acoustical quality of the wood. In particular, density, wave propagation speed and elastic modulus have been measured and calculated. Furthermore, considering these characteristics, it is possible to define other quantities useful to compare the acoustical properties of different materials: the quasi-longitudinal wave speed in beams, the characteristic impedance, the sound radiation coefficient and the loss factor were determined for three specimens. In order to compare the measured and the predicted data, propagation velocities and elastic modulus were measured in all of the directions, by using mechanical impulses and accelerometers: the equipment included a signal conditioner Brüel and Kjær NEXUS, two Piezoelectric charge accelerometer Brüel and Kjær 4371V, an instrumented impact hammer APTech AU02 (Bombay, India) and a calibrator Brüel and Kjær 4294. Figure 4 shows how the measurements were performed on the samples: two accelerometers were fixed at a reference distance d and impulses were generated with an impact hammer along the longitudinal axis (F_y), the tangential axis (F_x) and the radial axis (F_z). The signal acquired from the accelerometers and the impact hammer were therefore analyzed in order to calculate the wave propagation speed considering the time delay from the two signals. A Matlab code (The MathWorks, Boston, MA, USA) was then created to calculate the wave speed c_{ly} in the longitudinal direction with the impulse generated in the longitudinal direction. Similarly, the quasi-longitudinal wave speed c_{lx} in the tangential direction with the impulse generated in the tangential direction was determined. The values of the different wave speeds obtained with the method described are summarized in Table 5.

Table 5. Results of laboratory measurements on three wood specimens. ρ is the density, measured by weighting each specimen; c_{ly} is the quasi-longitudinal phase velocity measured in the y-direction; c_{lx} is the quasi-longitudinal phase velocity measured in the x-direction; c_{by} is the bending phase velocity in the y-direction; c_{bx} is the bending phase velocity measured in the x-direction; the elastic moduli in the longitudinal and tangential directions are, respectively, E_l and E_t, done by Equations (4) and (5); z is the modulus of characteristic impedance, done by Equation (2); R is the Radiation coefficient, done from Equation (3).

Beams	ρ kg/m^3	c_{ll} m/s	c_{tt} m/s	c_{lr} m/s	c_{tr} m/s	E_l GPa	E_t GPa	z 10^{3}kg/(m^{2}s)	R m^4/(kg s)
not-aged	390	3000	920	900	260	3.4	0.34	1170	7.57
two-year aged	400	4570	870	1000	340	8.1	0.30	1828	11.2
historical	450	5300	540	800	230	12.8	0.13	2385	11.85

The wave propagation speed along different axes was measured in order to compare the three samples' characteristics. With the obtained values, it was possible to determine the dynamic elastic modulus E from Equation (1), the characteristic impedance z and the sound radiation coefficient R from Equations (2) and (3), respectively. The sound velocity is defined as the root square of the ratio between elastic modulus E and density ρ:

$$c = \sqrt{E/\rho} \quad (m/s). \tag{1}$$

The characteristic impedance is defined as a product of sound velocity c and density ρ:

$$z = \rho c \quad (kg/(m^2 s)) \tag{2}$$

and, finally, the radiation coefficient R is defined as the square root of ration between the elastic modulus E and cubic density ρ^3:

$$R = \sqrt{E/\rho^3} \quad (m^4/(kgs)). \tag{3}$$

These quantities allowed for analyzing the acoustic properties of the different samples and comparing them with the ones found in other studies. From these values, it is possible to calculate the elastic modulus in longitudinal and tangential directions, respectively E_l and E_t [55]:

$$E_l = \rho c_l^2 \quad (GPa), \tag{4}$$

$$E_t = E_l \left(\frac{c_t}{c_l}\right)^2 \quad (GPa). \tag{5}$$

See Table 5 for the values measured in the specimens. The main result is that the radiation properties of the two-year aged specimen were quite similar to the historical one, in the first approximation of the used model.

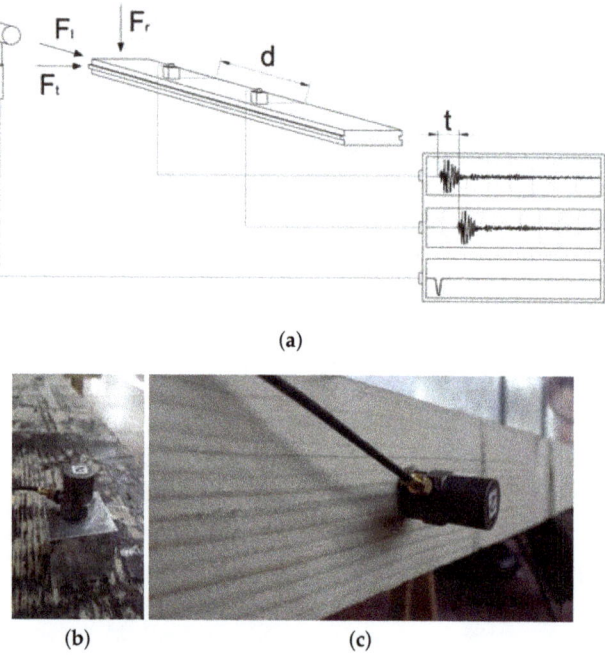

Figure 4. Procedure of laboratory vibro-acoustic measurements. Layout of the propagation velocity measurement (**a**). F_l, F_t e F_r are, respectively, longitudinal, tangential and radial mechanical impulses, forced by an instrumental hammer; accelerometers placed on the historical beam (**b**) and not-aged beam (**c**).

Moreover, vibration measurements were carried out on the stage of the Alighieri Theatre in November 2014, see Figure 5. Using the instrumented hammer as the source, impulse responses were recorded by accelerometers. Then, the measured impulse responses were evaluated two at a time, taking into account the disposition and the distance. The velocities extracted were statistically analyzed through histograms of occurrences: only the measurements with the receivers aligned are plotted in Figure 5c. The mean value of the measured velocities on the stage is about 770 m/s and this value is comparable with the laboratory measurement value of 800 m/s, done for historical fir in Table 5.

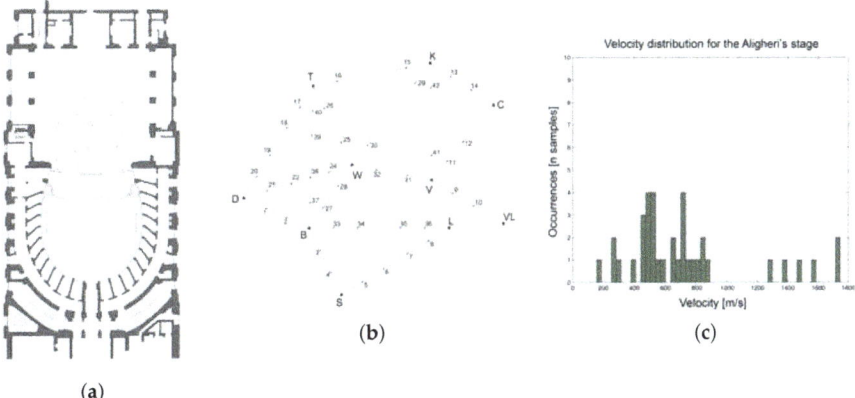

Figure 5. Procedure of in situ vibro-acoustic measurements. Plan of the stage of the Alighieri Theatre (**a**). Details on the source and receiver positions (**b**). Statistics of the structural sound velocities (occurrences vs. velocities) were measured on the stage for couples of aligned sources-receivers only (**c**).

The in situ measured vibro-acoustic impulse responses were also analysed in frequency. Figure 6 shows the amplitude of the modulus of the Fourier Transform of the measured velocities, which were normalised, with each curve plot a source–receiver couple. The maxima are in the range 70–100 Hz and this aspect was considered in the calibration process, by increasing the absorption of the wooden stage in the octave band of 125 Hz. Since the voice of a soloist shows a negligible contribution in this frequency range, the vibro-acoustic behaviour of the stage/proscenium may be considered negligible in all other aspects of the present study.

Finally, it is interesting to note that the acoustic properties of the two-year aged specimen are close to the ones of the historical specimen. It follows that only aged wooden boards should be used in the stage refurbishment; at the same time, the renewed stage can 'play' like the former one, from a vibro-acoustic point of view. It should be noted that the present works concerned fir wood only, which is the most used wood essence for the stage.

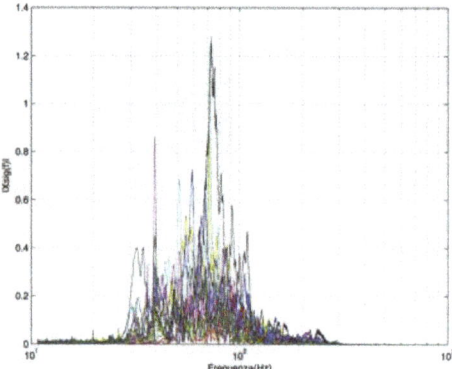

Figure 6. Frequency responses of the Alighieri Theatre's wooden stage, with each curve plot a source–receiver couple (see Figure 5b). Frequency on the abscissa axis (logarithmic scale), modulus of the Fourier Transform of the measured velocities on the ordinate axis (linear scale).

3. Results and Discussion

Table 6 compares the results of simulations, averaged for each sound source over three groups of receivers: stalls, boxes and gallery. A sound source with omnidirectional directivity was used in the first part of the simulations, in which sound strength at mid frequencies G_M and sound clarity $C_{50,3}$ values were simulated. A temporal threshold of 50 ms was used for sound clarity because of its meaning for HOH [56]. In the second part of the simulations, a directional sound source—corresponding to a "Soprano" directivity—was used in order to evaluate the speech transmission index STI.

Table 6. Measured values, spatially and spectrally averaged. The subscript "M" means the value was averaged over the octave bands of 500–1000 Hz. The subscript "3" means the value was averaged over the octave bands of 500–1000–2000 Hz. Speech Transmission Index (STI) values were simulated considering an infinite signal-to-noise ratio value.

		G_M (dB)	$C_{50,3}$ (dB)	STI
Present (without proscenium)	Stalls	7.6	6.0	0.65
	Boxes	1.5	2.9	0.63
	Gallery	4.4	−1.4	0.56
Original (with proscenium)	Stalls	10.0	1.2	0.61
	Boxes	2.5	0.8	0.57
	Gallery	4.7	−2.6	0.55

With the proscenium (original design), the sound of soloist is louder; in fact, the sound strength at mid-frequencies of the original configuration is higher than the present one: 10 dB instead of 7.6 dB in the stalls, 2.5 dB instead of 1.5 dB in the boxes; values in the gallery are quite similar. It means that the proscenium influences the direct sound and, to a lesser extent, the early reflections. In fact, the differences in stalls are higher than the ones in boxes, whereas, in the gallery, the direct sound contribution is negligible and the difference is due to early reflections only [10]. Sound strength values in the stalls are plotted in Figure 7. It is interesting to note that sound clarity values span from −1.4 to 6 dB in the present configuration, whereas it is from −2.6 to 1.2 dB in the original design. On one hand, it means that now the intelligibility is higher than the one in the original design, which is also confirmed by STI simulations. On the other hand, the timbre of the voice in the original design seems to be more 'balanced' in frequency than the present configuration. In an HOH, the sound clarity may mean, in first approximation, how the room can 'equalise' the voice. In fact, if the C_{50} value is high (e.g., 5–8 dB), the direct field is reinforced by the early reflections that have plenty of high frequencies, due to low absorption at high frequencies of the smooth wall surfaces surrounding the stalls, the ceiling and the proscenium arch. The voice timbre may be too 'brilliant' but poor of 'body'. In other words, it can be full of harmonics in the formant region (the octave band of 4000 Hz) but poor of energy in the octave band of 250 Hz and below. Otherwise, if the C_{50} value is close to 0 dB, the voice timbre can be appear to be more 'balanced': the energy of early reflections is attenuated and the 'body' of the voice is enhanced by the energy of the subsequent reverberation.

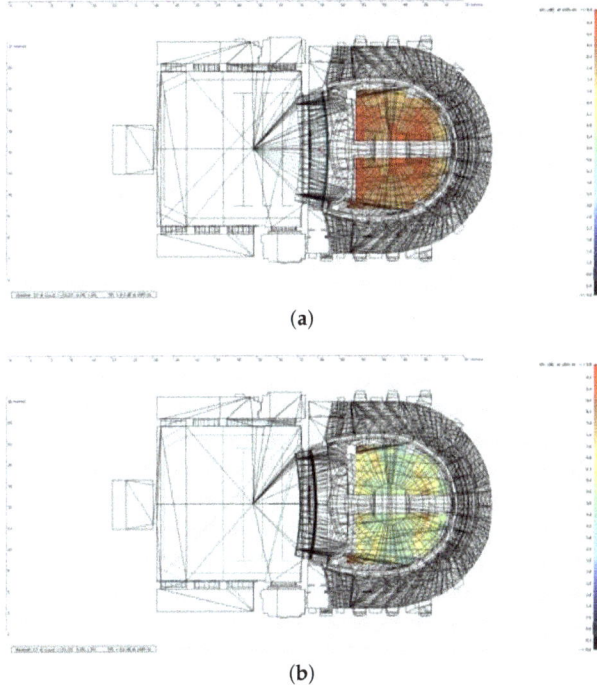

Figure 7. G values (in dB) at 1000 Hz simulated in the stalls with in the original (**a**) and in the present configuration of the theatre (**b**).

Furthermore, the analyses of spatial distribution of the sound strength in Figure 8 show further differences between the soloist on the proscenium of the original design, and the soloist in the fore-stage of the present layout. The Barron's "revised theory" of the semi-reverberant sound field [57] gives a relationship between the expected sound strength value at distance r from the source, $G(r)$ (in dB), the volume of the hall, V (in cubic metres) and the mean of the reverberation time measured in the hall, T (in seconds), as follows:

$$G(r) = 10 \log \frac{100}{r^2} + 31{,}200 \frac{T}{V} e^{-0.04r/T} \quad \text{(dB)}. \tag{6}$$

In the original Barron's work, the hall under study was the concert hall, without coupled volumes. In case of the opera house, there were several coupled volumes that should be taken into account: the volume of the main hall V_{hall}, the volume fo the fly tower $V_{flytower}$ and also the volume of the orchestra pit, the volume of the boxes, etc. Previous studies considered the volume of the hall only [11] or, more recently, the 'effective volume' V', which was preliminary studied in the case of churches [58] and opera houses [59].

Considering the receiver in stalls only, the spatial distribution of sound energy in the original design shows values near the Barron's Equation (6) with the effective volume V' is equal to:

$$V' = V_{hall}. \tag{7}$$

When the sound source is placed on the fore-stage, the values—in the stalls—are closer to the lowest curve, in which Barron's Equation (6) considered the effective volume equal to the whole volume of the opera house, i.e., the volume of the hall and the volume of the fly tower:

$$V'' = V_{hall} + V_{flytower}. \tag{8}$$

These results are in agreement with the ones of a survey of over eleven HOHs [59]: when the sound source is on the proscenium, only the volume of the hall contributes to the reverberation; when the sound source is moved back, i.e., when the proscenium was 'cut' from the stage, both the volume of the stage house and the volume of the hall may be taken into account, the weight of each depending on the sound source position between the two coupled volumes.

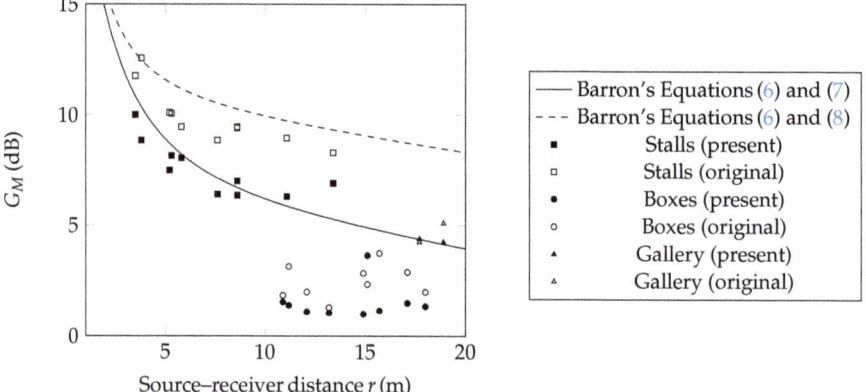

Figure 8. Simulated G_M values, considering the present layout and the original design. Receivers in stalls (squares), in boxes (circles) and in the gallery (triangles). Simulated values are compared with two analytical prediction curves: Barron's Equation (6) which takes into account the hall volume only V' (see Equation (7)); Barron's Equation (6), which takes into account the whole volume V'' (hall and fly-tower), see Equation (8).

The sound strength of receivers in the boxes did not vary with the distance because of the 'dominant' volume of the boxes. For the receivers in the boxes, the changes in sound strength values may depend on the distance only: higher values for the receiver in the central zones of the lower tiers of boxes, lower values for the receivers in the lateral zones of the highest tiers of boxes.

Figure 8 shows that, when the soloist was on the proscenium, the acoustics of the voice depended on the volume of the hall only. It follows that the audience felt a louder voice and a more balanced timbre.

4. Conclusions

In the early decades of the 20th Century, almost all the stages of the Italian theatres were modified in order to include an orchestra pit. The front of the stage, the so-called proscenium, was cut about 2–3 m, depending on the theatre. This fact drastically changed the acoustics of the theatre: the singers had to draw back from their position during the aria.

A virtual simulation of an HOH original configuration, the Alighieri Theatre in Ravenna, helps to quantify these differences. Acoustic measurements were done in the theatre and a simulation model was calibrated using room criteria extracted from these measurements. Another model of the theatre was simulated, including the proscenium and removing the orchestra pit. The analyses show that historical stage is quite similar to the two-year aged wood, so we can assume that the historical stage may be modeled as the present one, which was 'calibrated' with the acoustic measurements in the theatre. Results show that the proscenium influences the loudness of the voice, increasing the sound strength and enhancing the tonal balance, while decreasing the sound clarity. Further developments of this approach may be extended to subjective analyses, e.g., by MIMO (Multiple-Input-Multiple-Output) auralisations, using instrumental directivity and multitrack anechoic recordings [60].

Author Contributions: Conceptualization and methodology, D.D., M.G.; Investigation, D.D., Simulation, A.R.; Writing—Original Draft, D.D. and A.R.; Revision, D.D., A.R. and M.G.

Funding: This research received no external funding.

Acknowledgments: The authors would like to thank Antonio De Rosa, Davide Mazzavillani and the technical staff of the Alighieri Theatre.

Conflicts of Interest: The authors declare no conflict of interest.

References and Note

1. Vecco, M. A definition of Cultural Heritage: From the tangible to the intangible. *J. Cult. Herit.* **2010**, *11*, 321–324. [CrossRef]
2. Đorđević, Z. Intangible tangibility: Acoustical heritage in architecture. *Struct. Integr. Life* **2016**, *16*, 59–66.
3. Fausti, P.; Pompoli, R.; Prodi, N. Acoustics of opera houses: A cultural heritage. *J. Acoust. Soc. Am.* **1999**, *105*, 929. [CrossRef]
4. Tronchin, L.; Farina, A. Acoustics of the Former Teatro—La Fenice—In Venice. *J. Audio Eng. Soc.* **1997**, *45*, 1051–1062.
5. Iannace, G.; Ianniello, C.; Maffei, L.; Romano, R. Objective measurement of the listening condition in the old Italian opera house "Teatro di San Carlo". *J. Sound Vib.* **2000**, *232*, 239–249. [CrossRef]
6. Farina, A. Acoustic quality of theatres: Correlations between experimental measures and subjective evaluations. *Appl. Acoust.* **2001**, *62*, 899–916. [CrossRef]
7. Cammarata, G.; Fichera, A.; Pagano, A.; Rizzo, G. Acoustical prediction in some Italian theatres. *Acoust. Res. Lett. Online* **2001**, *2*, 61–66. [CrossRef]
8. Astolfi, A.; Bortolotto, A.; Filippi, A.; Masoero, M.; Pisani, R. Acoustical characterisation of small Italian opera house. In Proceedings of the Forum Acusticum, Budapest, Hungary, 29 August–2 September 2005; pp. 2301–2306.
9. Pompoli, R. (Ed.) *Proc. Teatri d'Opera dell'Unità d'Italia (Proceedings Opera houses of the Italian Unification)*; Acoustical Society of Italy (AIA): Venezia, Italy, 2011. (In Italian)
10. Garai, M.; Morandi, F.; D'Orazio, D.; De Cesaris, S.; Loreti, L. Acoustic measurements in eleven Italian opera houses: Correlations between room criteria and considerations on the local evolution of a typology. *Build. Environ.* **2015**, *94*, 900–912. [CrossRef]
11. Prodi, L.; Pompoli, R.; Martellotta, F.; Sato, S. Acoustics of Italian Historical Opera Houses. *J. Acoust. Soc. Am.* **2015**, *138*, 769–781. [CrossRef]
12. Weinzierl, S.; Sanvito, P.; Schultz, F.; Büttner, C. The acoustics of renaissance theatres in Italy. *Acta Acust.* **2015**, *101*, 632–641. [CrossRef]
13. Pompoli, R.; Prodi, N. Guidelines for acoustical measurements inside historical opera houses: Procedures and validation. *J. Sound Vib.* **2000**, *232*, 281–301. [CrossRef]
14. *ISO 3382-1: Acoustics—Measurement of Room Acoustic Parameters. Part 1: Performance Spaces*; ISO: Geneva, Switzerland, 2009.
15. Pisani, R.; Duretto, F. Il restauro ed i problemi di acustica dei teatri storici (The restoration and the acoustics problems in historical theatres). In Proceedings of the XXVII National Conference of the Italian Acoustics Association (AIA), Genova, Italy, 26–28 May 1999.

16. Prodi, N.; Pompoli, R. Acoustics in the restoration of Italian historical opera houses: A review. *J. Cult. Herit.* **2016**, *21*, 915–921. [CrossRef]
17. Prodi, N. From Tangible to Intangible Heritage inside Italian Historical Opera Houses. *Heritage* **2019**, *2*, 826–835. [CrossRef]
18. D'Orazio, D.; Nannini, S. Towards Italian Opera Houses: A Review of Acoustic Design in Pre-Sabine Scholars. *Acoustics* **2019**, *1*, 252–280. [CrossRef]
19. Wagner, R. The Opera House at Bayreuth. In *Selected from His Writings and Translated*; Wagner, R., Ed.; H. Holt and Company: New York, NY, USA, 1875; pp. 255–288.
20. Ledoux, C.N. *Architecture Considérée Sous le Rapport de l'Art des Moeurs et de la Législation*; Porroneau: Paris, France 1804.
21. Carnegy, P. *Wagner and the Art of the Theatre*; Yale University Press: New Haven, CT, USA, 2006.
22. Müller, K. Die Demokratisierung der Akustik (The democratization of acoustics). In *Das Richard Wagner Festspielhaus Bayreuth*; Kiesel, M., Ed.; Nettpress: Düsseldorf, Germany, 2007; pp. 174–197.
23. Stefani, D. L'orchestra del Teatro alla Scala nella prima metá dell'800 Organizzazione, Funzioni, Gerarchie. Ph.D. Thesis, Politecnico di Milano, Milan, Italy, 2015.
24. Lautenbach, M.R.; Vercammen, M.L.S.; Lorenz-Kierakiewitz, K.H. Acoustic Aspects of Stage and Orchestra Pit in Opera Houses. In Proceedings of the 38th German Acoustical Society (DAGA), Darmstadt, Germany, 19–20 March 2012.
25. Vercammen, M.L.S.; Lautenbach, M.R. Stage and Pit Acoustics in Opera Houses. In Proceedings of the International Symposium on Music and Room Acoustics (ISMRA), La Plata, Argentine, 11–13 September 2016.
26. Disposizione scenica per l'opera "Otello", dramma lirico in quattro atti, versi di Arrigo Boito, musica di Giuseppe Verdi, compilata e regolata secondo la messa in scena del Teatro alla Scala Milano, Ricordi, 1887. (In Italian)
27. Reggiani, L. Il Teatro Alighieri di Ravenna: Rilevazioni Acustiche e Modelli Numerici per la Conservazione, la Ricerca Musicologica e l'Auralizzazione di Futuri Allestimenti. Master's Thesis, University of Bologna, Bologna, Italy, 2011.
28. Rinaldi, M. Vibro-Acoustic Measurements of Wooden Stages in Historical Italian Theatres. Master's Thesis, University of Bologna, Bologna, Italy, 2015.
29. Vorländer, M. Computer simulations in room acoustics: Concepts and uncertainties. *J. Acoust. Soc. Am.* **2013**, *133*, 1203–1213. [CrossRef] [PubMed]
30. Vissilantopoulos, S.; Mourjopoulos, J. Virtual acoustic reconstruction of ritual and public spaces of ancient Greece. *Acta Acust. United Acust.* **2001**, *87*, 604–609.
31. Katz, B.F.G.; Wetherill, E. Fogg art museum lecture room: A calibrated recreation of the birthplace of room acoustics. In Proceedings of the Forum Acusticum, Budapest, Hungary, 29 August–2 September 2005.
32. Weinzierl, S.; Rosenheinrich, H.; Blickensdorff, J.; Horn, M.; Lindau, A. The acoustics of the concert hall in Leipzig's Gewandhaus. History, reconstruction and auralization. In Proceedings of the Fortschritte des Akustik, Berlin, Germany, 15–18 September 2010.
33. Takala, J.; Kylliäinen, M. Comparison of modelled performance of a vanished building with historical information on its acoustics. In Proceedings of the Forum Acusticum, Krakow, Poland, 7–12 September 2014.
34. Sü Gül, Z.; Xiang, N.; Çaliskan, M. Investigations on sound energy decays and flows in a monumental mosque. *J. Acoust. Soc. Am.* **2016**, *140*, 344–355. [CrossRef]
35. Sender, M.; Planells, R.; Perello, R.; Segura, J.; Gimenez, A. Virtual acoustic reconstruction of a lost church: Application to an Order of Saint Jerome monastery in Alzira, Spain. *J. Build. Perf. Simul.* **2017**, *11*, 369–390. [CrossRef]
36. Suárez, R.; Alonso, A.; Sendra, J.J.A. Virtual acoustic environment reconstruction of the hypostyle mosque of Cordoba. *Appl. Acoust.* **2018**, *140*, 214–224. [CrossRef]
37. D'Orazio, D.; De Cesaris, S.; Morandi, F.; Garai, M. The aesthetics of the Bayreuth Festspielhaus explained by means of acoustic measurements and simulations. *J. Cult. Herit.* **2018**, *34*, 151–158. [CrossRef]
38. Bo, E.; Shtrepi, L.; Pelegrin Garcia, D.; Barbato, G.; Aletta, F.; Astolfi, A. The accuracy of predicted acoustical parameters in ancient open-air theaters: A case study in Syrscusae. *Appl. Sci.* **2018**, *8*, 1393. [CrossRef]
39. Postma, B.N.J.; Dubouilh, S.; Katz, B.F.G. An archeoacoustic study of the history of the Palais du Trocadero. *J. Acoust. Soc. Am.* **2019**, *145*, 2810–2821. [CrossRef] [PubMed]

40. Đorđević, Z.; Novković, D.; Andrić, U. Archaeoacoustic Examination of Lazarica Church. *Acoustics* **2019**, *1*, 423–438. [CrossRef]
41. Meduna, T.; Meduna, G. *Il Teatro La Fenice in Venezia. . . (On the La Fenice Theatre. . .)*; Antonelli: Venezia, Italy, 1849. (In Italian)
42. Cocchi, A.; Garai, M.; Tronchin, L. *Influenza di cavità risonanti poste sotto la fossa orchestrale: Il caso del teatro Alighieri di Ravenna (The Influence of Resonating Cavities under the Orchestra Pit: The Case of the Alighieri Theatre in Ravenna)*; Nardini Editore: Firenze, Italy, 1997; 135–153. (In Italian)
43. Sarasini, F. Il teatro Alighieri di Ravenna. Storia e restauri dalla fondazione ad oggi. *Ravenna Studi e Ricerche* **2005**, *12*, 165–199. (In Italian)
44. D'Orazio, D.; De Cesaris, S.; Guidorzi, P.; Barbaresi, L.; Garai, M.; Magalotti, R. Room acoustic measurements using a high SPL. In Proceedings of the 140th Audio Engineering Society Convention (AES), Paris, France, 4–7 June 2016.
45. Guidorzi, P. Barbaresi, L.; D'Orazio, D.; Garai, M. Impulse responses measured with MLS or Swept–Sine signals applied to architectural acoustics: An in-depth analysis of the two methods and some case studies of measurements inside theaters. *Energy Procedia* **2015**, *78*, 1611–1616. [CrossRef]
46. Christensen, C.L.; Koutsouris, G.; Gil, G. *ODEON Room Acoustics Software, Version 12, User Manual*; ODEON A/S: Lyngby, Denmark, 2013.
47. Postma, B.N.J.; Katz, B.F.G. Perceptive and objective evaluation of calibrated room acoustic simulation auralizations. *J. Acoust. Soc. Am.* **2016**, *140*, 4326–4337. [CrossRef] [PubMed]
48. Vorländer, M. *Auralization: Fundamentals of Acoustics, Modelling, Simulation, Algorithms and Acoustic Virtual Reality*; Springer Science & Business Media: Berlin/Heidelberg, Germany, 2007.
49. Cox, T.J.; D'Antonio, P. *Acoustic Absorbers and Diffusers: Theory, Design and Application*, 3rd ed.; CRC Press: Boca Raton, FL, USA, 2004.
50. Koutsouris, G.; Norgaard, A.K.; Christensen, C.L. Discretisation of curved surfaces and choice of simulation parameters in acoustic modeling of religious spaces. In Proceedings of the 23rd International Congress on Sound and Vibration (ICSV), Athens, Greece, 10–14 July 2016.
51. Nakanishi, S.; Sakagami, K.; Daido, M.; Morimoto, M. Acoustic properties of a cavity backed stage floor: A theoretical model. *Appl. Acoust.* **1999**, *57*, 17–27. [CrossRef]
52. Augusztinovicz, F. Vibroacoustic analysis of the stage floor of a concert hall—A case study. *Appl. Acoust.* **2012**, *73* 648–658. [CrossRef]
53. Guettler, K.; Askenfelt, A.; Buen, A. Double basses on the stage floor: Tuning fork-tabletop effect, or not? *J. Acoust. Soc. Am.* **2012**, *1*, 795–806. [CrossRef] [PubMed]
54. Beranek, L.; Johnson, F.R.; Shultz, T.J.; Walters, B.G. Acoustics of Philharmonic Hall, New York, during its first season. *J. Acoust. Soc. Am.* **1964**, *36*, 1247–1262. [CrossRef]
55. Cremer, L.; Heckl, M.; Petersson, B.A.T. *Structure-Borne Sound—Structural Vibrations and Sound Radiation at Audio Frequencies*, 3rd ed.; Springer: Berlin/Heidelberg, Germany, 2005.
56. De Cesaris, S.; Morandi, F.; Loreti, L.; D'Orazio, D.; Garai, M. Notes about the early to late transition in Italian theatres. In Proceedings of the 22nd International Congress on Sound and Vibration (ICSV), Florence, Italy, 12–16 June 2015.
57. Barron, M.; Lee, L.J. Energy relations in concert auditoriums. *J. Acoust. Soc. Am.* **1988**, *84*, 618–628. [CrossRef]
58. D'Orazio, D.; Fratoni, G.; Garai, M. Acoustics of a chamber hall inside a former church by means of sound energy distribution. *Can. Acoust.* **2017**, *45*, 7–16.
59. Garai, M.; De Cesaris, S.; Morandi, F.; D'Orazio, D. Sound energy distribution energy distribution in Italian Historical Theatres. In Proceedings of the 22nd International Congress on Acoustics (ICA), Buenos Aires, Argentine, 5–9 September 2016.
60. D'Orazio, D.; De Cesaris, S.; Garai, M. Recordings of Italian opera orchestra and soloists in a silent room. *Proc. Mtgs. Acoust.* **2016**, *28*, 015014.

© 2019 by the authors. Licensee MDPI, Basel, Switzerland. This article is an open access article distributed under the terms and conditions of the Creative Commons Attribution (CC BY) license (http://creativecommons.org/licenses/by/4.0/).

Article

Historic Approaches to Sonic Encounter at the Berlin Wall Memorial

Pamela Jordan

Amsterdam Center for Ancient Studies and Archaeology (ACASA), Department of Archaeology, University of Amsterdam, 1012 XT Amsterdam, The Netherlands; p.f.jordan@uva.nl

Received: 1 April 2019; Accepted: 4 July 2019; Published: 16 July 2019

Abstract: Investigations of historic soundscapes must analyze and place results within a complex framework of contemporary and past contexts. However, the conscious use and presentation of historic built environments are factors that require more deliberate attention in historic soundscape analysis. The following paper presents a multimodal research methodology and promising preliminary results from a study at the Berlin Wall Memorial in Berlin, Germany. Here, the historic context from the Wall's recent past is presented within the surroundings of the contemporary unified capital city. The study approached the past soundscape and present site by combining historic and current-conditions research, linking archival research, conditions assessments via binaural recording and psychoacoustics analysis tools, and soundscape surveys rooted in standardized soundscape research practices. In so doing, archival textual and pictorial sources provided a rich source of primary information integrated within the study and are suggested as a resource for similar inquiries elsewhere. The investigation identified concerns specific to heritage sites that require critical consideration for historic soundscape research of the recent past—survey-participant composition and the problematized use of typical descriptors in soundscape surveys are the two concerns that are discussed. Some standardized soundscape terminology and research methodologies were found to be insufficient in historic contexts. Initial qualitative results from the research are presented as a proof of concept for the research approach with signposts for future analysis and developments.

Keywords: historic soundscapes; Berlin Wall; archives; soundscape survey; memorial; architectural conservation; sound mapping

1. Introduction

Historic soundscape research is expectedly intricate due to its focus on past, seemingly intangible or vanished sensorial conditions. But for many such projects, the investigation also requires an additional level of attention due to the official designation of the site as historic. When a research setting is recognized by a community or society as a location of important cultural heritage, due consideration must be paid to how it is and has been presented: through changes in the interpretation of physical fragments for visitors, in (changing) conservation decisions through time, and in the active presentation choices of parts or the whole as specifically representative of historic events, experiences, or cultural significances. The historic context of heritage sites should not be assumed to be a passive condition but rather the result of choices by a select group of stakeholders with specific results in mind, be they pedagogical or recreational, that are then upheld by a user group that buys into and supports historic designation. Historic sites do not simply "happen". The many factors that construct such locations all affect the built environment as well as the soundscape. Beyond simply recognizing changes in acoustic properties and sound sources over time, the influence of intention and expectation on perception is also an important factor for historic soundscapes [1,2]. Such physical, organizational and personal factors form an important contextual layer that is not addressed in standard soundscape investigation practices and is rarely addressed directly in soundscape research [3,4]. Sonic elements have been previously acknowledged as potentially defining and unifying for a community through time [5,6]; however, the particular agency of acknowledged historic spaces—and the vital layers of additional experiential, cultural and administrative context these deliver—requires more direct attention as soundscape investigation continues to be standardized.

One way to situate an historic level of consideration for soundscape purposes is to compare the research requirements at sites created in different time periods. Research may draw from multiple devices in the soundscape toolbox, including contemporary soundwalks, measurements, interviews with varying approaches, historic documentation sources, and even reconstructions of past physical conditions. The investigatory approach where existing building remains are supported by limited original contextual information (such as an ancient building complex [7,8]) would necessarily require alternate source material and methodologies from research at a site where living witnesses to significant history still are able to provide recollected testimony to past sonic conditions. Historic or event ancient documentation may exist and give relevant information concerning past sonic environments. Such evidence can be secondary to the document's original intentions; in such cases, the origin and author of the document must be interrogated as to their motivations, societal licenses and reliability. Such distinctions are particularly important towards acknowledging whom is arranging the pieces in constructing historic narrative of place and thus whom is being given voice in presenting information about "the" past [9]. Considering soundscape in historic locations also depends on the depth of historical information available concerning sonic aspects and what roles the soundscape played in the historic past.

The variability in contextual information sources, as well as the required research methodologies for investigating them, underscores the particular difficulty to researching soundscapes within overt cultural heritage contexts: a standardized approach based on contemporary settings alone may be insufficient. This was the starting position of the presented research project, which augmented a standardized soundscape research methodology at the Berlin Wall Memorial (hereafter: Memorial) with the addition of heritage-inflected components. The purpose was to test whether the heritage context would change the responses of soundwalk participants by exploring associative meaning related to history. Would participants link the soundscape of the present as somehow meaningful and/or representative of past conditions? One avenue of inquiry was to trace whether and how participant impressions changed with the introduction of historic contextual information and narratives. Participants were first asked to analyze the sensory environment at each study location without any input from the soundwalk guide to gather their first impressions; later, some information about the historic sonic environment was recounted, after which participants were again asked to note their impressions of the sonic environment.

It was also a goal of the investigation to observe how participants responded to the heritage concepts integrated as descriptors within the public soundwalk and survey portion of the study—such as "authentic" and "significant"—and to see if any patterns arose. The resulting observations were considered a proof of concept: namely, that historic context indeed plays a role in people's perceptions of the soundscape. As the paper will show, the complexity of this role remains open for more extensive study within soundscape research; the methods utilized at the Memorial illustrate some requirements and potentials of multimodal historic soundscape work with specific regard to memorials of recent history, that is, heritage locations of the recent past with an overtly pedagogical thrust to their presentation and management [10]. Such sites afford the opportunity to combine tools available to soundscape researchers with new approaches informed by historic and architectural research. The methodology explored at the Memorial specifically demonstrates the potential of integrating archival material as a primary source of historic soundscape information. It also validates the applicability of systematic psychoacoustic mapping of the current conditions as a way to corroborate qualitative analysis. Lastly, the work highlights the strong potential for participant-based field surveys while also demonstrating the need for augmenting such standardized tools depending on the historic context and suggestions for such changes.

2. Materials and Methods

2.1. The Memorial Today

From 1961 to 1989, the Berlin Wall encircled West Berlin, following a 140 km/87 mi path that wove through neighborhoods, fields, and dense urban fabric alike. Since reunification, most of the large Wall portions have been removed (though a close inspection reveals that much of the associated infrastructure and non-structural traces persist [11]). One of the most intact segments to survive still stands along Bernauer Straße, a two-lane residential thoroughfare that links northern neighborhoods with the main train station and the central portion of Berlin (see Figure 1). The portion of the Wall along Bernauer Straße and the interrelated grounds on the eastern side have been made into an outdoor national Memorial to German division, including original Wall remnants, fragments of buildings that were demolished for the Wall's construction, new interpretative and commemorative elements that occasionally include sound and video playback, a new chapel, and documentation and visitor centers. The outdoor portion of the site extends along approximately 1.4 km of Bernauer Straße and is maintained as an open, grass-laid zone within the mostly residential corridor of the city. A small portion of the Wall has been reconstructed to include the full militarized apparatus of the Wall, including the western concrete barrier wall, the border strip and patrol road known as the 'deathstrip' and eastern barrier wall. Within this segment, the reconstructed deathstrip remains physically inaccessible to the public as a silent memorial, though visitors can peer inside. The Memorial grounds adjoin the Sophien Parish cemetery to the southeast, an active cemetery with internments and distinctive gravesite tending. Bernauer Straße itself is highly transited, with electric trams passing every three to five minutes, regular vehicle, pedestrian and bicycle circulation, and steady private and public tour bus traffic, particularly during the summer season. Several smaller residential streets bisect or intersect the main road. Two underground metro stations lie within the grounds, providing dense nodes of local and tourist access as well. In 2017, approximately 956,000 individuals purposely visited the Memorial grounds, with over half from outside Germany [12]. The surrounding neighborhood has an average population density of approximately 35 individuals per square kilometer [13], contributing to substantial daily circulation throughout the study area.

Figure 1. Location of the Berlin Wall from 1961–1989, with portion along Bernauer Straße highlighted (© P. Jordan).

2.2. Sources and Methods of Historic Context Assessment

For the purposes of this study, the relevant historic soundscape conditions that once existed at the site focused on the history being honored and presented at the Memorial—this is centered almost exclusively on the period of German division (1961–1989) and directly contributing events before and since, such as the active period of "wallpecker" ("Mauerspechte") activity, when individuals chipped off souvenir pieces of the wall for months after the initial fall of the Wall in November 1989. The study aimed to find sources of information that would reveal sonic characteristics of the site that—whether momentary or consistent over time—were historically defining in some sense. Such information would include any details directly implicating the sonic environment of the past, such as physical or architectural changes contributing to site acoustics or patterns of use, or behavior that would result in certain sonic conditions becoming routine or expected. Particular moments that were indicative of broader historic sonic realities were also sought.

With the international profile of the Berlin Wall—and the Memorial specifically—there exists an abundance of material concerning local experience of the Wall during its operation as a militarized barrier, from news media to amateur photographs and personal letters. The chief source of information utilized was the archive of the Memorial Documentation Center, which holds photographic, audio-visual, and textual material related to the Wall's history with a local focus on the history of Bernauer Straße. A major contribution of primary information was also found in the living witness (Zeitzeuge) collection, which houses dozens of interviews with individuals who lived and/or worked along Bernauer Straße during division. The accounts of daily life and particular events therein were related during free-form interviews conducted primarily by employees of the Berlin Wall Foundation in the early 2000 s. Witnesses included residents on the West and East, guards and Eastern border personnel, and resident nuns in the Sophien Parish located in the West that looked out over the Wall and the Sophien Parish cemetery in the East. While not explicitly about the sonic past, valuable information was nonetheless gleaned from these interviews, such as the following excerpts (from the Zeitzeuge Collection of the Berlin Wall Foundation, translated from the original German):

> "Particularly on weekends it was such that the street [Bernauer Straße] was entirely a tourist mile. [...] The car horn choruses were awful. People would beep well into the night as a form of protest. For the locals who lived here and for the patients that I dealt with, it was, of course, catastrophic." ...

> ... "Because of this length of wall, because of this no-man's-land, the quiet allowed plants as well as animals to develop and settle in the cemetery. [...] There was also a nightingale that practically enjoyed total silence on the one side"—Nurse at Lazarus Hospital, lived on Bernauer Straße (Interview with G. Malchow, first paragraph on page 1.f, and page 24).

> "[Shouts] shattered the quiet that as a general rule prevailed. And when this quiet was interrupted by shouts, warning shots, dogs barking and the like, it was even more unusual" —Resident from eastern side of the Wall (Interview with R. Zausch, page 26).

> "[W]e couldn't see Bernauer Straße. We could only hear it and would ... go down to the street to see what was going on. Our police officers were standing down there with pistols drawn to give cover for people who were coming out [of their windows]"—Resident from western side of the Wall (Interview with K. De, page 18).

First-person narratives and perspectives like the above interviews were selected over, for instance, news broadcasts from the time or other forms of historic analysis to prioritize personal experiences. Some photographic and military documentation was occasionally used to corroborate accounts given in witness testimony (see for instance [14–16] for information and representations of the construction phases of the Wall and its surveillance elements). For instance, images of the guard dogs kept on wires, vehicles with speakers attached to the roof for protest broadcasts, crowds gathering to protest or help save fleeing East Berliners or tourists, and West Berliners gathering in large numbers to view the East on constructed platforms, all supported stories recounted in the living witnesses archive (see Figure 2).

Figure 2. Images of the Berlin Wall along Bernauer Straße from the Berlin Wall Foundation archive: (a) Guard dogs kept on wire runs within the deathstrip, 1979 (© Berlin Wall Foundation, "Hunde in Hundelaufanlage zwischen Signalzaun und Hinterlandmauer," Photo: Edmund Kaperski); (b) Lines of people using wooden platforms to peer over the Berlin Wall along Bernauer Straße into East Berlin (© Berlin Wall Foundation, "Schaulustige und Fotografen auf einer Aussichtsplattorm in der Bernauer Straße," Photo: Edmund Kaperski).

Detailed reports made by the Eastern patrol guards of Wall repairs, protest, or political incidents and escape attempts were also consulted (housed in the Federal Archives military collection (Bundesarchiv-Militärarchiv) in Freiburg). These served to expand the comparatively limited yet vital perspective of the guards who patrolled the death-strip, as well as substantiating the anecdotal accounts of certain behaviors from the Western residents' perspectives. The physical reality of the Wall could also be traced, from the initial barbed wire and hollowed-out original building facades to the large concrete slabs taken down in 1989. As evidenced in pictures and personal accounts in the archives, one of the most frequently cited and thus defining sonic elements of the Wall along Bernauer Straße was the cinching of communication between East to West. With families divided on either side, every iteration of the Wall increasingly restricted direct communication across its seized land.

Research revealed a sonic history along Bernauer Straße prior to the creation of the Memorial that could be divided into periods of essential activity:

- 1961–1962: the initial construction phase, when protests and escape attempts dominated and the initial barbed wire was continually "upgraded" until a West and East Wall enclosed what became known as the death strip;
- 1962–1965: escape attempts continued, protests were common and tourism became prevalent;
- 1965–1975: the sheared-off building facades functioning as the West Wall were replaced by concrete barriers, development on the West rendered the once commercial/residential mix into a purely residential neighborhood, and protest and tourism continued;
- 1975–1989: the "final" version of the Wall was installed, animal life flourished in the cemetery and tourism and protest diminished; and
- 1989–1990: when the Wall "fell" and wall-peckers slowly disintegrated the Wall's remains for souvenirs day and night.

Significant, defining events within the timeframes above were noted. Understanding the periods when certain sonic connections were severed and other expressions were introduced on Bernauer Straße proved vital to identifying the changing (soundscape) dynamics of this cleaved neighborhood and what role sound played as either a direct actor or an expression of those dynamics. The gathered information was then applied towards designing a soundwalk and survey with reference to as much of the historic sonic conditions as possible.

2.3. Current Context Assessment Approaches

2.3.1. Binaural Recording and Mapping

Concurrent with archival research, a survey of the current soundscape was conducted using binaural recording and psychoacoustic analysis tools as advised in standardized soundscape work [4] in order to gain a clear understanding of the average sonic conditions faced by Memorial visitors today. The equipment used for field recordings was chosen for ease of use in outdoor exposures over extended periods of time without a power source. Recordings were made using the handheld SQobold II recording system paired with binaural headset (binaural microphones worn as headphones) by HEAD acoustics GmbH; a GPS receiver was attached that linked each recording with geolocation data. Binaural recording technology logs a 360-degree fingerprint of the sonic environment that can be analyzed for physical and perceived qualities while additionally serving as a high-fidelity documentation mechanism for producing data that can serve future reference and analysis.

As the entire 1.4 km Memorial grounds are too large for comprehensive study, a portion was chosen that represented the region where the majority of tourists visit. This portion extends between the new Chapel of Reconciliation to the north and the Nordbahnhof metro station at the southern end of the Memorial (see Figure 3). Situated in the middle lies the reconstructed portion of the Wall and deathstrip, which is the primary destination for tourists and is observable from the ground or a viewing platform above.

Within this area surrounding the reconstructed deathstrip, fifty points were identified in a loose grid spread evenly across the terrain and just outside the Memorial grounds, including the adjacent cemetery, sidewalks and side streets (see Figure 4). Recordings of three-minute duration were made at each location every day for a week, always facing Bernauer Straße (depending on which side of the street the point was located, this could result in the researcher facing east or west). Over 350 recordings were made in total. Additionally, four images were made at each recording location, one image per cardinal direction, to maintain a visual record of the general contextual conditions of each recording point. Positions were confirmed via the GPS readings associated with each recording.

Figure 3. Overhead view of the study area within the Memorial. The Chapel of Reconciliation (round wooden building) sits at the top left, Nordbahnhof lies out of view to the far right and Bernauer Straße links these two along the bottom. In the center of the image, the conserved and reconstructed Wall and "deathstrip" with guard tower area can be seen. (Image © P. Jordan).

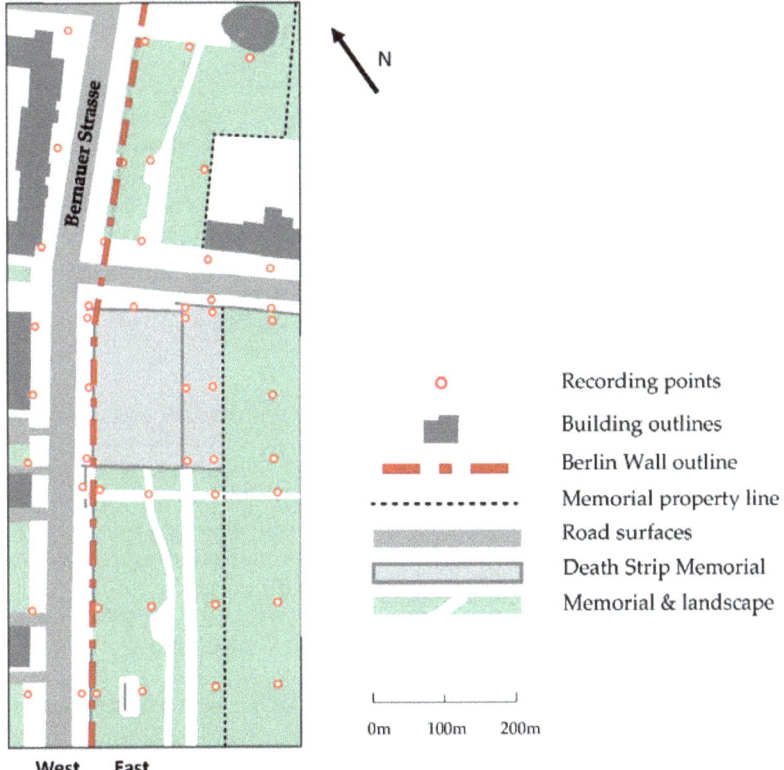

Figure 4. Sitemap of the Memorial study area with soundwalk study points highlighted. The conserved portion of the Wall and deathstrip assembly is shaded in gray at the center (© P. Jordan).

The recordings were analyzed for certain properties—an LA_{eq} analysis provided continuously measured dB(A) values at each point and a psychoacoustic Specific Loudness analysis provided the continuously measured perceived loudness in Sone (GF) at each point as well [17,18]. The values were then averaged within each recording and then across the seven days of recordings to provide representative values of a typical summer day (high visitor season). The results were mapped over a simplified rendering of the Memorial landscape to observe any unusual sonic pockets or relationships within the landscape that might have historic significance, depicted and discussed in detail in Section 3.

2.3.2. Soundwalks with Custom Surveys

The information derived from the archival research and the mapping study informed the design of a customized soundwalk and survey to investigate whether and how visitors connected the current soundscape to the history of the place. Five study points were selected that reflected, as much as possible, a diverse set of current and historical conditions, including perspectives from West residents, East residents, and guards. The following positions were identified for study (see Figure 5a–e for representative images):

a. "Mundt": Located on the west side of the Wall, today in an open area dominated by traffic noise from the adjacent intersection. Adjacent Wall fragments are not conserved and appear as they looked after the Wallpeckers finished their work in 1990. The Eastern landscape can be seen easily. Historically, this location was a bend in the original Wall where the killing of Ernst Mundt in 1962 (as he tried to escape from the East) became a political flashpoint. It was later a spot where Western residents attempted suicide by crashing their cars into the concrete wall.

b. "Deathstrip": Located on the East side of the Wall in the midst of the large grassy expanse that dominates the interpreted deathstrip landscape. Generally visually isolated from the West by the original Wall remnants at the street. Historically a part of the original Sophien Parish cemetery that was converted into a denuded area with guard dogs, tripwires, sirens and constant vehicle patrols by the East German guards.

c. "West Wall": Located on the West, today directly in front of the conserved portion of the concrete West Wall that faces the documentation center on the other side of Bernauer Straße, a major crossroads for tourist circulation and dominated by dense vehicle traffic noise refracting between non-porous surfaces (effectively creating a semi-closed urban canyon). No portion of the East can be seen except the tops of trees and buildings. Historically resembles the Wall as it appeared from 1975 onward. Part of the path along which tourist buses and cars honking in protest would travel while the Wall was in use.

d. "East Wall": Located on the East, today in the back area of the conserved portion of the Wall, dominated by the sound of crushed stone underfoot and echoes from Bernauer Straße. No part of the Western streetscape can be seen except the tops of buildings; the top of a guard tower is visible within the deathstrip area and the deathstrip itself can be viewed through gaps left in the Eastern Wall. Historically resembles the eastern portion of the Wall as it appeared in 1965 (though the aforementioned gaps are a modern intervention) and sits adjacent to the Sophien Parish cemetery that was also present throughout the Wall's historic use.

e. "Chapel": Located on the eastern edge of the Wall next to the grassy expanse of the interpreted deathstrip and facing the newly constructed (2000) Chapel of Reconciliation. The West can easily be seen and heard through a modern "Wall" interpretation. Located within the grounds of the Church of Reconciliation foundations (demolished in 1985), next to buildings where people attempted to escape from their windows in the first year of the Wall's construction and where building facades functioned as the Wall itself for many years before being replaced with concrete slabs in the 1970s.

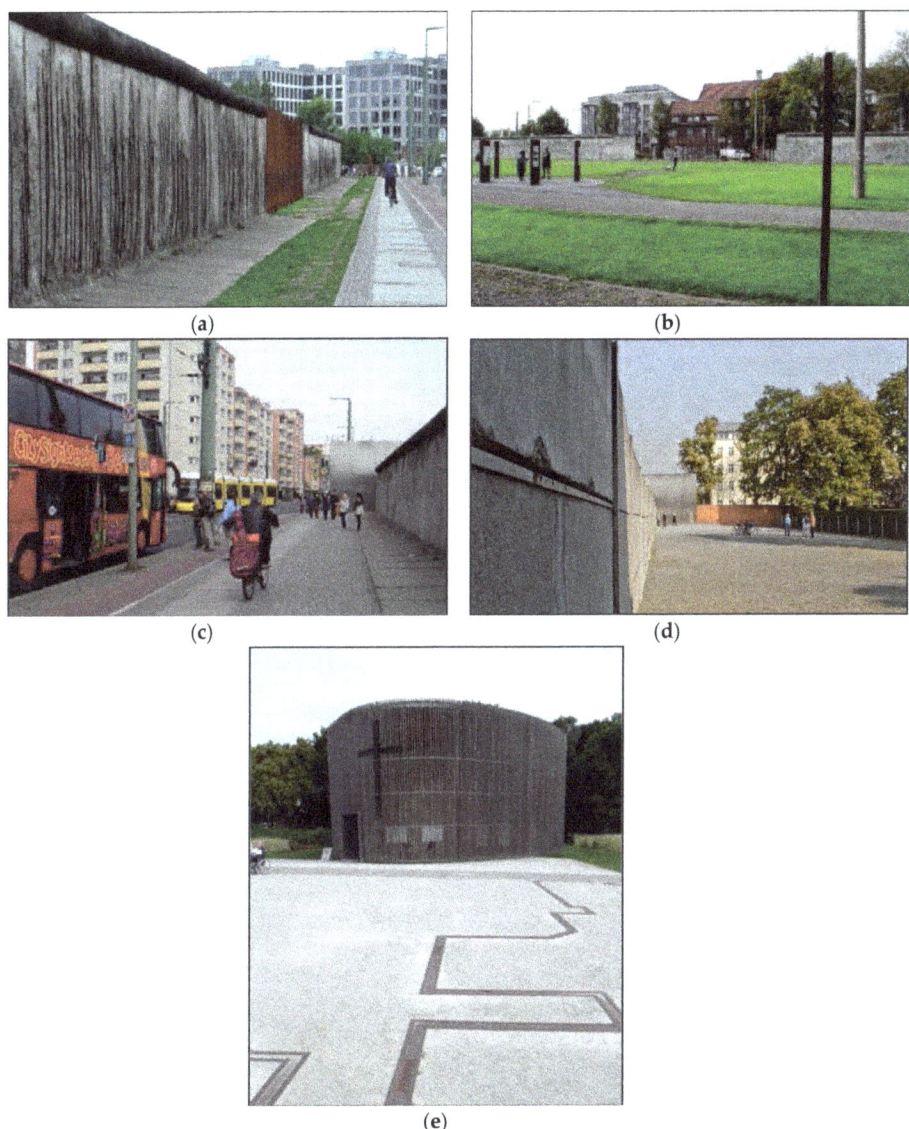

Figure 5. Images of soundscape study points (© P. Jordan): (**a**) "Mundt"; (**b**) "Deathstrip"; (**c**) "West Wall"; (**d**) "East Wall"; (**e**) "Chapel".

As opposed to the predisposition of many soundscape studies to focus on preference assessments [19,20], the focus of the work at the Memorial was to delve into associative meaning related to history. Categorizing soundscapes according to appropriateness continues to be explored within soundscape studies [21,22] and was implemented here also, such as the integration of descriptors such as authenticity and significance [23]. With the study locations established, the survey was designed to approach these inquiries by using a simple, standard format of twelve, five-point semantic scales using descriptor pairs [4,24]. The chosen pairs derived from earlier research on the topic of soundscape descriptors [23] and are shown in Table 1.

Table 1. Descriptor pairs used in the semantic scales of the survey.

Acceptable	Unacceptable
Appropriate	Inappropriate
Authentic	Altered
Clear	Confusing
Comfortable	Uncomfortable
Constant	Changing
Dense	Sparse
Meaningful	Meaningless
Natural	Artificial
Old	New
Pleasant	Unpleasant
Significant	Insignificant

The semantic scales were presented on an individual sheet for each study point; the rows depicted in Table 1 were shuffled on each sheet so that no two sheets appeared the same. Additionally, the columns in Table 1 were switched for return visits. The scrambling of the scales was designed to subvert any possible primacy and recency effects during the study [25].

To the same end, participants were not told or shown the path of the soundwalk. The walk began with a short introduction about soundwalks and soundscape research, instructions about filling in the survey, and a research and confidentiality agreement. The soundwalk was conducted in English. Participants were encouraged to adopt a "holistic" hearing approach to the sonic environment rather than identifying individual sound sources [26,27]. When assessing their experience, they were encouraged to mark the sheets freely, skipping scales they felt they could not use while annotating the form with any comments and concerns. Giving participants the freedom to skip scales entirely highlighted terms that participants felt unsure employing—any change in usage during return visits (after historic context dissemination) was of interest. Requiring participants to answer every scale, as is generally done in soundscape surveys, would not reveal such usage dynamics. In a similar mode of inquiry, a free-form question was also included at the bottom of each sheet that asked simply, "What else is on your mind"?

After the introduction given by the soundwalk guide, participants were led to point 1 and asked to listen for two minutes and then record their responses on the survey sheet. When everyone had completed her or his survey form, the group was led to point 2 and the process repeated identically until point 5. No questions related to the historic context were answered by the guide until this time. At point 5, the study was reoriented with a briefing on the concepts underlying historic soundscapes, their evolution through time and what they can embody that physical fabric alone might miss. Specific stories and historic sonic descriptions related to point 5 were shared with the group to demonstrate the kind of information they could expect to hear from that point on and all questions were answered by the guide. Participants were then asked to listen to the soundscape for two minutes and fill out the associated survey sheet. Point 5 acted as a position of conceptual transition within the study, a setting physically and acoustically removed from the other locations in order to introduce another layer of awareness to the perspective of participants. No comparative data derived from location 5. Once finished, the group was led to point 4 and the process repeated with historic context added until point 1. In this way, points 1–4 were surveyed twice. Each group naturally broke into discussion between survey points and at the conclusion of the soundwalk, resulting in each walk lasting approximately one and three-quarters of an hour.

The study was structured so as to gain insight into participants' use of descriptors without any historic contextual information (capturing their impressions based on the immediate sensory environment alone) followed by whether the introduction of historic information and narratives would dramatically alter their impressions and subsequent responses. To examine these dynamics, the soundwalk was designed for participants to assess each study point twice: first without any information provided for an initial appraisal of present conditions, and then again later in the soundwalk after learning about the historic soundscape from the soundwalk guide. Information provided was specific to each location whenever possible and included historical events, descriptions and stories from the archives related to sound in some direct capacity (such as the examples in Section 2.2).

The composition of participants was a concern for this historic soundscape study. A broad and even spread among age, gender, professional background and nationalities was desired to provide a representative population sample. Given the international profile and educational mission of the Memorial, three groups in particular were sought for inclusion as well: first-time visitors to the site, long-term residents of the area and/or living witnesses, and caretakers and managers of the Memorial itself. These categories were identified to capture the impressions of the primary interest groups related to the Memorial grounds, both as an historic reality and a contemporary interpretation. In total, 27 individuals participated in the study.

3. Results

3.1. Soundscape Mapping

Analysis of the soundscape maps in Figure 6 points to Bernauer Straße as the dominant sonic component to the current Memorial landscape. Comparing the LA_{eq} to the perceived loudness in sone does not reveal any striking differences in the objective acoustic qualities of the space with what was perceived—perceptual indicators were consistent with physical sound behavior. However, the maps do make tangible the wide range of conditions within the Memorial, particularly the differences between east and west sides of the Wall. Along Bernauer Straße, the LA_{eq} and sone were found to be consistently high, while the measurements taken in the interpreted deathstrip landscape were often similar to those found in the cemetery beyond. Unsurprisingly, research point 4 ("East Wall") was the most sonically isolated with the lowest measured values.

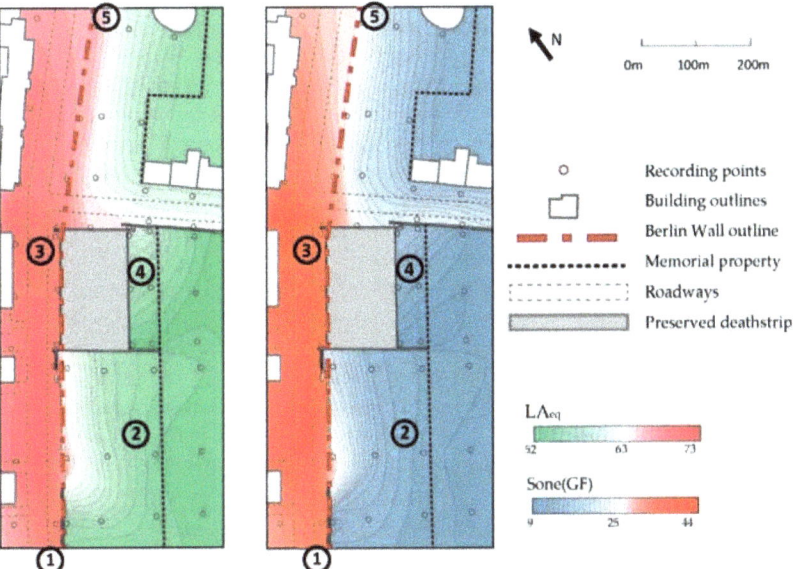

Figure 6. Soundscape maps of the Memorial on a typical summer day, measurement points in small circles, soundwalk study points in large numbered circles (© P. Jordan): (**left**) LA_{eq} values derived from FFT (Fast Fourier transform) vs. Time analyses over 3 min; (**right**) averaged sone (GF) values derived from Specific Loudness analyses over 3 min.

The maps also clearly show two key architectural elements that enforce such different sonic environments on the East and the West of the Wall to this day. First is the Wall itself, an effective barrier that appears to absorb and/or deflect sound from Bernauer Straße. South of the preserved Wall and deathstrip area, the original concrete Wall is mostly intact. Here, the difference in measurements immediately on either side of the Wall can be large; at the southern-most measuring points, (located near point 1), a reading of 69 dB(A) on the West is answered by only 54 dB(A) on the East.

The second architectural element to note is the open grassy expanse that was once the deathstrip, which forms the majority of the Memorial landscape today (on the maps, this is the region between the red line demarcating the Berlin Wall and the dark dotted line indicating the Memorial property boundary). This region appears to have a significant mitigating impact on the sound emanating from the street, as can be seen by comparing the northern portion of the study area (top portion of the maps, where no fragment of the Wall remains) to the southern portions of the study area (bottom portion of the maps, where the original Wall remains). Without the tall concrete wall structure, a full spectrum of direct sounds from Bernauer Straße penetrates the open landscape but is absorbed through ground attenuation as one gets farther away from the street [28]. The concrete slabs and open landscape are both designed elements that are physically remnant of the original Wall construction during its later years. Their preservation provides authentic visual continuity of the site as whole for visitors, though the grass is materially different from the original groomed sand that was utilized by guards. As the maps illustrate, the soundscape conveys some of the most powerful messaging of the original militarized architecture, which can still be experienced at the Memorial today; the Wall and deathstrip areas continue to cut off sonic connection between East and West and produce two distinct sonic environments. How these differences manifest throughout the Memorial and are subsequently understood by visitors requires a detailed study of the soundwalk survey results.

3.2. Soundwalk and Surveys

The information gathered in the surveys illustrates a few key findings concerning the applicability of descriptors for historic soundscapes. At the Memorial, several descriptor pairs proved problematic for consistent use by some participants and were skipped altogether. Table 2 shows the total number of participants who did not use a semantic scale on one or both visits to a study point by each participant. The pair *altered—authentic* proved the most consistently avoided term across all study points.

To search for possible reasons behind these findings, instances of skipping a semantic scale were broken up between first and second visits. If a semantic scale was skipped both on first and second visits to a study location, it could indicate that the terms themselves were not found to be clear or applicable at all for participants. Table 3 shows the percentage of participants throughout the entire soundwalk who skipped a semantic scale per visit to each study point.

In Table 3, percentages are calculated based on the number of participants in each gender group across the four points studied. This results in 152 data points per visit for female participants (19 participant responses at each of four study locations per visit), 24 data points for male participants and eight data points for those who identified in the category of "other". Looking at the semantic pair Meaningful—Meaningless, for instance, the scale was left empty twice across the 152 visits by female participants on the first visit, resulting in a 2.9% skip rate for first visits throughout the walk. This scale was subsequently used by all female participants on the second visit, resulting in a 0% skip rate for second visits.

It is important to recognize the difference between percentage results given the varied number of participants in each gender group. Nonetheless, the findings point out several semantic scales that some participants chose not to apply on both first and second visits: Natural—Artificial, Altered—Authentic, Old—New, Appropriate—Inappropriate and Acceptable—Unacceptable. These choices do not indicate the value of the terms, since they were applied for the majority of the time throughout the entire study. Rather, the findings suggest ambiguities in how to conceive and apply them; several individuals voiced their confusion about the context under which authenticity should be judged, for instance. Authentic to whom, to what time period? Such ambiguity also may be a reason for the increase in skipping the semantic pair Natural—Artificial, a point returned to in more detail below.

Table 3 reveals a pattern concerning the use of each scale according to gender groups: an evident contrast of descriptor use exists between female and male participants. The combined female participants skipped (both on first and second visits to a study point) the five semantic scales noted above at least 10% of the time. This suggests the possibility that gender could have an influence on the determination of artificiality, authenticity, age, appropriateness and acceptability of historic soundscapes. Previous research has suggested the possibility that semantic-based scales may play a role in such assessment differences [29].

Table 2. Number of total participants (out of 27) who did not use semantic scales on the first and/or second visit to each study point.

Semantic Scale [1]		Mundt	Deathstrip	West Wall	East Wall
Constant	Changing	0	0	0	0
Natural	Artificial	3	3	5	3
Altered	Authentic	7	7	6	6
Old	New	4	5	6	4
Appropriate	Inappropriate	6	5	4	3
Insignificant	Significant	2	3	0	1
Comfortable	Unsettling	0	0	1	0
Dense	Sparse	2	2	1	0
Unacceptable	Acceptable	3	2	3	2
Pleasant	Unpleasant	1	0	0	0
Meaningless	Meaningful	1	1	0	2
Confusing	Clear	1	1	0	0

[1] Order of descriptors were as shown or reversed, as described in Section 2.3.2.

Table 3. Percentage of participants that skipped each semantic scale on first and second visits to all four study points, organized by self-identified gender groups. Results over 10% are highlighted.

		Female (19 Total)		Male (6)		Other (2)	
Semantic Scale [1]		1st Visit	2nd Visit	1st Visit	2nd Visit	1st Visit	2nd Visit
Constant	Changing	0%	0%	0%	0%	0%	0%
Natural	Artificial	**10.5%**	**17.1%**	4.2%	0%	0%	0%
Altered	Authentic	**23.7%**	**18.4%**	4.2%	0%	**37.5%**	**37.5%**
Old	New	**15.8%**	**10.5%**	**12.5%**	4.2%	0%	0%
Appropriate	Inappropriate	**17.1%**	**15.8%**	4.2%	0%	0%	0%
Insignificant	Significant	8.8%	0%	0%	0%	0%	0%
Comfortable	Unsettling	0%	0%	0%	4.2%	0%	0%
Dense	Sparse	2.9%	0%	0%	**12.5%**	0%	0%
Unacceptable	Acceptable	**13.2%**	**13.2%**	4.2%	0%	0%	0%
Pleasant	Unpleasant	1.5%	0%	0%	0%	0%	0%
Meaningless	Meaningful	2.9%	0%	0%	4.2%	0%	0%
Confusing	Clear	1.5%	0%	4.2%	0%	0%	0%

[1] Order of descriptors were as shown or reversed, as described in Section 2.3.2.

When the semantic scales were employed, shifts were evident from the first and second visits in a number of cases. To analyze aggregated responses, the scales were assigned numerical values for each point on the scale. For instance, in the Constant—Changing scale, Constant was assigned a value of 1 and Changing a value of 5. Participant responses were then assigned a number according to where they marked on the scale between these two descriptors, with 3 being a neutral answer. The results were aggregated for each study point and responses between the first and second visits were compared (see Figure 7a–d).

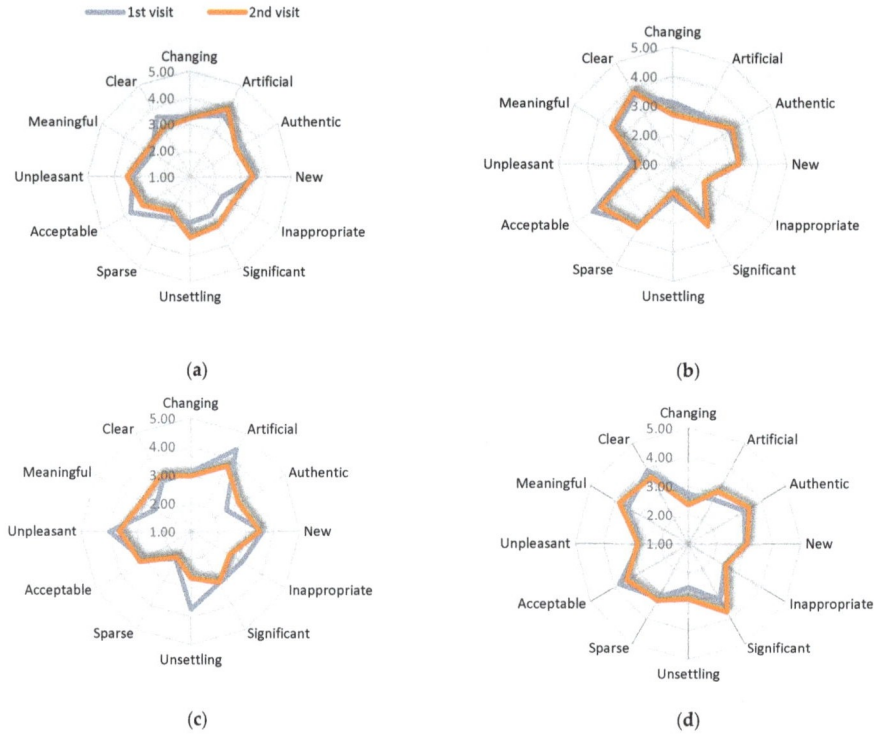

Figure 7. Web charts of average ratings given by all study participants on first (blue) and second (orange) visits to study points, where 1 is the center-point of each web (© P. Jordan): (**a**) "Mundt"; (**b**) "Deathstrip"; (**c**) "West Wall"; (**d**) "East Wall".

From the comparison charts, it is first evident that, averaged across all participants, the ratings given to the Deathstrip (b) and East Wall (d) study points shifted very little between the visits, whereas the shifts in perception of the West Wall point (c) generally tended to move similarly between participants along particular scales, notably the Calm—Unsettling, Natural—Artificial and Authentic—Inauthentic scales. One participant (a visitor) analyzed her/his impression this way on the survey sheet: "After getting information about how it was here in the past, for me it seems to be more authentic. The traffic is now more part of the history as well as the wall itself." This response was regularly echoed by other participants; it suggests that the sonic history provided the foundation for participants to re-contextualize the unpleasant, intense sonic environment as related or similar to certain historic conditions. This could explain why the average rating on the second visit found the West Wall to be less unsettling, more authentic and indeed more "natural".

In order to determine whether the relatively stable responses at (b) and (d) above were due to similarities in response or the averaging of disparate responses, the results were analyzed per individual and by degree of perception shift: whether the values given on a scale on the second visit shifted by 25% or more, between 0–25% or not at all. Table 4 examines how often participants shifted at least six of the twelve scales at each study point.

Table 4. Table showing the number of instances that perceptions shifted between the first and second visit at each study point (of 27 participants).

The Value of at Least Half of Semantic Scales:	Mundt	Deathstrip	West Wall	East Wall
Shifted by 25%+	14	12	10	12
Shifted by 0–12%	5	7	9	6
No change	4	1	6	1

The results of Table 4 illustrate that, while the West Wall indeed was a position of noticeable shifts in perception by participants, it was not the most extreme. Noticeable shifts across several scales took place for participants throughout the soundwalk: at every study point, at least 10 participants changed six or more of their answers by at least 25%. Perceptions appear to have shifted dramatically at the Deathstrip and East Wall study points in particular. However, participants did not always agree in the direction of such shifts, leading the averaged results in Figure 7 to appear rather stable. This observation was exemplified by a conversation between two participants after the conclusion of a soundwalk, in which their perception of the quieter locations (i.e., East Wall and Death Strip) shifted as a result of the soundwalk but in opposite directions: one found the quiet and park-like settings more soothing after learning about the history, while the other participant read them as disjointed and even disturbingly misrepresentative in a place of historical trauma.

Future validation of these results would also be able to trace whether subtler patterns of perceptive shifts exist in conjunction with specific factors—undeniably, multiple aspects contribute to such shifts. When investigating heritage sites, the participant's direct knowledge of its history may be a critical factor. There are multiple paths towards measuring such pre-existing participant knowledge; linked professional expertise (such as history, politics or city planning) and personal background (such as being part of a group directly affected by the associated history of the site) are two examples that could be easily probed within a survey. At the Berlin Wall, partly due to its international notoriety, it is also still possible to divide participants between those with living memory of the Wall's fall and those without by looking at their age. Consequently, the participant responses were compared between those 35 years old and older (i.e., individuals who were at least 10 years old when the Wall fell) and those 34 years old and younger to see if age played a noticeable role in any shifts.

The results in Figure 8 point to some possible differences between age groups. The groups vary in their descriptions of each point according to the scales—the most similar are Mundt and the Deathstrip points, where the second visit webs look very similar. The responses by the older group generally show very similar impressions on the first and second visits, while the younger group shows somewhat more varied responses on the second visit. The exception to this dynamic is the West Wall point, where some perceptions shifted in both groups. Thus, it appears that age may provide useful nuance when reading the survey results—a larger sample size is required to confirm and outline such connections meaningfully. At this stage, the results point to the possible influence of age on participant perception as well as the influence of historic information in less pleasant study points.

The open question at the bottom of each sheet proved helpful in capturing some details of the shifts in perception that some participants experienced during the soundwalk. Numerous individuals noted a connection between people's perception of sound joined with historic conditions:

"The information about the sonic history of the site influences the listening to the actual sounds. The noise today seems indifferent; other than the 'historical noise' with a certain intention/meaning." (Memorial staff)

"There's a gap between my acoustic imagination (gun shots, church bells, protests ...) and the actual sounds I can hear. The combination is interesting." (Memorial staff)

"The wall is silent; the wall produces silence in passersby. It is a powerful thing to sense the wall in such proximity. It is unsettling that it still shapes the space in such powerful ways." (visitor)

"[I am] more aware of hum of park–wind in trees, and birds. ... I have a very different attention now that I have more narratives about the space." (visitor)

A fuller understanding of the scalar survey results was possible using the information provided in the open question responses, supporting the utility of such free-form options in historic soundscape surveys of this type.

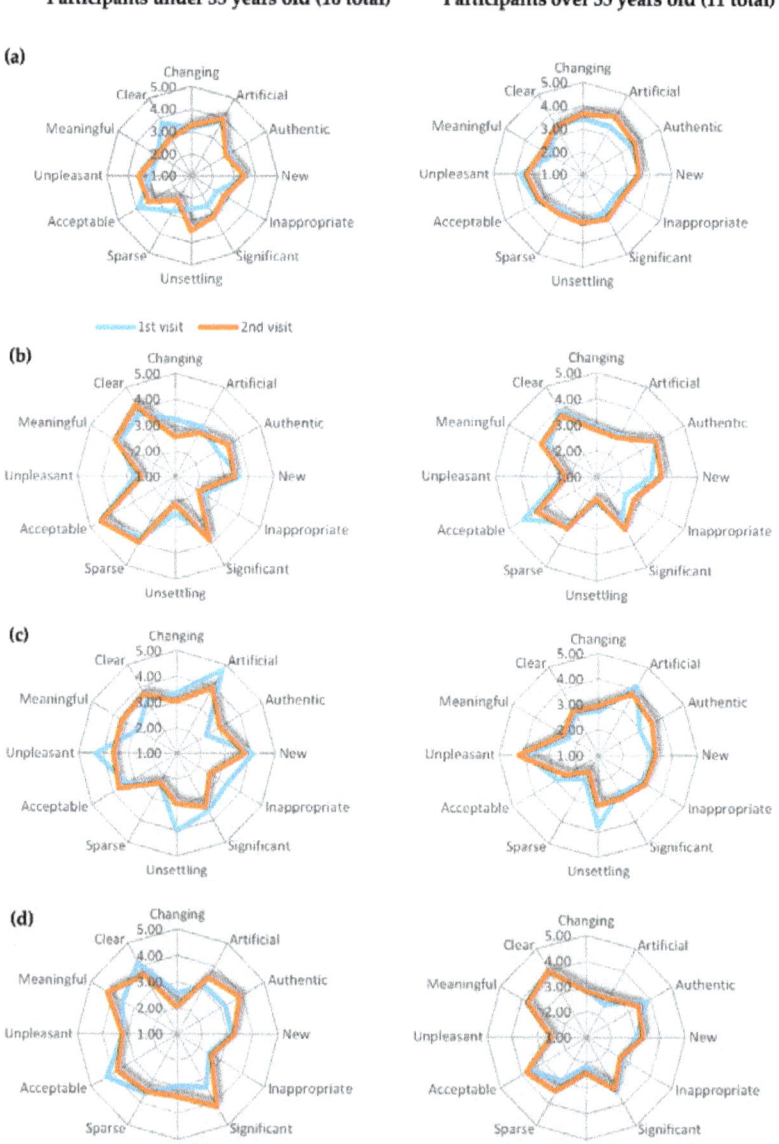

Figure 8. Web charts of average ratings given by study participants divided by age groups on their first (blue) and second (orange) visits to study points, where 1 is the center-point of each web (© P. Jordan): (**a**) "Mundt"; (**b**) "Deathstrip"; (**c**) "West Wall"; (**d**) "East Wall".

4. Discussion

The survey results indicated that learning about past soundscape conditions can influence perception of historic environments. The noted shifts throughout the soundwalk should not be interpreted as a suspect distortion of results, a mere product of conditioning participants through suggestion and prompting. First, the variability in how participants shifted their perception suggests an internalization of the terms used and an effective grappling with varying conditions. Respondents did not simply mark every study point as more meaningful or significant after hearing historical information. Moreover, in a pedagogical, commemorative context of national memorialization, the process of "re-viewing" (or in this case, "re-hearing") the site as a result of added historical context is a preferred if not characteristic result of a visit. Shifts in awareness about historic realities is a function of memorials. The soundwalk's focus on historic experience easily melded with typical visitor patterns of information assimilation; the soundwalk study results cannot be understood in isolation from the site and standard modes of interaction.

The choice to focus on qualitative analysis was informed by the predominantly textual nature of the data as well as the limited number of participants, with the intent that results could function as preparatory or generalization models for further quantitative analysis and conclusions [30,31]. Looking at the findings of the survey specifically, the gathered data exposed a problematized dialogue between positive perceptions and historic soundscapes that challenges the research trends of the field over time. One of the early concerns of the field was noise, particularly people's adverse reactions to it within urban contexts and how to conceive "better" soundscape experiences. In the past twenty-five years since the reprinting of Schafer's original soundscape work [6], the language and concepts that undergird many soundscape studies have been bound by ideas of human preference (see References [32–38]). For example, research approached the study of noise with an implicit connection between quiet conditions and feelings of pleasantness [23]. A recent shift in the field, owing much to the COST Action TD-0804 Soundscape of European Cities and Landscape project and the introduction of the ISO soundscape standards [3,4,37,39], refocused research attention towards considering all sound as a potential asset rather than simply investigating in opposition to noise. This reframing is important towards broadening the applicability of research results generally and for the Memorial research especially.

The survey results suggested the possibility that the meanings and values associated with historic sites could sway typical soundscape interpretations. The seemingly dependable application of standard soundscape descriptors, such as the terms natural and appropriate, could be destabilized when applied in certain historic scenarios at the Memorial. The terms could even be rendered unusable. Thus, the Memorial study offers a particular challenge to the implicit bias of framing noise in opposition to perceived-pleasant conditions. Despite the location's ostensible similarities to urban parks—for instance, large open spaces dominated by grass crossed by gravel pathways, proximity to urban density, indications of respite such as benches, streetscape barriers (walls) and audible birdsong—it was critical not to extend the assumptions behind urban park soundscape research directly to the Memorial landscape [38,40]. This is because the motivations behind visiting historic sites are not solely guided by the search for a pleasant or enjoyable experience [39,41]. Such a visit could seek out education and leisure, involving emotions ranging from duty to curiosity, animosity, patriotism, trauma, pride or shame. The respite offered by a bench in the Memorial setting may not be from the intensity of the urban environment but from the emotional demands of traumatic history recall. Typical assumptions concerning the impact of basic physical and sonic elements must be reexamined at a memorial. It stands to reason that the subsequent interpretation of the landscape—and soundscape—may be influenced by these factors as well, which would affect one's classification of a positive experience in historic contexts. An experience involving incredulity, shame, and leisure simultaneously might be positive while not necessarily pleasant. Moreover, the motivations in creating and presenting a memorial are likewise multifaceted and require consideration in historic soundscape analysis; as discussed earlier, they are active constructs of visitor experience of the site and the resulting soundscape. Therefore, applying established soundscape descriptors derived from investigations of positive or negative experience has the potential of being inadequate or even misleading.

The West Wall study point provides a good example of the complexities in applying standard terminology in historic soundscapes. As shown in Figures 5c and 7c, the soundscape at this location was rated on average more natural/ less artificial on the second visit. Typical use of the term natural in soundscape research denotes a relationship with the natural world: animal life, plant growth and favorable meteorological conditions. Describing the environment as natural is striking given the hardscaping of stone-slab sidewalk bordered by a tall concrete wall, paved busy road and multi-story stuccoed buildings. But discussions with participants plus notations on surveys revealed that some participants found the soundscape more "naturally historic" rather than constructed or imposed. Natural and artificial were interpreted and applied differently in the historic setting of the Memorial, particularly after more historic information was conveyed. This was not true at all study points, where associations with nature (particularly birdsong) dominated participant impressions and use of the term. This raises the possibility of descriptors hosting starkly different interpretations by various participants throughout an entire study or even fluctuating between different points by the same participant. It is also important to consider the results in Table 3 concerning the use of the Natural—Artificial scale, where female participants skipped the term slightly more on the second visit than the first throughout the study. If the historic context of the study site destabilized the term natural for participant perception, particularly female respondents, one option to avoid confusion may be to define terms as concretely as possible in advance of soundwalks. Alternatively, using different semantic pairings on the same survey, such as Natural—Artificial, Natural—Unnatural and Natural—Humanmade, might be a way of honing the meaning of terms more specifically. Generally speaking, the West Wall example underscores the necessity of reexamining standard descriptive terms in settings that could evoke multiple associations – consistency with research results gathered across different site environments should not be assumed.

It appears important to factor in the historic background of study sites, both as a general contextual layer and as potential influence on participants' personal experience during the study. The results underscore how important it is for soundscape researchers to constantly revisit the terminology used in survey-based soundscape studies and include ways to capture and interpret the more flexible approach to soundscape interpretation held by some participants. The free-response questions included in this study present only the first step, as not every participant took the opportunity to reflect directly on their experience—some simply described their hunger levels or their competing thoughts. There may be some value in adding a question that prompts participants to reflect on their own results more directly, either at each study point or at the very end of the soundwalk.

Another point of consideration is raised by looking at the differences between age groups depicted in Figure 8. Here, the impressions of respondents with a possible living memory of the Wall did not seem to change with added historic information in comparison to the younger group. The exception, again, is the West Wall point, though the impressions of the older age group only changed mildly. It is too simplistic to suggest that age defines one's historic experience and awareness and thus is the only factor here; however, these results point to an interplay of age and historic interpretation that may be specific to heritage spaces.

Lastly, the results of the Memorial study must be understood within the limits of the participant mix, both in terms of number and background characteristics. Many more participants would be needed to identify cross-demographic trends in the survey data. Living witnesses in particular would require more robust representation, in turn requiring a thoughtful translation of the soundwalk and survey into German rather than English. With more survey responses to scrutinize, biases related to confirmation, valorization and expectation would require examination to contextualize shifts between first and second visits. But taken as a whole, the current study points out discreet hypotheses on the connection of historic soundscapes and soundscape analysis that are open to future comparative research. The intricate relationship of soundscape and historical context cannot be simplified or easily categorized solely by pre-existing soundscape practices. As cultural heritage begins to embrace embodied (sonic) experience as vital to historic context, it is equally important that soundscape study begins to incorporate more than the present tense directly in its contextual investigations.

5. Conclusions

The study at the Memorial demonstrated a practical methodology for using archival sources to learn about historic soundscape conditions over time as well as integrating this information into participant-based field research via soundwalks and surveys. The mapping study of present conditions helped to visualize the varied sonic conditions that define the primary soundscape relationships throughout the Memorial. Both of these research efforts provided information used directly in the composition of the soundwalk and survey, which sought to identify and explore relationships between historic sonic realities and current historic experience. Further, the project probed whether the soundscape tools commonly employed across soundscape research were readily applicable to investigations within a designated (officially recognized) heritage location. The results of the semantic scalar survey were analyzed qualitatively to identify any patterns in participant use of provided descriptors. This information was cross-referenced for corroborative context with informal conversations and observations from the soundwalks as well as the free-responses and annotations provided by participants on their survey sheets.

The results indicate that historic realities and events can form a vital layer of contextual information that require directed integration into soundscape studies. Moreover, it was shown that the terms and approaches proposed for standardized soundscape investigations are not necessarily equipped to assess historic contextual influence. Soundwalks structured with historic context in mind, like the case study presented here, may prove vital for studying historic soundscapes and how people conceive of history embodied by sound; but they also hold potential as a mechanism of education for the Memorial and other historic places as well. This overlap suggests a partial answer to the challenge of expanding the impact of soundscape study into other fields [32,33,40,41]. But in order to be a viable tool in both soundscape research and heritage applications, the composition of the survey and/or soundwalk must be tuned specifically to the site, its history and current intent.

Funding: This research was funded by the Alexander von Humboldt Foundation and the HEAD Genuit Foundation.

Acknowledgments: The author wishes to thank the Berlin Wall Foundation and Berlin Wall Memorial for research and support assistance, particularly Lydia Dollmann, Axel Klausmeier and Manfred Wichmann. Translation of Zeitzeuge transcripts and research assistance was also provided by Tessa Smith. Brigitte Schulte-Fortkamp, Klaus Genuit, and André Fiebig provided essential suggestions for soundscape analysis. Lastly, the anonymous participants and their dedicated interest throughout each soundwalk made the research both possible and consistently rewarding, for which the author is especially grateful.

Conflicts of Interest: The author declares no conflict of interest. The funders had no role in the design of the study; in the collection, analyses, or interpretation of data; in the writing of the manuscript, or in the decision to publish the results.

References

1. Bruce, N.S.; Davies, W.J. The effects of expectation on the perception of soundscapes. *Appl. Acoust.* **2014**, *85*, 1–11. [CrossRef]
2. Liu, A.; Wang, X.L.; Liu, F.; Yao, C.; Deng, Z. Soundscape and its influence on tourist satisfaction. *Serv. Ind. J.* **2018**, *38*, 164–181. [CrossRef]
3. International Organization for Standardization. *ISO 12913-1:2014—Acoustics—Soundscape Part 1: Definition and Conceptual Framework*; International Organization for Standardization: Geneva, Switzerland, 2014.
4. International Organization for Standardization. *ISO 12913-2:2018—Acoustics—Soundscape Part 2: Data Collection and Reporting Requirements*; International Organization for Standardization: Geneva, Switzerland, 2018.
5. Truax, B. Sound, listening and place: The aesthetic dilemma. *Organ. Sound* **2012**, *17*, 193–201. [CrossRef]

6. Schafer, R.M. *The Soundscape: Our Sonic Environment and the Tuning of the World*, 2nd ed.; Destiny Books: Rochester, VT, USA, 1994.
7. Jordan, P. Soundscapes in historic settings—A case study from ancient Greece. In Proceedings of the INTER-NOISE 2016—45th International Congress and Exposition on Noise Control Engineering: Towards a Quieter Future, Hamburg, Germany, 21–24 August 2016. [CrossRef]
8. Holter, E.; Muth, S.; Schwesinger, S. Sounding out Public Space in Late Republican Rome. In *Sound and the Ancient Senses*, 1st ed.; Butler, S., Nooter, S., Eds.; Routledge: London, UK, 2019; pp. 44–60.
9. Stoever-Ackerman, J. Splicing the Sonic Color-Line. *Soc. Text* **2010**, *28*, 59–85. [CrossRef]
10. Schulte-Fortkamp, B.; Jordan, P. When soundscape meets architecture. *Noise Map.* **2016**, *3*, 216–231. [CrossRef]
11. Klausmeier, A.; Schmidt, L. *Wall Remnants—Wall Traces: The Comprehensive Guide to the Berlin Wall*, 1st ed.; Westkreuz-Verlag Gmbh: Berlin/Bonn, Germany, 2005.
12. Gedenkstätte-Berliner-Mauer, Stiftung Berliner Mauer Zieht Positive Bilanz. 2017. Available online: https://www.berliner-mauer-gedenkstaette.de/de/presse-17,250,16.html (accessed on 20 January 2019).
13. SenStadtWohn, 06.06 Population Density. 2016. Available online: https://www.stadtentwicklung.berlin.de/umwelt/umweltatlas/ekm606.htm (accessed on 20 January 2019).
14. Baker, F. The Berlin Wall: production, preservation and consumption of a 20th-century monument. *Antiquity* **1993**, *67*, 709–733. [CrossRef]
15. Verheyen, D. Commemorating a Vanishing Monument. In *United City, Divided Memories? Cold War Legacies in Contemporary Berlin*; Lexington Books: Lanham, MD, USA, 2008; pp. 219–232.
16. Gröschner, A.; Messmer, A. *Aus Anderer Sicht/The Other View*; Hatje Cantz: Berlin, Germany, 2011.
17. Stevens, S.S. A scale for the measurement of a psychological magnitude: loudness. *Psychol. Rev.* **1936**, *43*, 405–416. [CrossRef]
18. Zwicker, E.; Fastl, H. *Psychoacoustics: Facts and Models*, 1st ed.; Springer: Berlin/Heidelberg, Germany, 1990.
19. Brown, L.; Brown, A.L. Soundscapes and environmental noise management. *Noise Control Eng. J.* **2010**, *58*, 493–500. [CrossRef]
20. Brown, A.L.; Gjestland, T.; Dubois, D. Acoustic Environments and Soundscapes. In *Soundscape and the Built Environment*, 1st ed.; Kang, J., Schulte-Fortkamp, B., Eds.; CRC Press: Boca Raton, FL, USA, 2017; pp. 1–16.
21. Xiao, J.; Aletta, F. A soundscape approach to exploring design strategies for acoustic comfort in modern public libraries: a case study of the Library of Birmingham. *Noise Map.* **2016**, *3*, 264–273. [CrossRef]
22. Nielbo, F.L.; Steele, D.; Guastavino, C. Investigating Soundscape Affordances through Activity Appropriateness. In Proceedings of the Meetings on Acoustics (ICA), Montreal, QC, Canada, 2–7 June 2013; Volume 19, pp. 1–8. [CrossRef]
23. Jordan, P. Valuing the soundscape—Integrating heritage concepts in soundscape assessment. In Proceedings of the INTER-NOISE 2017—46th International Congress and Exposition on Noise Control Engineering: Taming Noise and Moving Quiet, Hong Kong, China, 27–30 August 2017; pp. 5694–5702.
24. Toepoel, V.; Das, M.; Van Soest, A. Design of web questionnaires: The effect of layout in rating scales. *J. Off. Stat.* **2009**, *25*, 509–528. [CrossRef]
25. Murdock, B.B., Jr. The serial position effect of free recall. *J. Exp. Psychol.* **1962**, *64*, 482–488. [CrossRef]
26. Botteldooren, D.; Andringa, T.C.; Aspuru, I.; Brown, A.L.; Dubois, D.; Guastavino, C.; Kang, J.; Lavandier, C.; Nilsson, M.E.; Preis, A.; et al. From Sonic Environment to Soundscape. In *Soundscape and the Built Environment*, 1st ed.; Kang, J., Schulte-Fortkamp, B., Eds.; CRC Press: Boca Raton, FL, USA, 2017; pp. 17–41.
27. Raimbault, M. Qualitative Judgements of Urban Soundscapes: Questionning Questionnaires and Semantic Scales. *Acta Acust.United Acust.* **2006**, *92*, 929–937.
28. Rasmussen, K.B.B. Sound propagation over grass covered ground. *J. Sound Vib.* **1981**, *78*, 247–255. [CrossRef]
29. Wirth, M.; Horn, H.; Koenig, T.; Stein, M.; Federspiel, A.; Meier, B.; Michel, C.; Strik, W. Sex Differences in Semantic Processing: Event-Related Brain Potentials Distinguish between Lower and Higher Order Semantic Analysis during Word Reading. *Cereb. Cortex* **2007**, *17*, 1987–1997. [CrossRef] [PubMed]
30. Fiebig, A. Cognitive Stimulus Integration in the Context of Auditory Sensations and Sound Perceptions. Ph.D. Thesis, Technische Universität Berlin, Berlin, Germany, 2015.
31. Hsieh, H.F.; Shannon, S.E. Three Approaches to Qualitative Content Analysis. *Qual. Health Res.* **2005**, *15*, 1277–1288. [CrossRef] [PubMed]
32. Ismail, M.R. Sound preferences of the dense urban environment: Soundscape of Cairo. *Front. Archit. Res.* **2014**, *3*, 55–68. [CrossRef]

33. Cassina, L.; Fredianelli, L.; Menichini, I.; Chiari, C.; Licitra, G. Audio-Visual Preferences and Tranquillity Ratings in Urban Areas. *Environments* **2017**, *5*, 1. [CrossRef]
34. Sasaki, M. The preference of the various sounds in environment and the discussion about the concept of the soundscape design. *J. Acoust. Soc. Jap.* **1993**, *14*, 189–195. [CrossRef]
35. Miller, N. Understanding Soundscapes. *Buildings* **2013**, *3*, 728–738. [CrossRef]
36. Carles, J.L.; Barrio, I.L.; de Lucio, J.V. Sound influence on landscape values. *Landsc. Urban Plan.* **1999**, *43*, 191–200. [CrossRef]
37. Kang, J.; Chourmouziadou, K.; Sakantamis, K.; Wang, B.; Hao, Y. *(Eds) COST Action: TD0804—Soundscape of European Cities and Landscapes*; Soundscape-COST: Oxford, UK, 2013.
38. Filipan, K.; Boes, M.; De Coensel, B.; Lavandier, C.; Delaitre, P.; Domitrovic, H.; Botteldooren, D. The Personal Viewpoint on the Meaning of Tranquility Affects the Appraisal of the Urban Park Soundscape. *Appl. Sci.* **2017**, *7*, 91. [CrossRef]
39. Poria, Y.; Butler, R.; Airey, D. Links between Tourists, Heritage, and Reasons for Visiting Heritage Sites. *J. Travel Res.* **2004**, *43*, 19–28. [CrossRef]
40. Kang, J.; Aletta, F.; Gjestland, T.T.; Brown, L.A.; Botteldooren, D.; Schulte-Fortkamp, B.; Lercher, P.; van Jamp, I.; Genuit, K.; Fiebig, A.; et al. Ten questions on the soundscapes of the built environment. *Build. Environ.* **2016**, *108*, 284–294. [CrossRef]
41. Kang, J.; Aletta, F. The Impact and Outreach of Soundscape Research. *Environments* **2018**, *5*, 58. [CrossRef]

© 2019 by the author. Licensee MDPI, Basel, Switzerland. This article is an open access article distributed under the terms and conditions of the Creative Commons Attribution (CC BY) license (http://creativecommons.org/licenses/by/4.0/).

MDPI
St. Alban-Anlage 66
4052 Basel
Switzerland
Tel. +41 61 683 77 34
Fax +41 61 302 89 18
www.mdpi.com

Acoustics Editorial Office
E-mail: acoustics@mdpi.com
www.mdpi.com/journal/acoustics

www.ingramcontent.com/pod-product-compliance
Lightning Source LLC
LaVergne TN
LVHW070421100526
838202LV00014B/1501